Chemical Oxidations
with Microorganisms

OXIDATION IN ORGANIC CHEMISTRY
A Series of Monographs

Series Editor

JOHN S. BELEW

Baylor University
Waco, Texas

Other volumes in preparation

Chemical Oxidations
with Microorganisms

GUNTHER S. FONKEN / ROY A. JOHNSON

Cancer Research
The Upjohn Company
Kalamazoo, Michigan

Experimental Chemistry Research
The Upjohn Company
Kalamazoo, Michigan

MARCEL DEKKER, INC., New York   1972

MARCEL DEKKER, INC.
95 Madison Avenue, New York, New York 10016

LIBRARY OF CONGRESS CATALOG CARD NUMBER 70-148597

ISBN 0-8247-1211-0

PRINTED IN THE UNITED STATES OF AMERICA

PREFACE

This book is intended to provide for the organic chemist a brief introduction to preparative oxidative reactions as they are performed by microorganisms. It is not intended to be an exhaustive treatise and has been deliberately restricted in its scope to those areas that seem likely to be useful in the laboratory. It is intended to allow the average chemist, with average access to equipment and supplies, to add microbial oxidation to his armamentarium of reagents. If a complex synthetic problem reaches earlier or more satisfying resolution by the use of microorganisms, the chemist will have had the pleasure of broadening his scientific experience, and perhaps gained a modicum of unusual status among his less diversified brethren. If the use of microbial reagents attracts the chemist away from his more conventional tools, luring him into full-time study of the capabilities of the "wee beasties," he will find an abundance of interesting reactions and phenomena, most of them only lightly explored, not to mention a greater abundance of almost completely unexplored areas. He will find a challenge in the methodology of microbiology, much of which retains the simple elegance of Pasteur's day, but which is ripe for the attentions of a clever organic chemist. And he will respond to the fascinations of the more fastidious, complex, tricky organisms as his capability in their manipulations grows.

Since this book is directed toward the organic chemist rather than the microbiologist, some liberties have been taken with microbiological terminology. The Glossary should make this clear. Following the customary practice, with which the chemist may be unfamiliar, a microorganism is named by its genus and species names, *e.g.*, *Sporotrichum sulfurescens*, when it first appears in the text. In subsequent appearances in the same section or chapter, the genus name is abbreviated, *e.g.*, *S. sulfurescens*. We have not been rigorous, but

have tried to avoid ambiguity by occasionally introducing microbiological redundancy. Other small intricacies of microbiological usage
are to be found in Chapter 14, "Practical Experimental Methods," and
it would probably be a good idea for the reader to begin with at
least a cursory perusal of that chapter.

The organization of the available material into chapters by reaction type presented a number of problems, some of which were resolved by rather arbitrary and loose accommodation of examples, with
occasional duplications in different sections. Purists may object to
such mismatching as the inclusion of certain C_{19} compounds in the diterpene category, but we felt that the alternative would have been a
less coherent discussion.

Recognizing that the yield of a reaction means "product-in-a-
bottle" to the chemist, although it may only mean "spectrophotomet-
ric units" to a biochemist, we have tried to show yields of isolated
pure product, specified simply as (%) yield. In some cases, where the
product was isolated but the yield was not specified, this is indicated: (isol.). The same symbol following an assay yield indicates
that at least some of the product was isolated. There are undoubtedly some errors in this approach, of which the reader is hereby fore-
warned!

Although scientific literature published as recently as mid-1970
has been included, the coverage subsequent to 1968 is highly selective. Even for the earlier years, only a single citation may be given
for an example that represents the net result revealed in a series of
papers, and the reader is urged to use it as a leading reference.

It is a pleasure to acknowledge John S. Belew's catalytic influence. We are deeply indebted to The Upjohn Company for making available the time and resources needed for the preparation of this book,
and we are particularly grateful to Donna M. Shoup, for coping with
early drafts of varying degrees of intelligibility, and to Elizabeth
Clark, for the preparation of text material in a final form suitable
for photoreproduction.

CONTENTS

*Chemical Oxidations
with Microorganisms*

Chapter 1
NONACTIVATED CARBON HYDROXYLATION

Many microorganisms are able to oxidize organic compounds at C-H bonds to produce alcohols. The reaction may proceed either by direct introduction of gaseous oxygen or, less commonly, it may involve dehydrogenation, followed by addition of water to the double bond. In either case, the end result is an oxidized compound that may be difficult to obtain by more conventional chemical methods.

Further oxidations of the alcohol to ketones, aldehydes, or carboxylic acids may be catalyzed by dehydrogenation reactions, and are considered in Chapter 9. This chapter will consider only the oxidation of nonactivated C-H bonds. Allylic systems, aromatic structures, and other activated structures are dealt with elsewhere.

As yet, little is known about the intimate details of the enzymatic process responsible for oxidation of an unactivated C-H bond. Two generalizations seem warranted, however. First, the oxygen atom has been shown to be derived from molecular oxygen in a number of cases. Secondly, the replacement of the C-H bond is stereospecific in those cases studied, the hydroxyl group replacing the proton without inversion of configuration. This stereospecificity extends to the introduction of chirality into an achiral substrate. A second type of stereospecificity is that of oxidation of enantiomeric substrates at different positions. A number of attempts at predicting microbiological oxidations on the basis of current experimental results have been made. These are mentioned briefly in the context of the results upon which they have been based.

Steroids, isoprenoids, alkaloids, hydrocarbons, and other types of molecules have been oxidized successfully by microbial "reagents." For want of any better classification system, this chapter has been divided into these classes of compounds for purposes of discussing hydroxylation reactions.

1

STEROIDS

Owing to their great commercial importance,[1] steroids have been the most actively studied group of compounds, and a number of excellent reviews and monographs on steroid oxidations by microorganisms are available. Since the subject has been so exhaustively reviewed, our presentation is designed to give a brief overview with special emphasis on those reactions that appear to offer synthetic utility either because they afford high yields or because they give essentially a single product that is relatively inaccessible by other means. Many low yielding steroid oxidations may well lend themselves to fermentation development studies, with vastly improved yields as the consequence. Should this be of interest to the reader, the indicated reviews and monographs will serve well to illuminate the field.

For the purposes of discussion, the steroids have been further subdivided into three arbitrary and rather loosely defined structural categories, which in no case imply any biological activity of the types suggested by the category name: Progestins and Corticoids (these are compounds with a 17β, two-carbon side chain; progestins, unlike corticoids, have no hydroxyl group(s) at C-17 and/or C-21, and hence would be expected to be somewhat less susceptible to side-chain disruption by chemical oxidative reagents than would corticoids); Androgens and Estrogens (compounds lacking a carbon side-chain at C-17); and Miscellaneous Steroids (includes sterols, sterol antibiotics, bile acid-related compounds, cardenolides, bufadienolides, and steroidal alkaloids).

Progestins and Corticoids. The first demonstration of a microbial hydroxylation of a nonactivated methylene group -- the conversion of progesterone to 11α-hydroxyprogesterone by *Rhizopus arrhizus* in 10% yield -- triggered two decades of intensive effort that shows no signs of abating. With extensive commercial development, the production of 11α-hydroxyprogesterone can now be effected in 90% or better yields, with several microorganisms (*Rhizopus nigricans* and *Aspergillus ochraceus* are outstanding),[2-8] and is, in fact, carried out

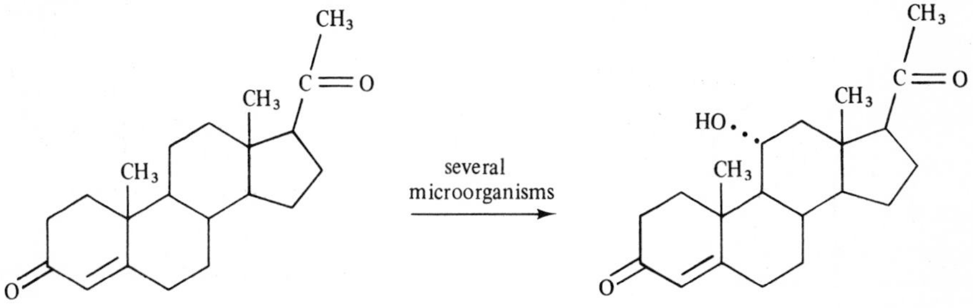

Progesterone

on a huge scale to produce raw material for medicinally important
steroids. A few examples of other high yield 11α-hydroxylations are
illustrated, almost all using *Rhizopus nigricans*. (The presence in
these substrates of groups sensitive to many chemical oxidative reactions is noteworthy.)

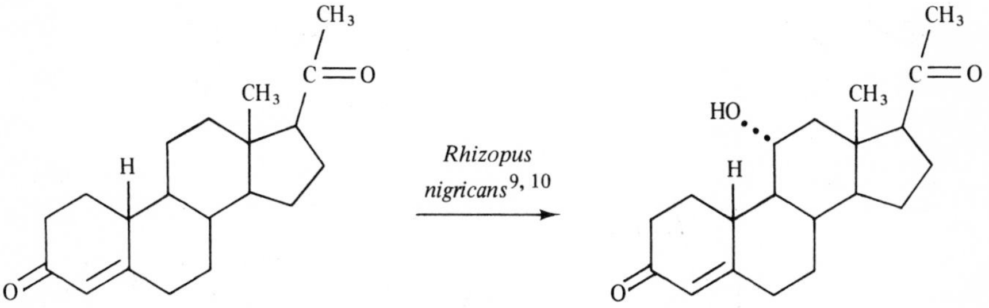

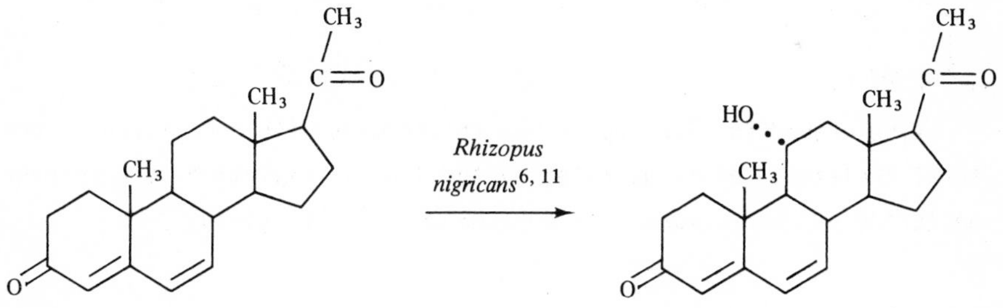

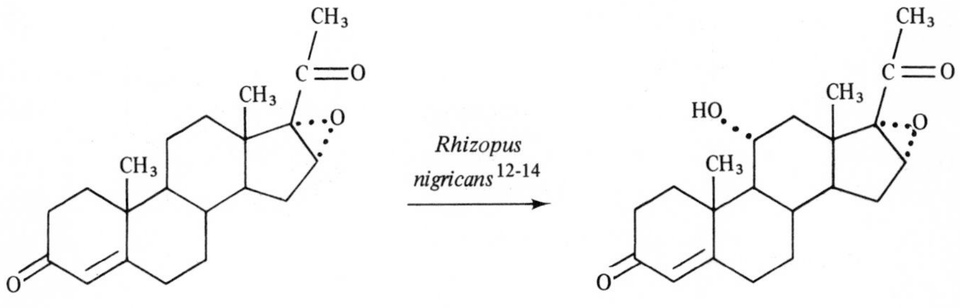

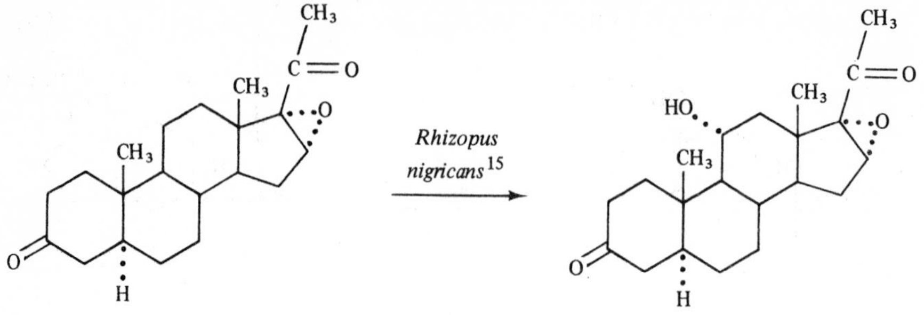

Rhizopus nigricans[15]

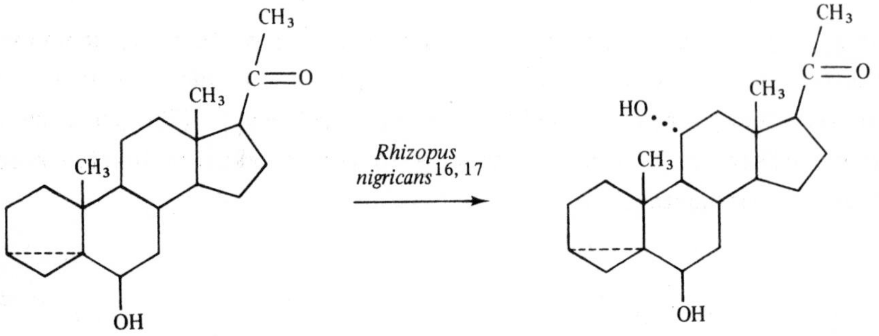

Rhizopus nigricans[16, 17]

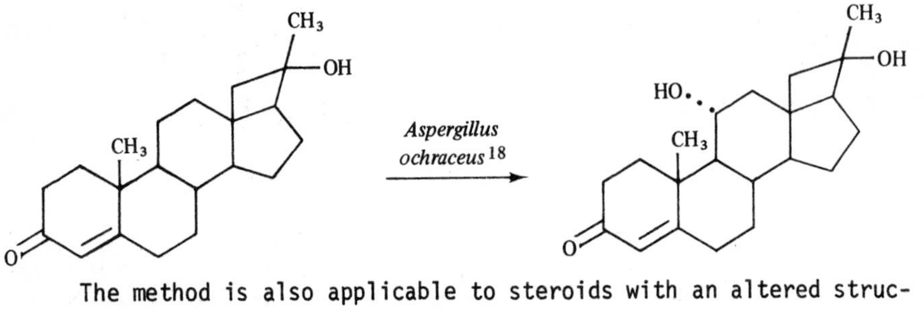

Aspergillus ochraceus[18]

The method is also applicable to steroids with an altered structural skeleton, as illustrated by the 16α-hydroxylation of A-norprogesterone by *Streptomyces roseochromogenus* in 59% yield.[19,20]

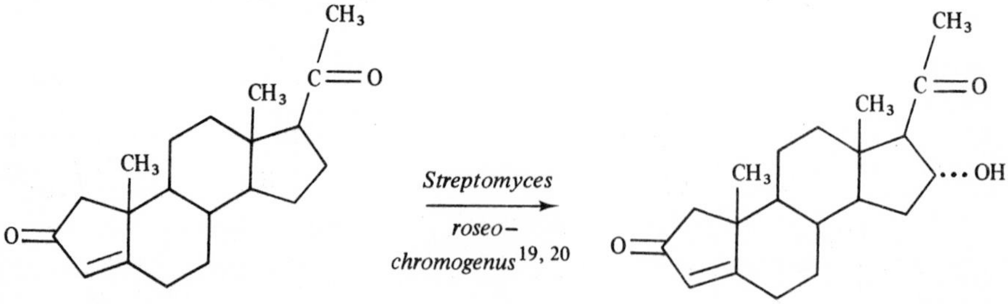

A-Norprogesterone

Streptomyces roseo-chromogenus[19, 20]

The lovely pink fungus *Calonectria decora*, an efficient 15α-hy-droxylator as exemplified by the conversion of 11α-hydroxy-5α-preg-nane-3,20-dione to 11α,15α-dihydroxy-5α-pregnane-3,20-dione in 60%

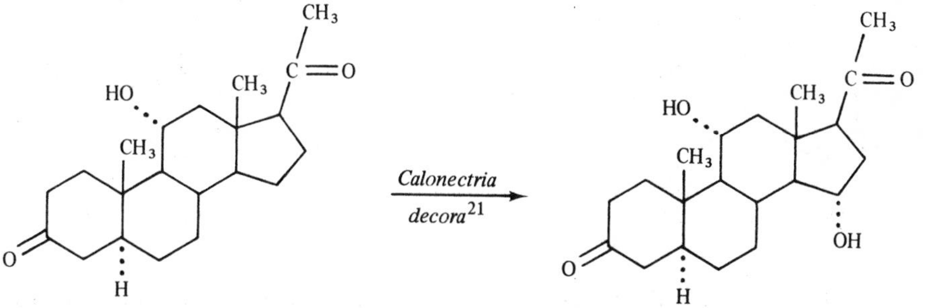

11α-Hydroxy-5α-
pregnane-3,20-dione

yield,[21] introduces two hydroxyls when confronted with a substrate like progesterone, to give a 77% yield of 12β,15α-dihydroxyprogester-one.[22-24] Many other microorganisms also carry out dihydroxylations

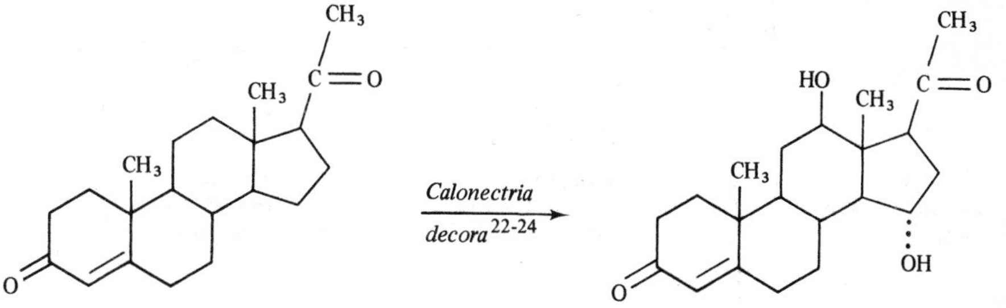

Progesterone

on appropriate substrates, 6β,11α-dihydroxylation being common.[25] In the 11α-hydroxylation of progesterone by *Aspergillus ochraceus* it is possible to suppress the formation of the unwanted 6β,11α-dihydroxy derivative by the use of spores of a mutant "albino" culture, in which 11α-hydroxylation efficiency is unimpaired.[26]

A number of hydroxylating microorganisms also have the capabili-ty to cleave the 17-acetyl side chain, a reaction discussed more ex-tensively in the chapter on microbiological Baeyer-Villiger type re-actions. Mixed reactions may thus be seen, such as the oxidation of progesterone to 11α-hydroxytestosterone by several species of *Sporo-trichum* in about 50% yield.[27]

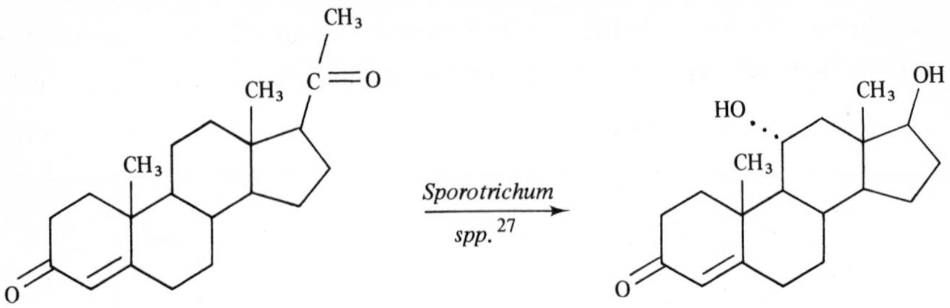

Progesterone   11α-Hydroxytestosterone

Thus far our discussion has dealt exclusively with microbial oxidation of natural *d*-steroids, *i.e.*, optically active materials of
skeletal configuration 8β, 9α, 10β, 13β, 14α. As modified *retro* or

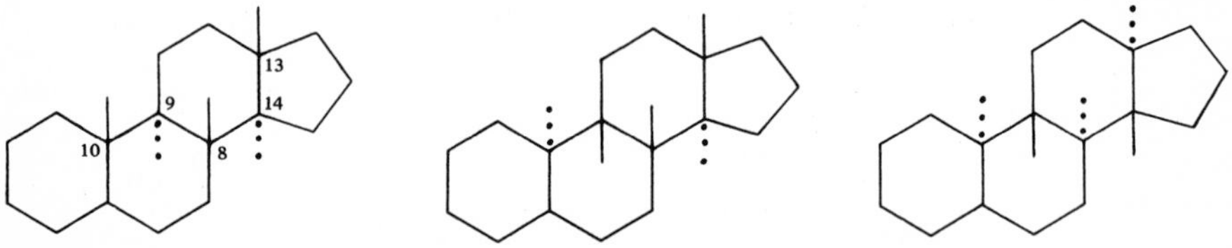

d-steroid   *retro*-steroid   *l*-steroid
(8β, 9α, 10β, 13β,   (8β, 9β, 10α, 13β   (8α, 9β, 10α,
14α)   14α)   13α, 14β)

racemic steroids have become available as a result of total synthesis
work, it is not surprising that attention has been directed to the
microbial oxidations of these compounds.

Microbial hydroxylations of *retro* (9β,10α-) progestins have been
studied by two groups, one Dutch[28,29] and one Swiss.[30] *retro*-Proges-
terone with *Colletotrichum gloeosporioides* gives the 15α-hydroxy de-

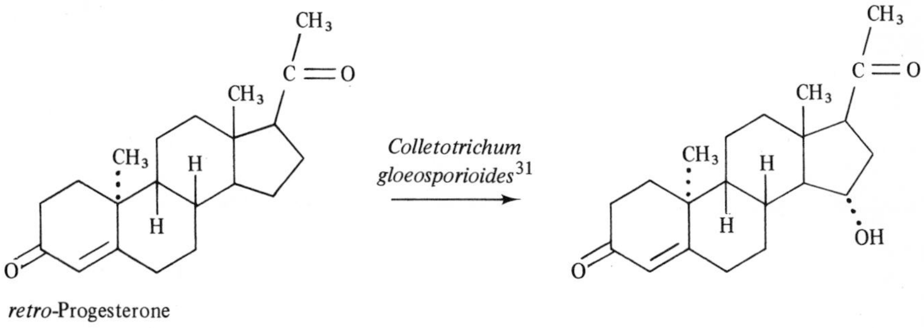

retro-Progesterone

rivative in 29% yield[31] while *Cephalothecium roseum,*[30] *Curvularia lunata,*[30] *Aspergillus ochraceus,*[28] and *Sepedonium ampullosporum*[29] give the 9β-, 8β-, 11α- (70%), and 16α- (71%) hydroxy derivatives, respectively. The corresponding reaction with the 6-dehydro analog (9β,10α-pregna-4,6-diene-3,20-dione) has been carried out on a kilogram scale (at least) in 81% yield! Interestingly, the yield of 16α-hydroxypro-

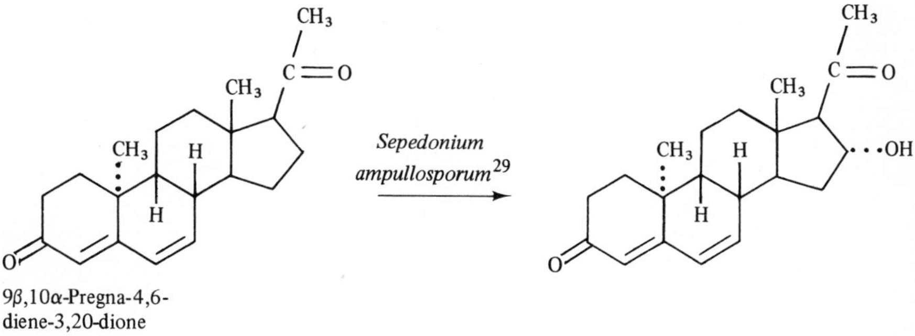

9β,10α-Pregna-4,6-
diene-3,20-dione

gesterone from progesterone under the same conditions is only about 10%.[29]

Hydroxylation reactions of the kinds outlined for the progestins are also observed with corticoids as shown in the following examples:

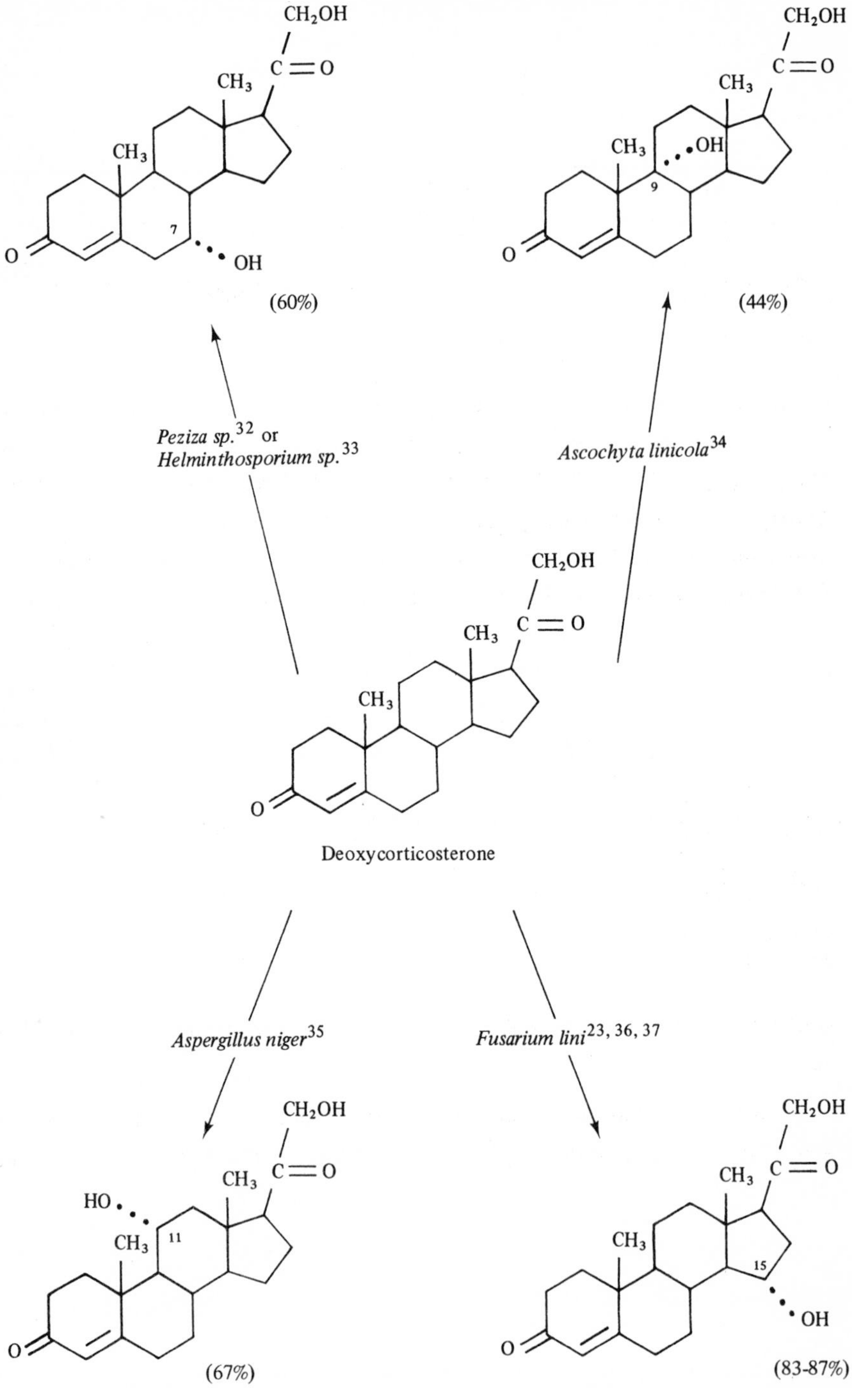
CH₂OH
C=O
CH₃
CH₃
7
OH
O
(60%)
CH₂OH
C=O
CH₃
CH₃
OH
9
O
(44%)
Peziza sp.32 or
Helminthosporium sp.33
Ascochyta linicola34
CH₂OH
C=O
CH₃
CH₃
O
Deoxycorticosterone
Aspergillus niger35
Fusarium lini23, 36, 37
CH₂OH
C=O
CH₃
HO
CH₃
11
O
(67%)
CH₂OH
C=O
CH₃
CH₃
15
OH
O
(83-87%)

Reichstein's Substance S, the immediate precursor to hydrocortisone requiring only 11β-hydroxylation, has served as a substrate for many microorganisms as shown.

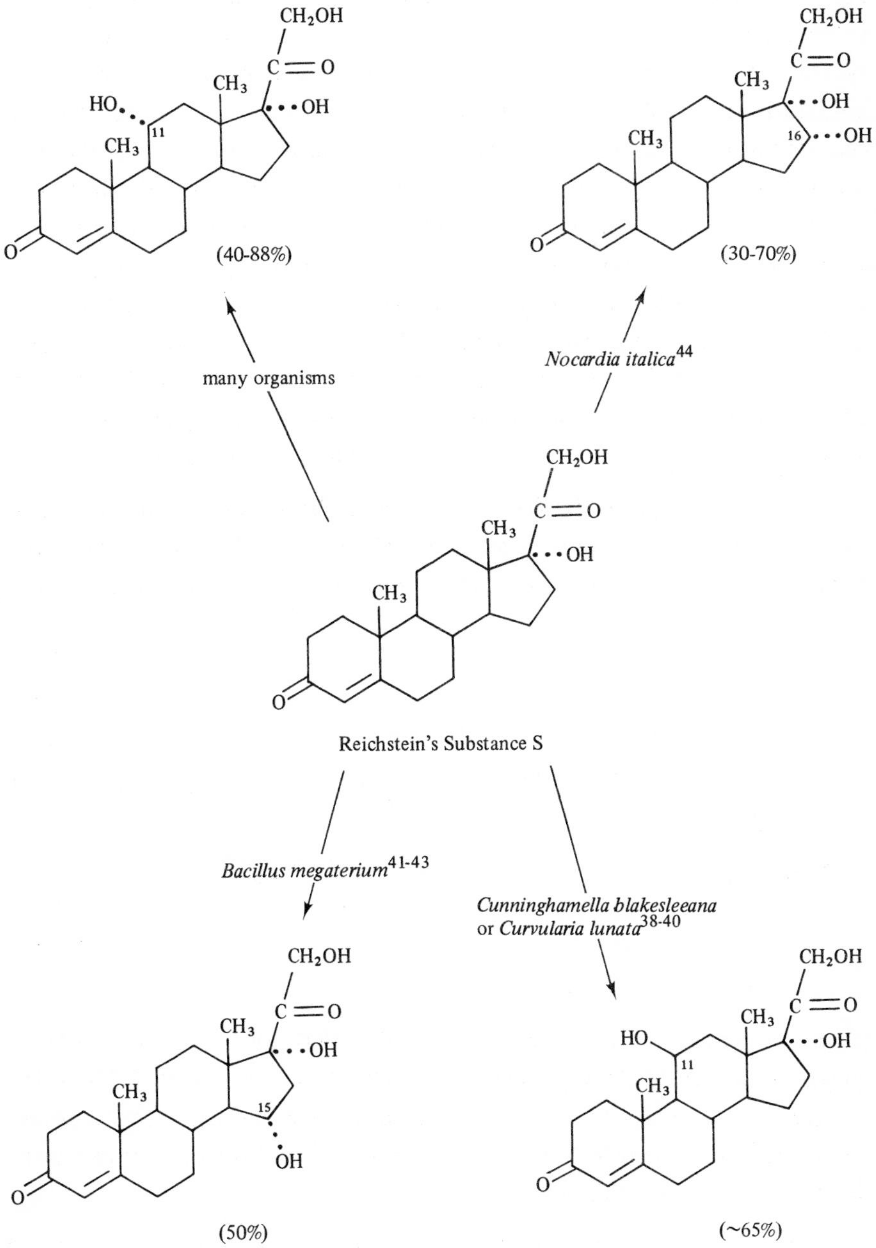
CH₂OH
C=O
CH₃
HO
11
CH₃
OH
O
(40-88%)
CH₂OH
C=O
CH₃
OH
16
OH
CH₃
O
(30-70%)
many organisms
Nocardia italica44
CH₂OH
C=O
CH₃
OH
CH₃
O
Reichstein's Substance S
Bacillus megaterium41-43
Cunninghamella blakesleeana
or Curvularia lunata38-40
CH₂OH
C=O
CH₃
OH
CH₃
15
OH
O
(50%)
CH₂OH
C=O
CH₃
OH
HO
11
CH₃
O
(~65%)

The isolation of products from microbial fermentation media may be difficult, making desirable simplified systems retaining oxygenating capabilities.   The desirability of the conversion of Reichstein's Substance S to hydrocortisone has stimulated two approaches to this problem.   The first is the preparation of a cell-free enzyme system obtained from *Curvularia lunata* that is capable of catalyzing the 11β-hydroxylation.   Exacting conditions are required to obtain consistent results by this method.[39a] The second approach is the inclusion of *C. lunata* mycelial cells in a polyacrylamide gel in which the cells are held entrapped but through the pores of which the substrate and product may pass.[39b] The gel is prepared in granular form and is easily separated from the aqueous medium.   Under optimum conditions, this method can transform 2 mg of substrate to product per hour per gram of dry polymer on a batch basis.   The functional life of such a gel preparation is not known, but the gel granules apparently can be reactivated several times following storage.[39b]

The use of spores (rather than vegetative mycelium) of *Aspergillus ochraceus* for the oxidation of 6α-fluoro-16α,17α-dihydroxyprogesterone to the 11α-hydroxy derivative in 85-98% yield offers the addi-

6α-Fluoro-16α,17α-dihydroxyprogesterone

tional advantage, not widely exploited, that the "reagent" may be produced and accumulated at a convenient time, and stored for later instant use as needed. Furthermore, conditions need not be rigorously aseptic when spores are used, permitting the use of simple equipment for the fermentation.[45]

As is the case for other steroid categories, hydroxylations of modified steroids have been carried out in the corticoid series, too. Examples are given for A-nor and for *retro*-steroids.

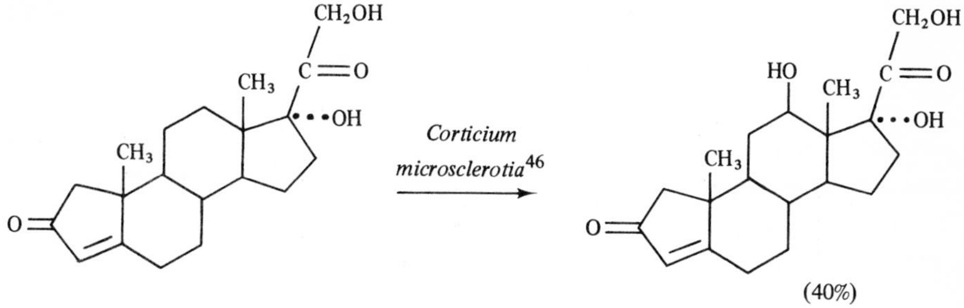

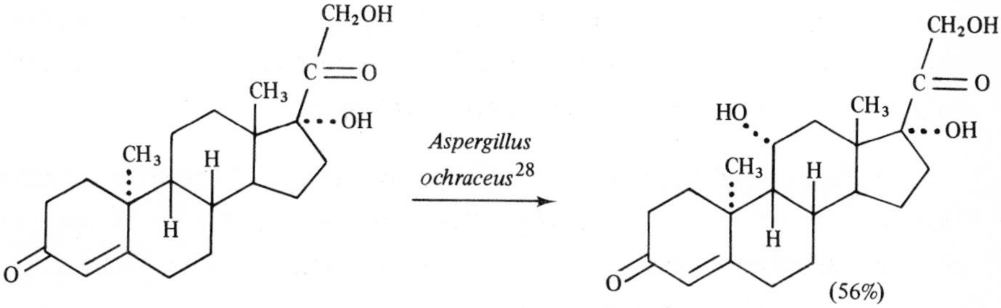

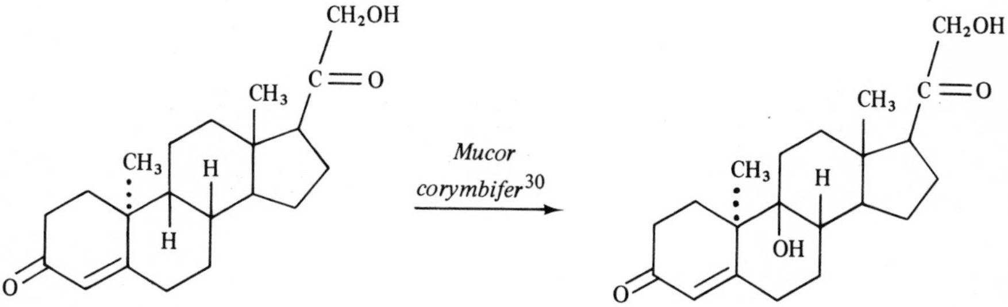

CH_2OH
$C = O$
CH_3
$\cdots OH$

Corticium microsclerotia[46]
$\longrightarrow$

CH_2OH
HO CH_3
$C = O$
$\cdots OH$

(40%)

CH_2OH
$C = O$
CH_3
$\cdots OH$

Aspergillus ochraceus[28]
$\longrightarrow$

CH_2OH
HO CH_3
$C = O$
$\cdots OH$

(56%)

CH_2OH
$C = O$
CH_3

Mucor corymbifer[30]
$\longrightarrow$

CH_2OH
CH_3 $C = O$
OH

Similarly, mixed reactions may occur. The *Helicostylum piriforme* transformation below undoubtedly involves microbial oxidation of

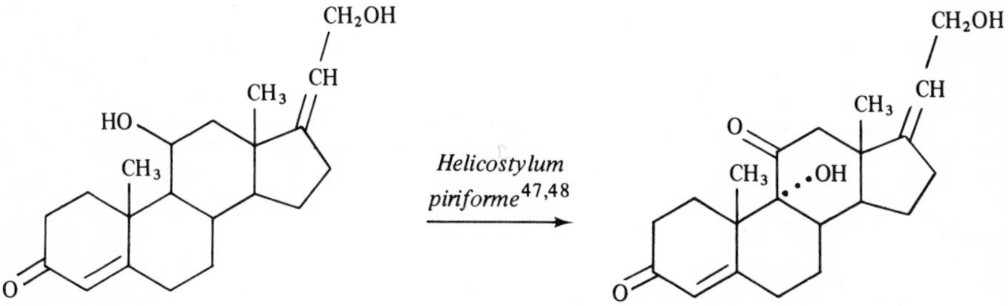

CH_2OH
CH
CH_3
HO CH_3

Helicostylum piriforme[47,48]
$\longrightarrow$

CH_2OH
CH
CH_3
O CH_3 $\cdots OH$

the 11β-hydroxyl to ketone, in addition to the introduction of the

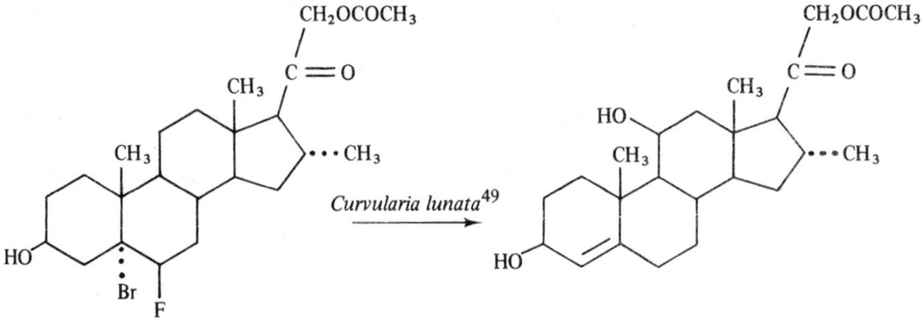

Curvularia lunata[49]

9α-hydroxyl group. However, the *Curvularia lunata* oxidation-dehydro-halogenation-hydrolysis above probably represents microbial action only for the oxidation and the ester hydrolysis.

Androgens and Estrogens. 11α-Hydroxylation of steroids of this type occurs with high yields regardless of the presence or absence of the angular methyl group at C-10. The oxidation of estr-4-ene-3,17-dione to 11α-hydroxyestr-4-ene-3,17-dione by *Aspergillus ochraceus* in

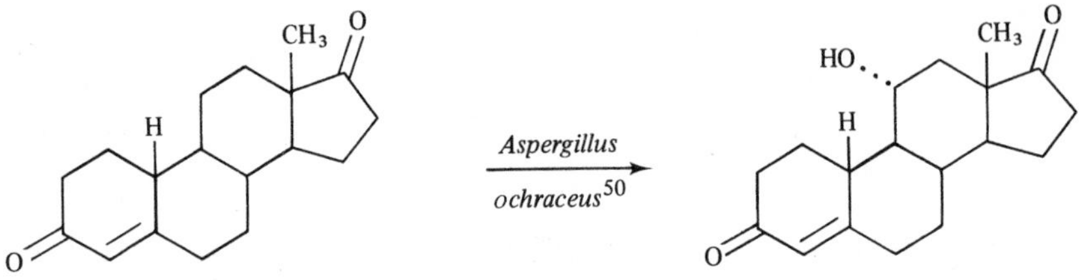

Aspergillus ochraceus[50]

Estr-4-ene-3,17-dione

69% yield[50] finds its counterpart in the oxidations of 17β-hydroxy-androsta-1,4-dien-3-one and of 17-methyltestosterone by *Sporotrichum*

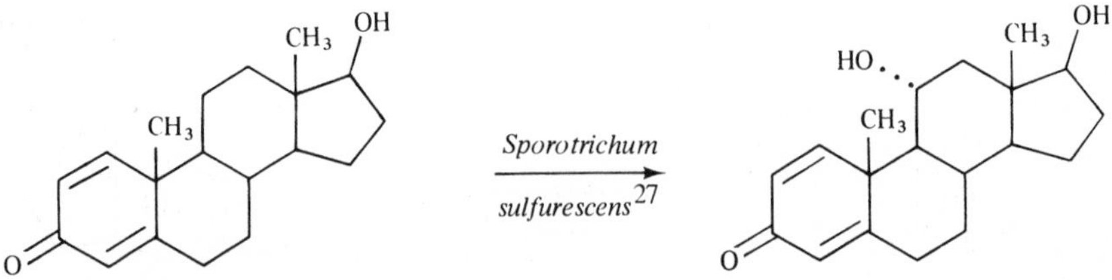

Sporotrichum sulfurescens[27]

17β-Hydroxyandrosta-
1,4-dien-3-one

sulfurescens[27] to the corresponding 11α-hydroxy compounds in 59 and

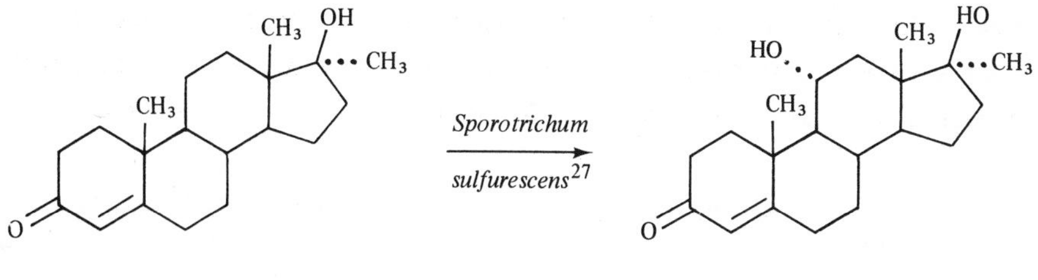

17-Methyltestosterone

70% yields, respectively. The aromatic substrate estrone is effici-
ently converted to 15α-hydroxyestrone in about 75% yield by *Gibber-*
ella fujikuroi.[51]

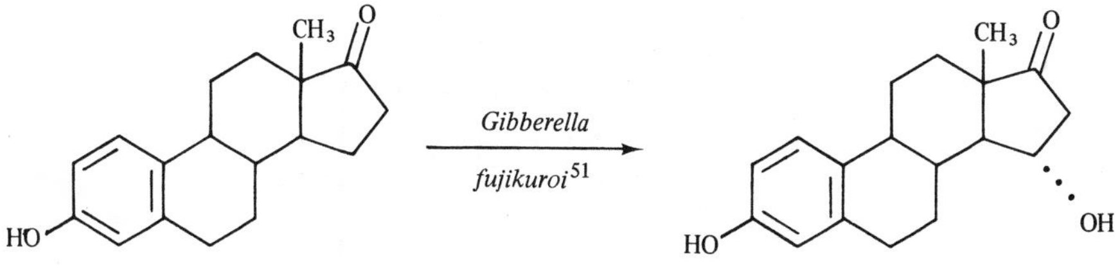

Estrone

  Although hydroxyl groups, both those present in the substrate
and those introduced by fermentation, may be retained as such, oxidation to ketone may occur (see Chapter 9). Mixed reactions are common, and the oxidation of 17β-hydroxy-A-norandrost-3-en-2-one (A-nortestosterone) to 9α-hydroxy-A-norandrost-3-ene-2,17-dione in 76%
yield by *Nocardia restrictus* illustrates the point for an unusual
steroid.[52]

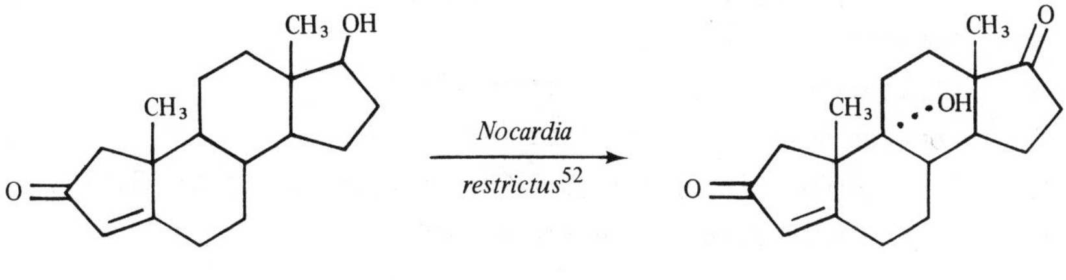

A-nortestosterone

It is often possible to predict the position on a steroid molecule that will be hydroxylated by the use of a particular microorganism.   This predictability is largely based on the experience gained
from studies that have been carried out on steroid substrates with
oxygen substituents at positions 3 and 17 (or 20, in the case of pro-

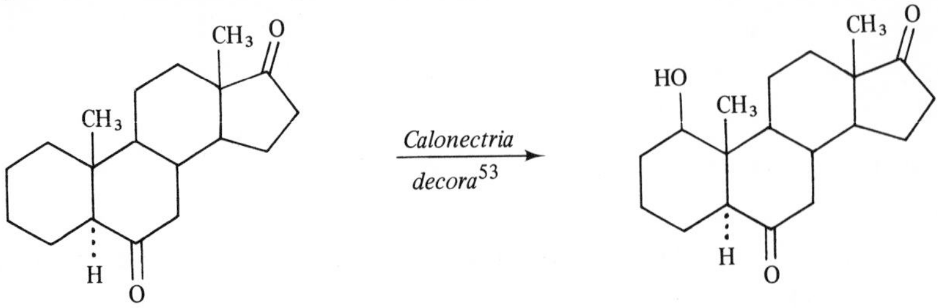

Calonectria
decora[53]

5α-Androstane-6,17-dione

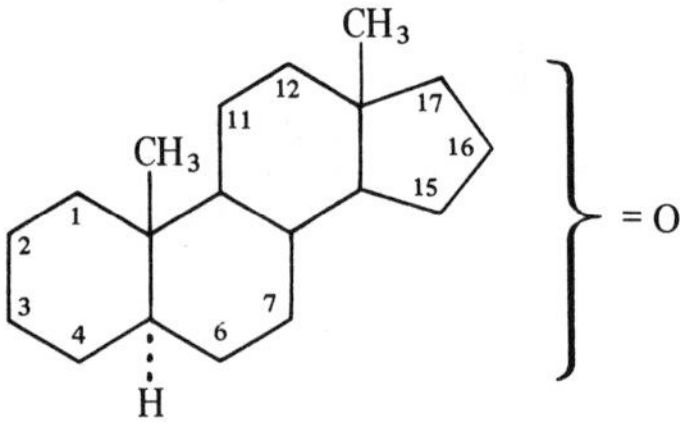

5α-Androstan-x-one

Table 1. 5α-Androstan-x-one Oxygenation by *Calonectria decora*

X	Hydroxylations (Assay yields, %)
1	Complex mixture + substrate (75%)
2	6α,12β-(OH)$_2$ (23%) + 6α,11α-(OH)$_2$ (11%) + substrate (13%)
3	12β,15α-(OH)$_2$ (40%) + 3β,12β,15α-(OH)$_3$ + substrate (22%)
4	12β,15α-(OH)$_2$ (41%) + 11α,15α-(OH)$_2$ (40%)
6	Substrate (90%)
7	12β,15α-(OH)$_2$ (4%) + substrate (72%)
11	Complex mixture + substrate (40%)
16	6α,11α-(OH)$_2$ (26%) + 1β,6α-(OH)$_2$ (7%) + substrate (31%)
17	1β,6α-(OH)$_2$ (42%) + substrate (39%)

gestins and corticoids), but a British group has undertaken a study of the microbiological oxidation of steroids with more unusually placed oxygen functions. This latter fundamental research is expected to yield insights into the role that variously positioned oxygen groups in the substrate play in determining the location of the newly introduced oxygen. For example, although *Calonectria decora* oxidizes progesterone in excellent yield (77%) to 12β,15α-dihydroxyprogesterone,[22-24] the same organism acting on 5α-androstane-6,17-dione, gives a 56% yield of 1β-hydroxy-5α-androstane-6,17-dione,[53] and converts androst-5-en-7-one to 12β-hydroxyandrost-5-en-7-one and 3β,12β-dihydroxyandrost-5-en-7-one. Comparison (Table 1) of a series of 5α-an-

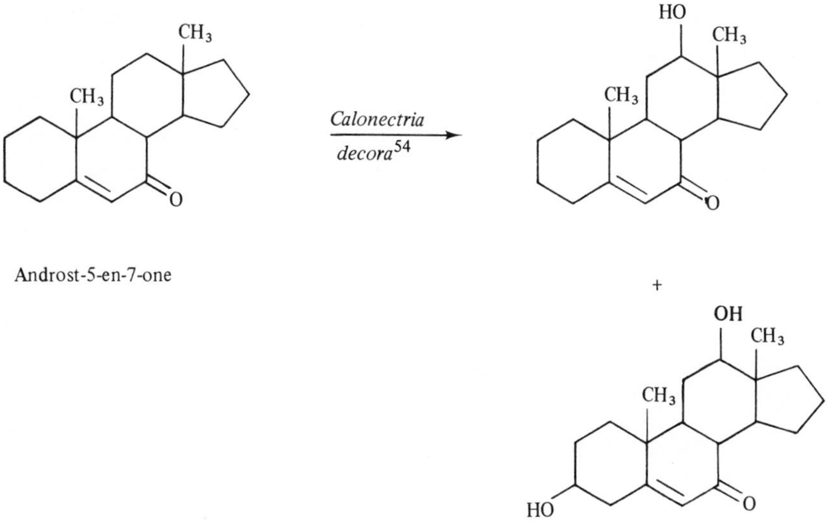

Androst-5-en-7-one

Calonectria decora[54]

+

drostan-x-ones subjected to *C. decora* shows a remarkable variety of products,[55] including the unusual (for microorganisms) 6α-hydroxylation.

Just as studies of this type may be expected to lead to better predictability, so studies aimed at controlling unwanted side reactions may be expected to lead to better control over product type and yield. The 9α-hydroxylation of androst-4-ene-3,17-dione by *Mycobacterium phlei*[56] ordinarily serves as prelude to further degradation of the molecule by ring cleavage between positions 9 and 10, but the ad-

dition of a chelating agent such as 8-hydroxyquinoline to the fermentation blocks the cleavage and affords a 17% yield of 9α-hydroxyan-
drost-4-ene-3,17-dione.

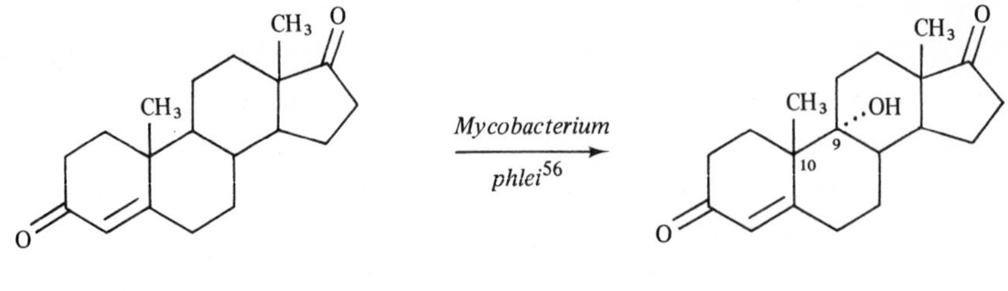

Androst-4-ene-3,17-dione

The two groups that studied oxidations of *retro*-progestins also
carried out microbial hydroxylations of *retro*-androgens.[28-30] Not
unexpectedly, *retro*-testosterone is converted to the 11α-hydroxy derivative (29% yield) by *Aspergillus ochraceus*[28] and to the 16α-hy-
droxy derivative (55% yield) by *Sepedonium ampullosporum*.[29] The

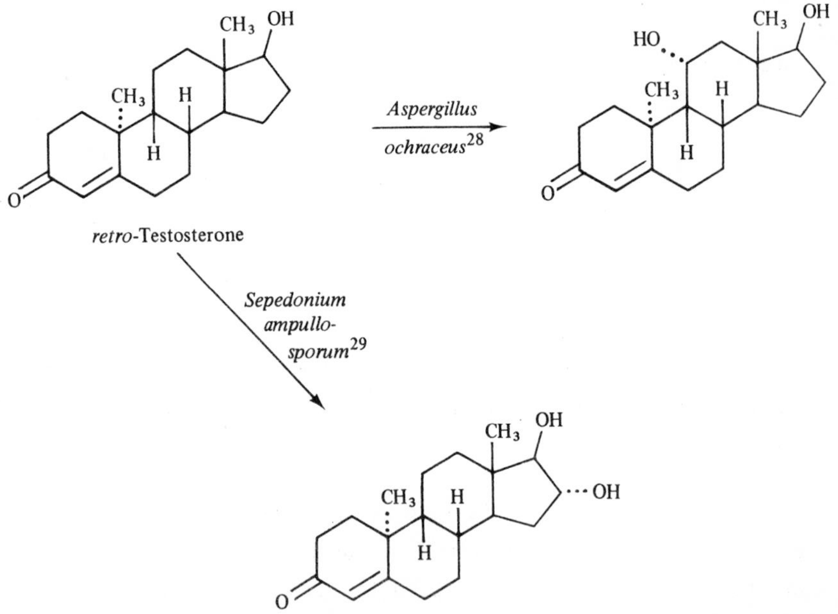

retro-Testosterone

transformations of the same substrate to the 8β-hydroxy derivative by
Aspergillus nidulans[30] and to the 9β-hydroxy derivative by *Choaneph-*

ora circinans,[30]   as well as of Δ^6-*retro*-testosterone   to its 9β-hy-
droxy derivative by *Nocardia lurida*[30]   may represent   deviations from
normal hydroxylation   patterns attributable to the altered   substrate
skeleton.

Δ⁶-*retro*-Testosterone

The hydroxylations of *d*- and *dl*-19-nortestosterone by *Curvularia
lunata*[57]   (which is used   commercially   for 11β-hydroxylation   but is
known to hydroxylate   a multiplicity of other positions   in steroids)
are compared in Table 2. (The 6β- and 10β-hydroxylations are allylic,
but are included for completeness.)

Table 2.   19-Nortestosterone Oxygenation by *Curvularia lunata*

Products from: *d*-19-nortestosterone	*dl*-19-nortestosterone
d-10β-OH (52%)	*d*-10β-OH (18.9%)
d-11β-OH (2.6%)	*dl*-11β-OH (6.1%)
d-14α-OH (9.5%)	*d*-14α-OH (0.5%)
--	*l*-12α-OH (1.4%)
d-6β-OH (1.4%)	*dl*-6β-OH (4.3%)
d-10β,11β-(OH)₂ (0.5%)	*dl*-10β,11β-(OH)₂ (1.3%)

A similar   experiment   using   *dl*-17β-hydroxy-13β-ethylgon-4-en-3-
one gives a slightly different   product pattern   (Table 3) from   that

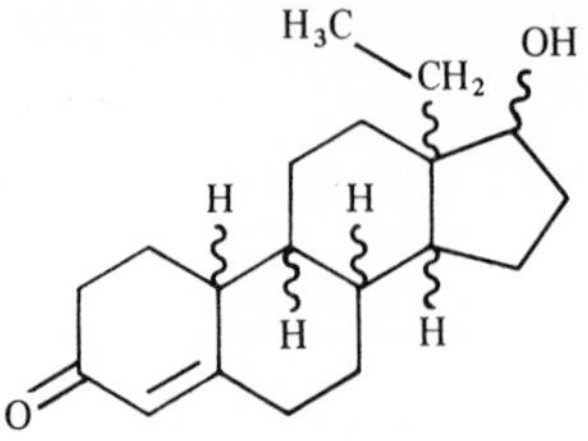

dl-17β-Hydroxy-13β-
ethylgon-4-en-3-one

Table 3. Oxygenation of *dl*-17β-Hydroxy-13β-ethylgon-
4-en-3-one with *Curvularia lunata*

Products
dl-10β-OH (16.7%)
l-12α-OH (14.3%)
d-14α-OH (4.8%)
dl-6β-OH (0.8%)
d-6β,10β-(OH)$_2$ (0.3%)

seen with the 13-methyl analog just described. Although it appears
likely (from this and similar work using *Aspergillus ochraceus*[58])
that enantiomers may respond differently to microorganisms, the data
now available do not provide sufficiently definitive evidence.
<u>Miscellaneous Steroids</u>. In contrast to the abundance of oxida-
tions implied in the preceding two sections, such reactions are rare
with sterols. Cholesterol, although largely oxidized to cholestenone
by a *Mycobacterium* species, affords a trace of 27-hydroxycholest-4-
en-3-one.[59]

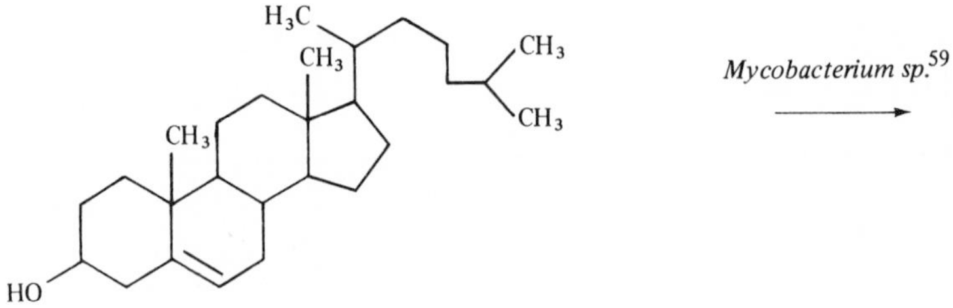

Mycobacterium sp.[59]

Cholesterol

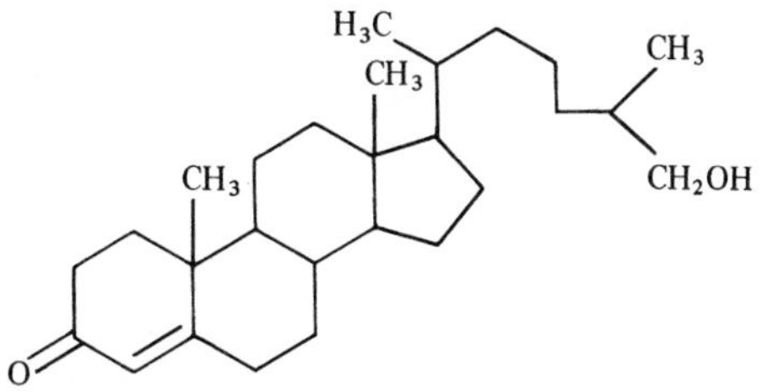

No hydroxylations of intact bile acids are known, but several related bisnor compounds are hydroxylated by microorganisms.

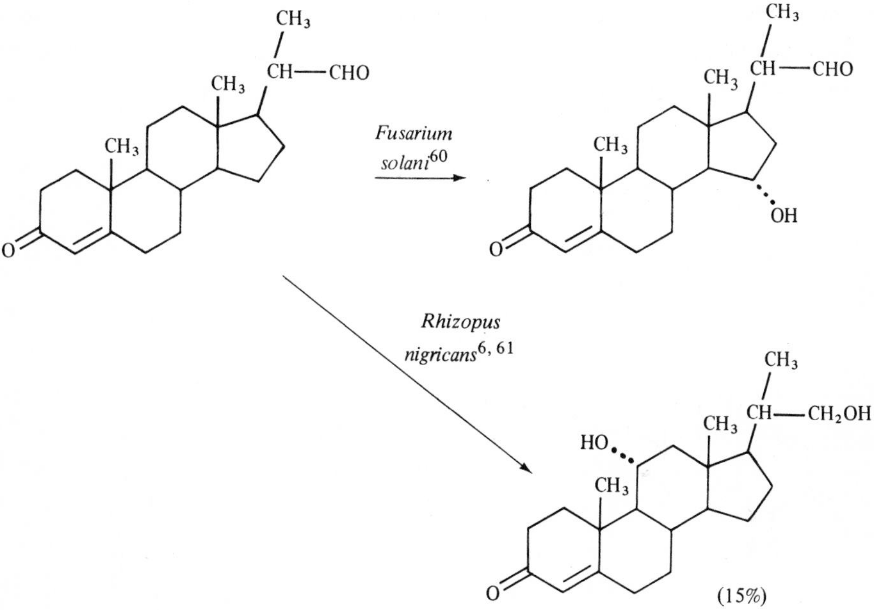

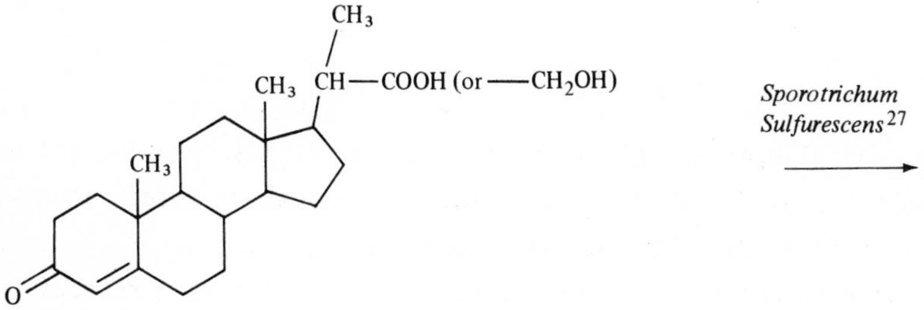

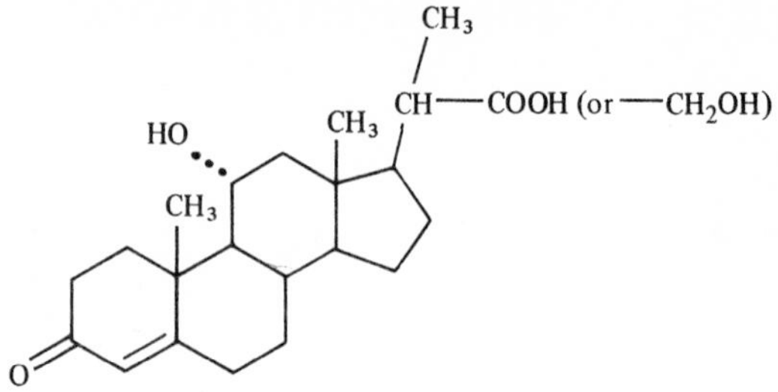

The sterol antibiotic fusidic acid undergoes 6α-hydroxylation by
Acrocylindrium oryzae, with some oxidation to the 6-keto compound.
The latter also arises from 11α-hydroxylation of 7-deacetoxy-1,2-di-
hydrohelvolic acid by *Fusidium coccineum*.[62]   Similar reactions occur
in the biosynthesis of helvolic (and fusidic) acid by *Cephalosporium
caerulans*.[63]

*Cephalosporium
caerulans*[63]

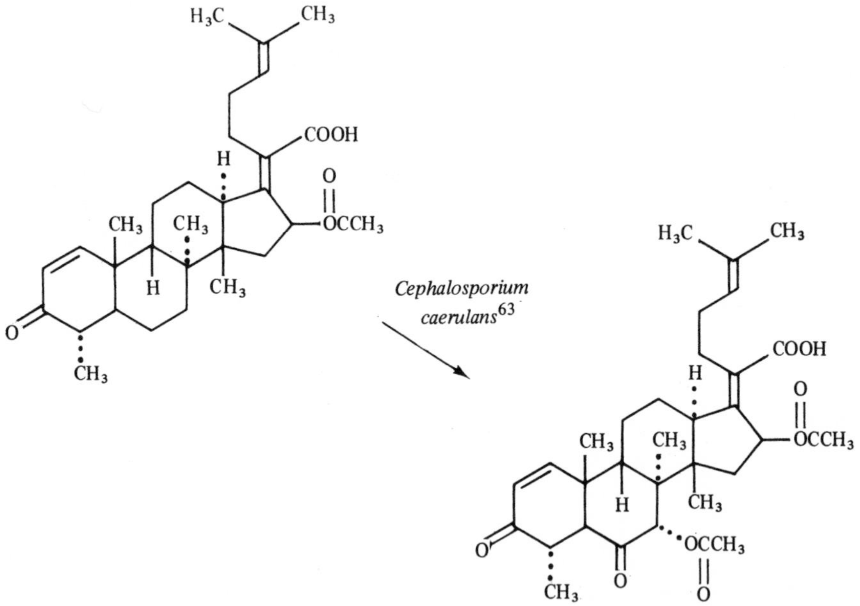

Helvolic acid

Microbial hydroxylations of steroids of the cardenolide and bu-
fadienolide types most frequently involve oxidative attack at positions 7β and/or 12β, although oxidations at 1β, 5β, 11α, 12α, and 16β
have also been observed.   The illustrative examples show respectable

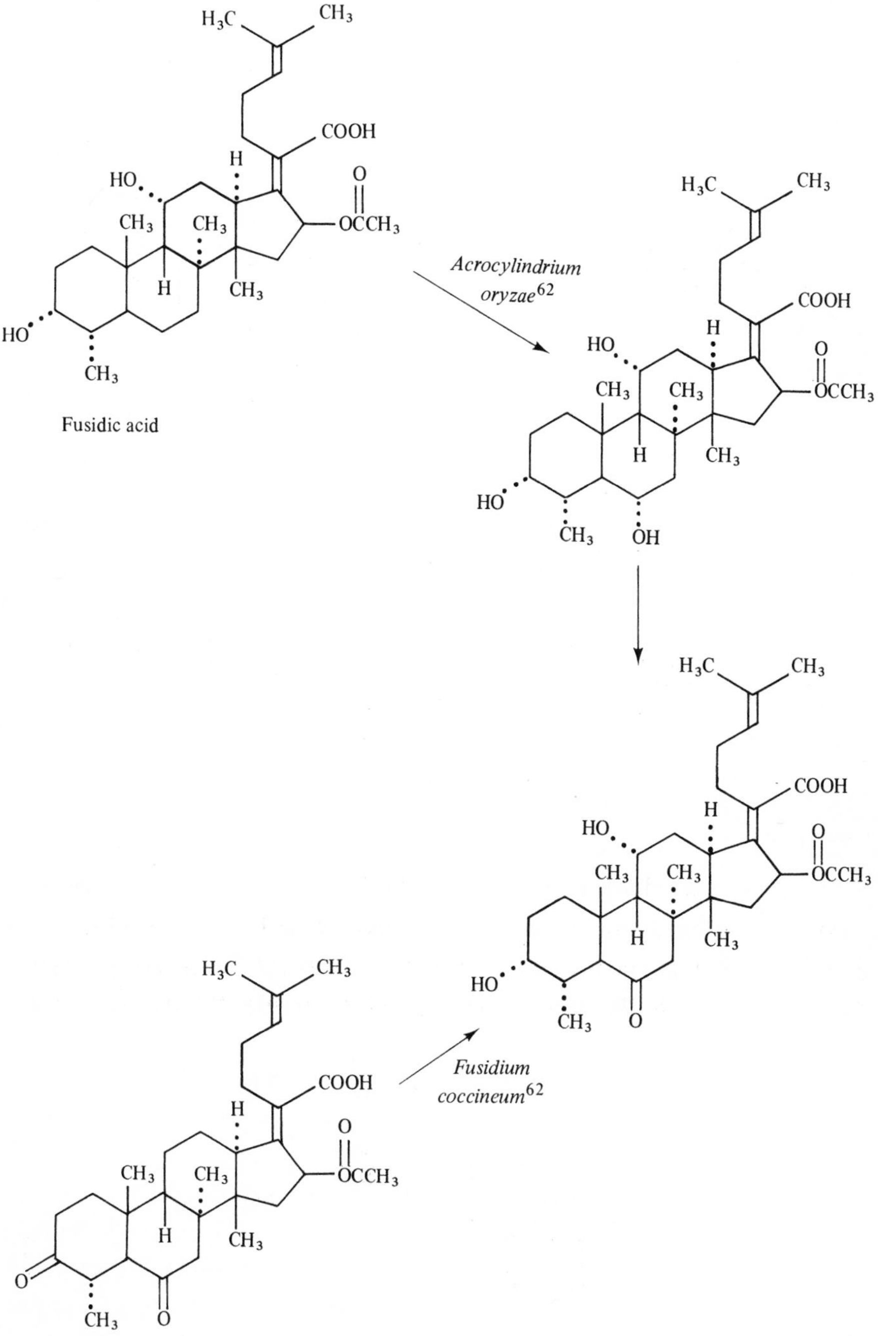

Fusidic acid

Acrocylindrium oryzae[62]

Fusidium coccineum[62]

7-Deacetoxy-1,2-dihydrohelvolic acid

yields, but in most other instances, the yields have been quite low.

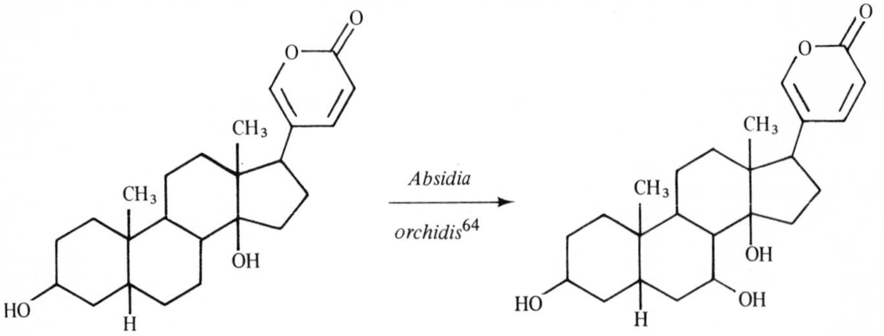

Bufalin   (~65%)

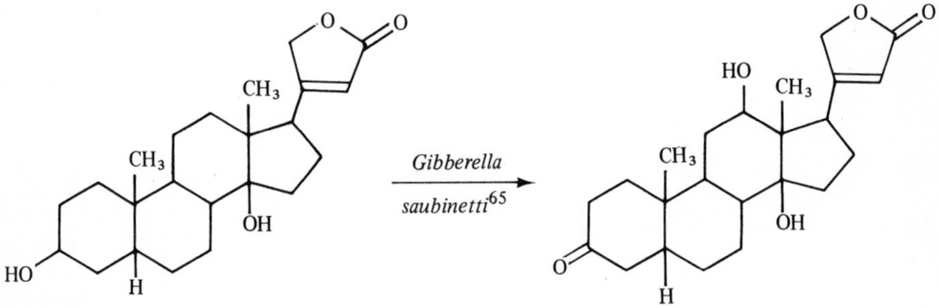

Digitoxigenin   (50%)

Several steroidal alkaloids may be hydroxylated using microorganisms. Solanidine, in a very slow oxidation by *Helicostylum piriforme*, gives the 11α-hydroxy derivative in about 10% yield,[66] while solasodine, with the same organism, gives largely the 9α-hydroxy de-

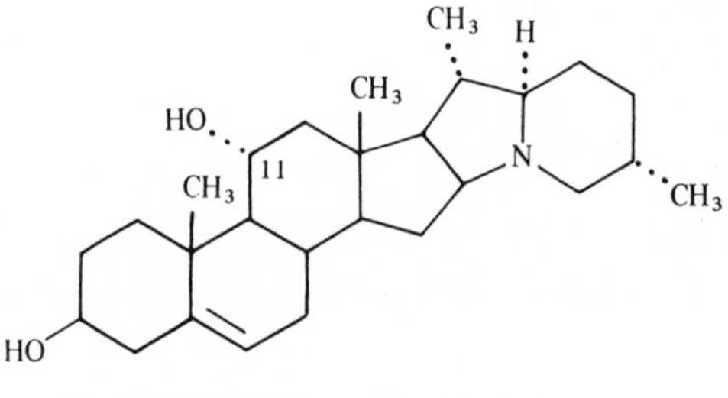

11α-Hydroxysolanidine

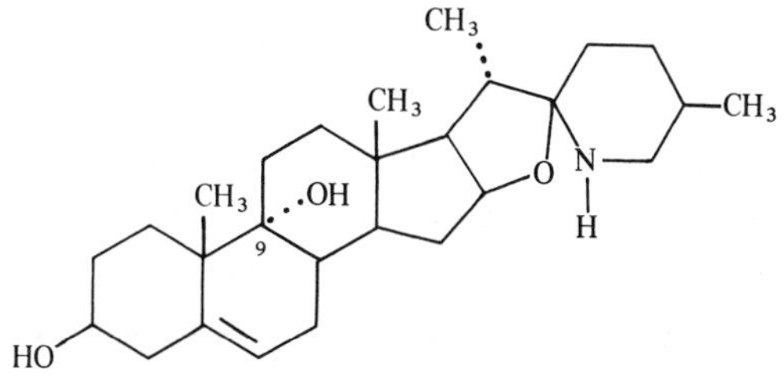

9α-Hydroxysolasodine

rivative in 30-35% yield.[67] Tomatidine undergoes 7α-hydroxylation in
5% yield, also with the same organism.

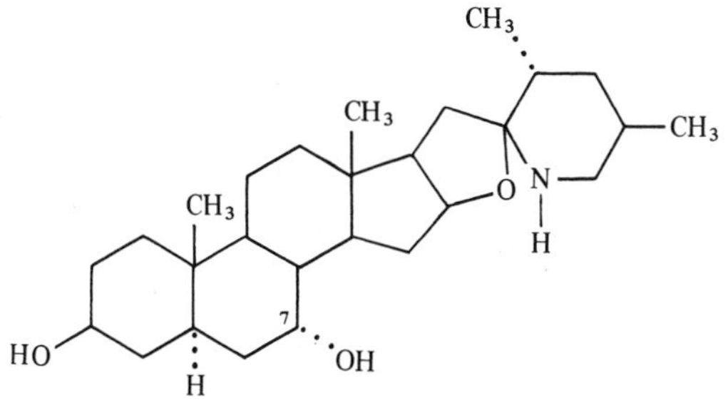

7α-Hydroxytomatidine

ISOPRENOIDS

In contrast to the almost monumental efforts expended on the microbial oxygenations of steroids, studies of terpenoid oxygenations
have been extremely hit-or-miss.   Metabolism of some compounds, such
as camphor, has been studied in considerable detail.   Terpenes such
as the pinanes, limonene, and menthane are oxygenated at numerous positions (either olefinic or allylic) by various microorganisms, but
yields are discouraging from a synthetic point of view.   Some well-
known terpenoids, such as abietic acid, the gibberellins, and fenchone have received more recent, but hardly extensive attention. Other
terpenes such as caryophyllenes, cedranes, and the diterpenes (of
which some forms are commercially available), other than abietic
acid, have received little attention.   The following discussion is

divided into the well-known terpene classes based on the number of isoprene units.

   Terpenes (C$_{10}$). The pathways of microbiological degradation of camphor have been studied in some detail at the University of Illinois. One organism, *Pseudomonas putida* (strain C$_1$), isolated from sewage sludge, oxidizes (+)-camphor to a mixture containing 5-*endo*- and 5-*exo*hydroxycamphor, 2,5-diketocamphane, and other hydroxyketones oxidizable to 2,5-diketocamphane.[68] Strain C$_1$ also oxidizes (-)-cam-

(+)-Camphor   2,5-Diketocamphane   5-*endo*-Hydroxycamphor

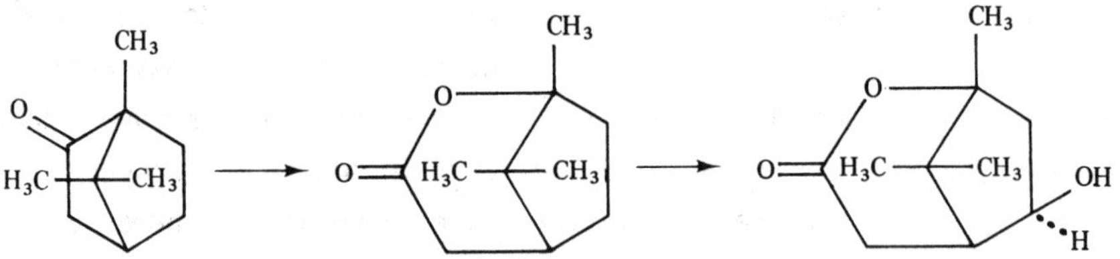

phor to 2,5-diketocamphane. Another *Ps. putida* (strain C$_5$) also degrades camphor but by another route involving 5-*exo*-hydroxylation of the initially formed 1,2-campholide.[69] Unlike the C$_1$ strain, the C$_5$ strain fails to give any 5-*endo*-hydroxylation.

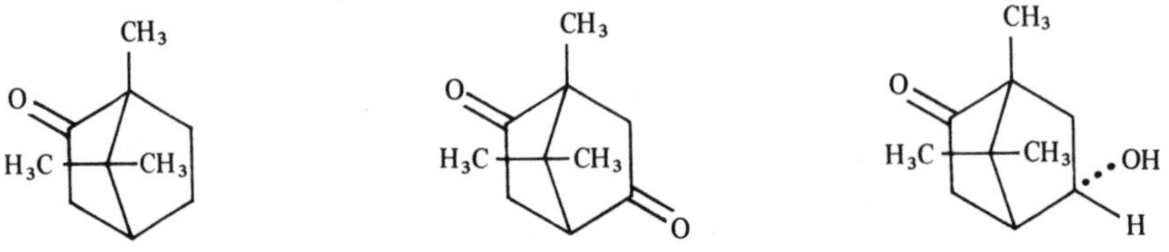

Camphor   1,2-Campholide   5-*exo*-Hydroxy-1,2-campholide

   Still another microorganism, a soil diphtheroid designated as strain T$_1$, degrades (+)-camphor by first producing 6-*endo*-hydroxycamphor and 2,6-diketocamphane.[70]

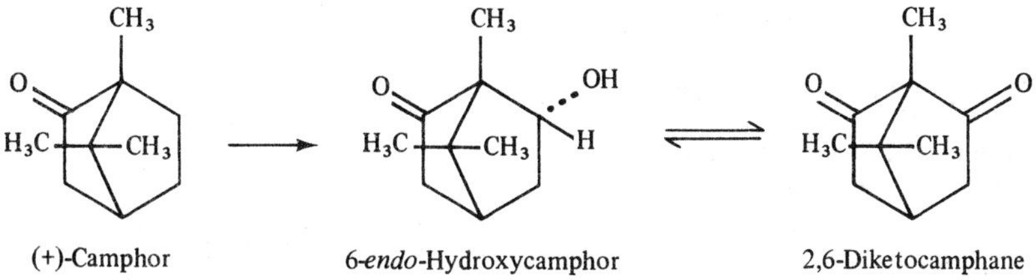

(+)-Camphor · 6-*endo*-Hydroxycamphor · 2,6-Diketocamphane

Although the products of the microbiological oxidations have, in most cases, been isolated and characterized, there have been no yields reported. Judging from some comparative analytical gas chromatographic data,[71] the yields are quite low.

For a convenient overview, the over-all pathway, as summarized by Gunsalus *et al.*,[72] is shown in Fig. 1.

Very similar hydroxylations of the enantiomers of fenchone and isofenchone can be obtained using *Absidia orchidis*, an organism that fails to oxidize camphor.[73,74] Optical integrity is maintained during the hydroxylations, which yield the 5-*exo*- and 6-*exo*-hydroxy derivatives of fenchone, and the 6-*exo*- and 6-*endo*- derivatives of isofenchone, each in about 8% yield.

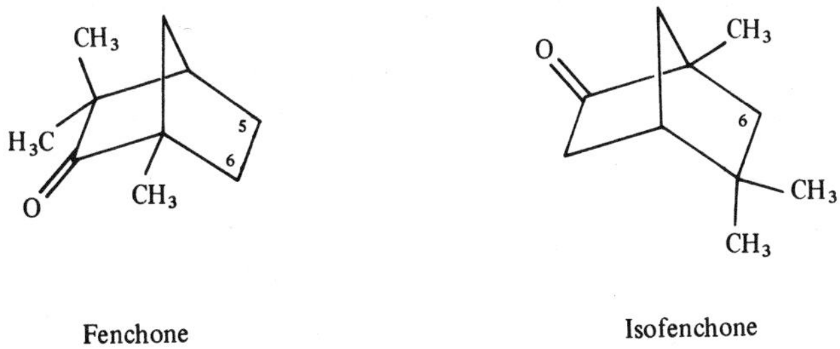

Fenchone · Isofenchone

Sesquiterpenes (C$_{15}$). A most unusual approach to the problem of the structure of guaioxide, a sesquiterpene found in guaiac wood oil, was taken by investigators at the Shionogi Research Laboratory. The oxygenation of guaioxide by *Mucor parasiticus*[75] gives three hydroxylated compounds, isolated and characterized as 4α-hydroxyguaioxide, 8α-hydroxyguaioxide, and 4α,8α-dihydroxyguaioxide, whose further

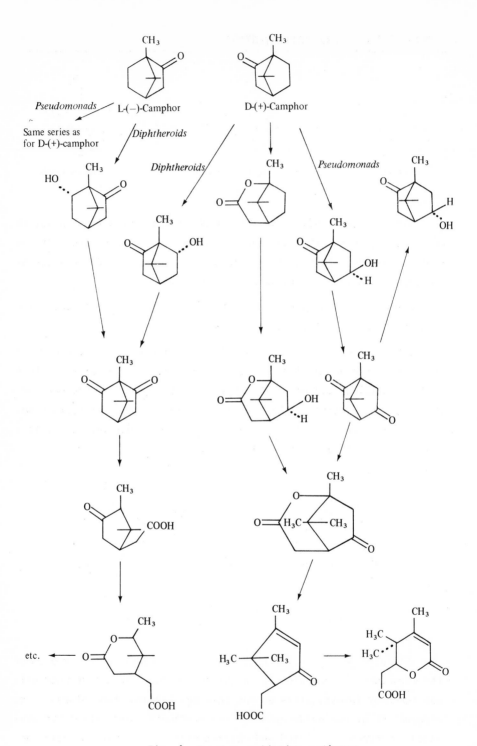

Fig. 1. Camphor oxidation pathways.

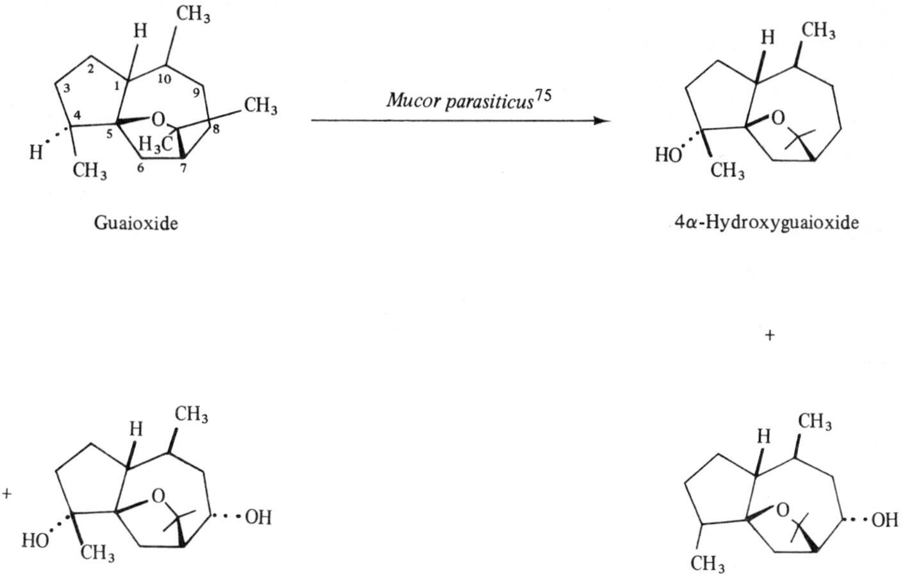

Guaioxide   *Mucor parasiticus*[75]   4α-Hydroxyguaioxide

+

4α,8α-Dihydroxyguaioxide

8α-Hydroxy-
guaioxide

chemical manipulation and appropriate correlation with related compounds of known structure results in the confirmation of the indicated structure for guaioxide.

Japanese valerian species contain the sesquiterpenoids α-kessyl
alcohol and kessyl glycol. While it is possible to convert kessyl
glycol to α-kessyl alcohol chemically, it is extremely difficult to
carry out the reverse transformation selectively by chemical means.
A Tohoku University group has shown that several microorganisms can
convert α-kessyl alcohol to a mixture of kessyl glycol (kessane-2β,
8α-diol) and kessane-2β,7-diol.[76] The indicated ratios of kessyl gly-

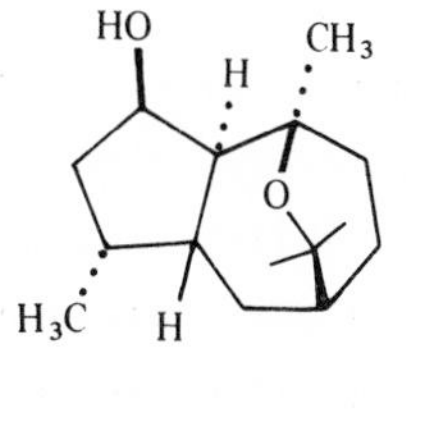

α-Kessyl alcohol 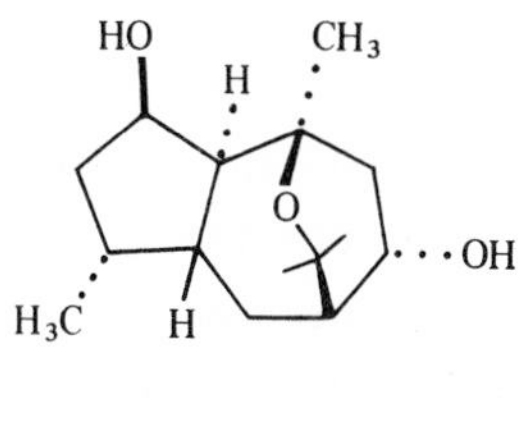   Kessyl glycol 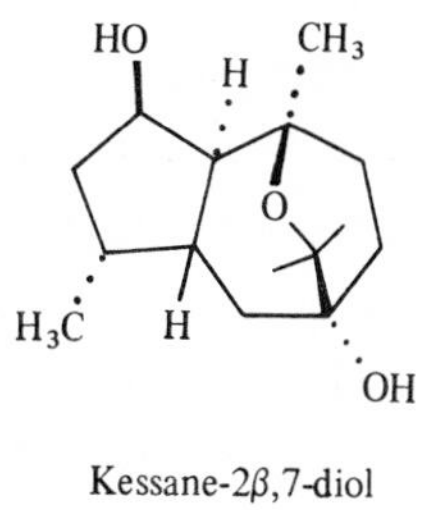   Kessane-2β,7-diol

col: kessane-2β,7-diol (measured by gas chromatography) are obtained with these microorganisms: *Cunninghamella blakesleeana* 35:65; *Corticium sasakii* 50:50; *Corticium centrifugum* 45:55; *Streptomyces aureofaciens* 87:13. The α-kessyl alcohol disappears from the fermentation after a few days (six or less), but the oxidation products are resistant to further attack, either in the same fermentation or when introduced to a fresh culture. The yields of the glycols are apparently quite low.

Kessane itself is oxygenated to a mixture of seven products, in unspecified yields, by *Cunninghamella blakesleeana*.[77]   The products include the 7-ol, 8β-ol, 2α-ol, 3ξ-ol, 4β-ol, 9ξ-ol, and the 3ξ,7-diol.

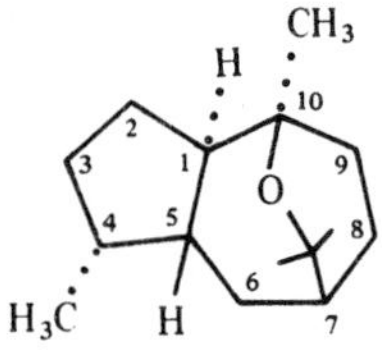

Kessane

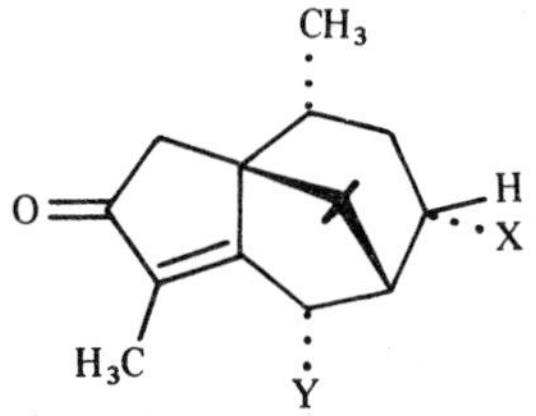

Cyperotundone   X = Y = H
Sugeonol   X = H, Y = OH
Isopatchoul-4-en-3-on-8α-ol   X = OH, Y = H

The Tohoku University group has also shown that the sesquiterpene, cyperotundone, a constituent of a Japanese nutgrass, is converted to two monohydroxylated products, sugeonol (13%) and isopatchoul-4-en-3-on-8α-ol (18%) by the organism *Corticium sasakii*.[78]

Diterpenes (C_{20}). A microorganism (*Flavobacterium resinovorum*, isolated from the soil of a French pine forest) that may be responsible for the biodegradation of oleoresins in nature oxidizes dehydroabietic acid to a mixture of trace products, of which one, representing hydroxylation at C-3 followed by oxidation to the ketone and decarboxylation, is obtained in 0.06% yield.[79] Since a variety of resin

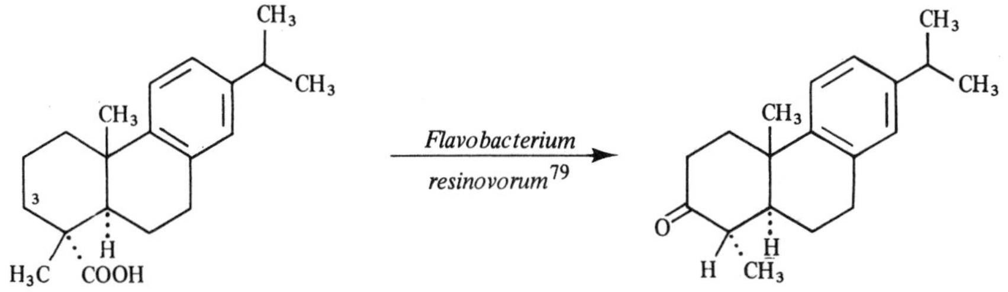

Dehydroabietic acid

acids and noncarboxylic related substances can be utilized as nutrients both by the *Flavobacterium* and by *Pseudomonas resinovorans*,[80] it seems likely that other oxidative products could be found in this series. Indeed, the incubation of methyl dehydroabietate with *Corticium sasakii*[81] gives methyl 3β-hydroxydehydroabietate (12% yield) and the diol methyl 3β,7β-dihydroxydehydroabietate (25% yield). The dihydroxy

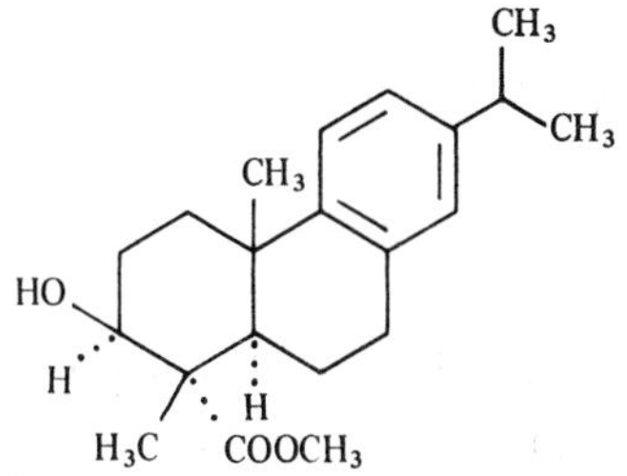

Methyl 3β-hydroxydehydroabietate 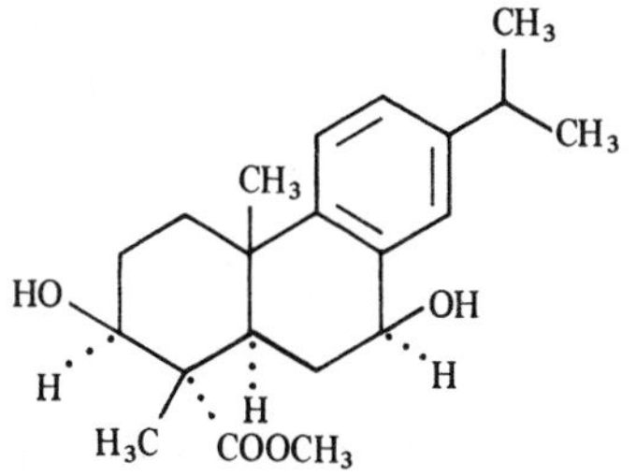   Methyl 3β,7β-dihydroxydehydroabietate

ester arises from further oxygenation of the monohydroxy ester and also from the microbiological oxygenation of methyl 7β-hydroxydehydroabietate. The carbomethoxy group remains intact in these transformations, but the possibility that some of the uncharacterized polar products that are formed are in fact the carboxylic acids has not been excluded. Since fungi usually have very effective esterases, this retention of the ester function illustrates an interesting selective reactivity of a microorganism with regard to a particular substrate.

Abietic acid itself is oxidatively attacked by an *Alcaligenes* species from soil, that converts it in infinitesimal yield to 5α-hydroxyabietic acid and other products.[82]

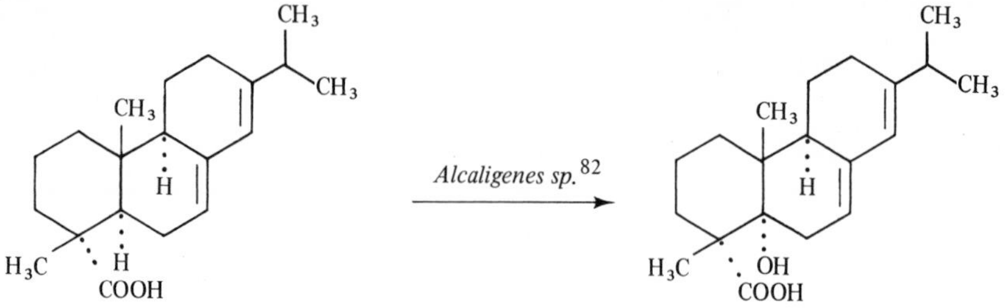

Abietic acid

Gibberellic acid, one of a class of nor-diterpenes obtained from *Gibberella fujikuroi*, causes an unusual growth stimulation of some higher plants. Because of the potential applicability to agriculture, and because of the inherently interesting structure of these compounds, extensive studies of its biosynthesis have been carried out. Although the yields of individual oxidative steps in this complex series of oxidations are quite low, and in fact are generally expressed in the form of percentage radioisotope incorporation, the reactions are potentially useful from a synthetic point of view, either in this series of compounds or in a related series, particularly if strains of microorganisms could be found that, by virtue of an appropriate biochemical characteristic, permitted accumulation of a desired product. An alternate approach, namely modification of the substrate to enhance the efficiency of its microbiological oxygenation, may also be considered. For example, although the conversion of gibberellin A_{12} to gibberellic acid proceeds to the extent of only 0.7%, the de-

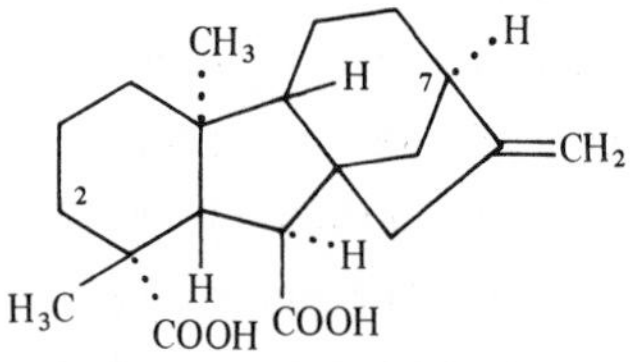

Gibberellin A_{12} 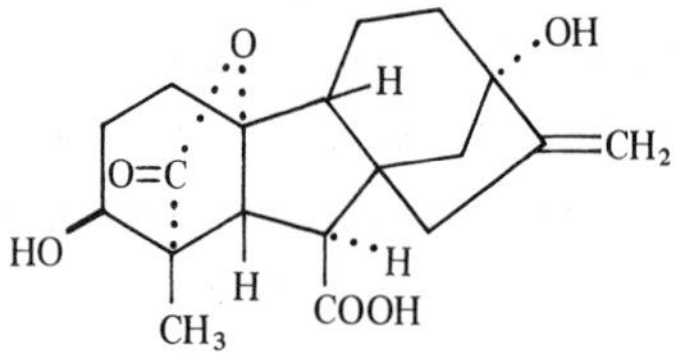   Gibberellic acid

rived diol (obtained by lithium aluminum hydride reduction of gibberellin A_{12}) is incorporated into gibberellic acid to the extent of 7.5%.[83] Since gibberellin A_{13} is obtained in both fermentations, it

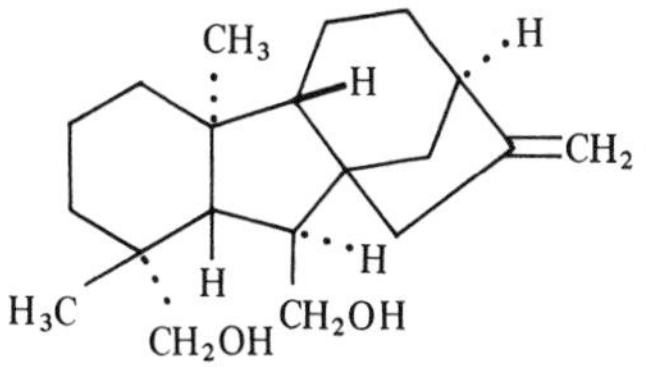

Diol

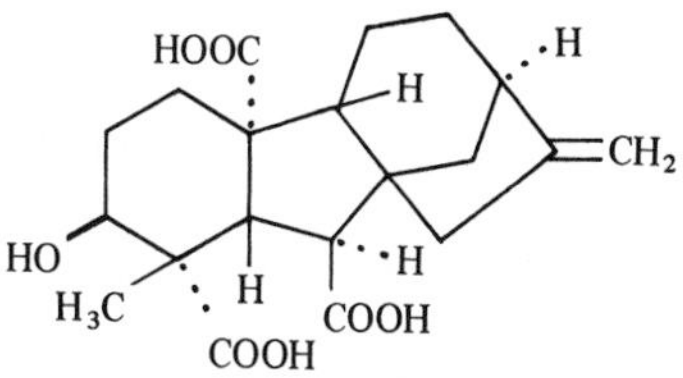

Gibberellin A_{13}

appears that in this complex series of oxidations, the introduction of the hydroxyl at C-2 and the oxidation of the angular methyl group to carboxyl precede the hydroxylation at C-7, the dehydrogenation at C-3 and C-4, and the oxidative decarboxylation at the angular position. Evidently the oxidation of the hydroxymethylene groups of the diol occurs very early in this sequence.

In a somewhat more drastic series of oxidative stages, kaurene undergoes dissimilation by *G. fujikuroi* to gibberellic acid.[84] The

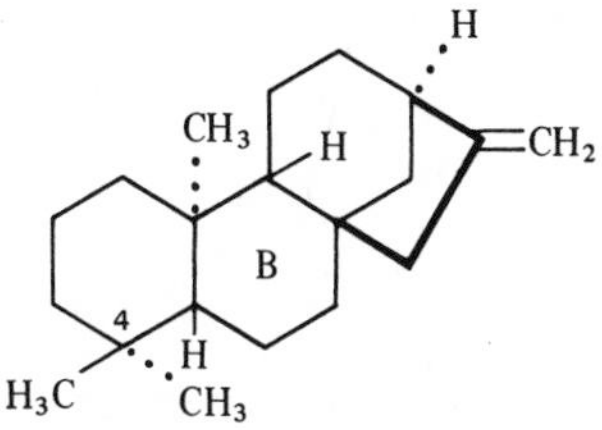

Kaurene

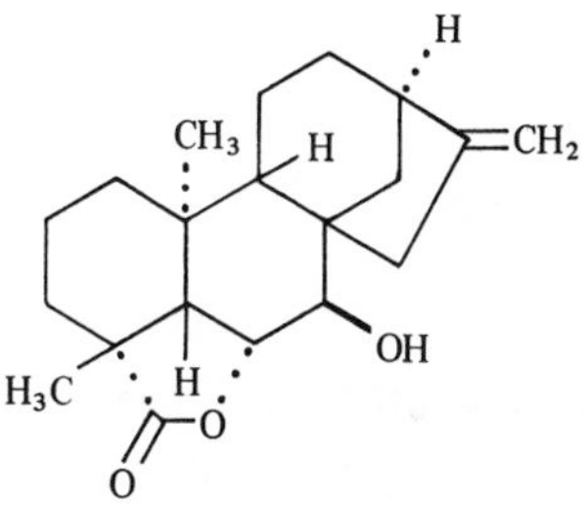

7-Hydroxykaurenolide

same hydroxylations are seen as with gibberellin A_{12}, and in addition the B-ring undergoes oxidative rearrangement, while both of the methyl groups at C-4 are oxidized. The isolation of 7-hydroxy kaurenolide and 7,17-dihydroxykaurenolide (0.05 and 0.44% radioisotope incorporation, respectively) from the same fermentation exemplifies at least three microbial hydroxylations. As the sole example of a potentially reasonably useful (*i.e.*, fair yielding) synthetic process in this series, 7-hydroxykaurenolide itself, when added to a *G. fuji-kuroi* fermentation, is hydroxylated in 43% yield (radioisotope incorporation) to 7,18-dihydroxykaurenolide.

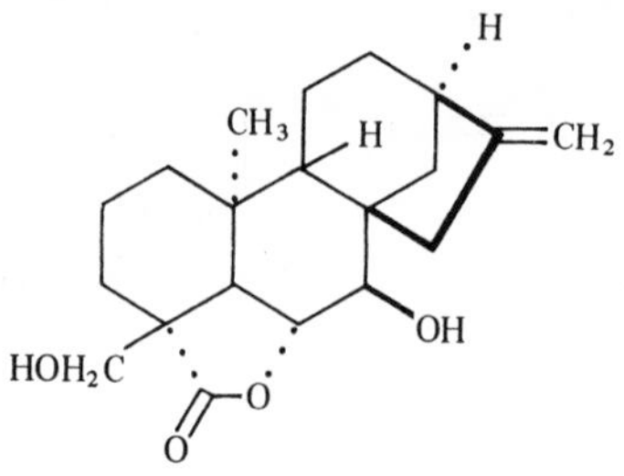

7,18-Dihydroxykaurenolide

Steviol, a diterpene closely related structurally to kaurene and the gibberellins, is hydroxylated at the 6α- and 7β-positions by *G. fujikuroi* to give the indicated lactone in 7% yield.[85]

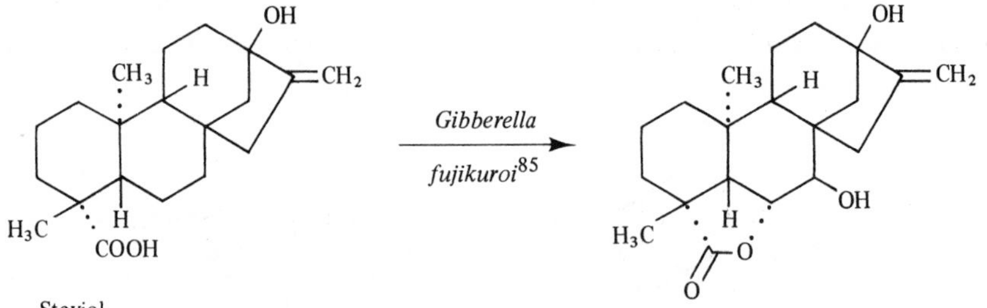

Steviol

Gibberella

fujikuroi[85]

Lactone

Judging from the successes of biotransformations of the tetracyclic diterpenes of the kaurene and gibberellin series by *G. fuji-kuroi*, one might expect this genus of organisms to be a fruitful source of microbial oxygenations of the resin acids.

Sesterterpenes (C$_{25}$). Studies[86,87] of the biosynthesis of the ophiobolins, which constitute the majority of the known members of this small class of natural terpenes made up of five isoprene units, have shown (by $^{18}O_2$ incorporation experiments) that the 14-hydroxyl and the carbonyl oxygens of ophiobolin A, produced *de novo* by the fungus *Cochliobolus heterostrophus*, or by *C. miyabeanus*, comes from gaseous oxygen, and that the other oxygens arise from the water in the medium. When either ophiobolin C or B is incubated with the culture, ophiobolin A is produced (apparently in minute yield), suggesting that 14α-hydroxylation of ophiobolin C to ophiobolin B is the pathway. It is worth noting that hydroxylation of a methine carbon

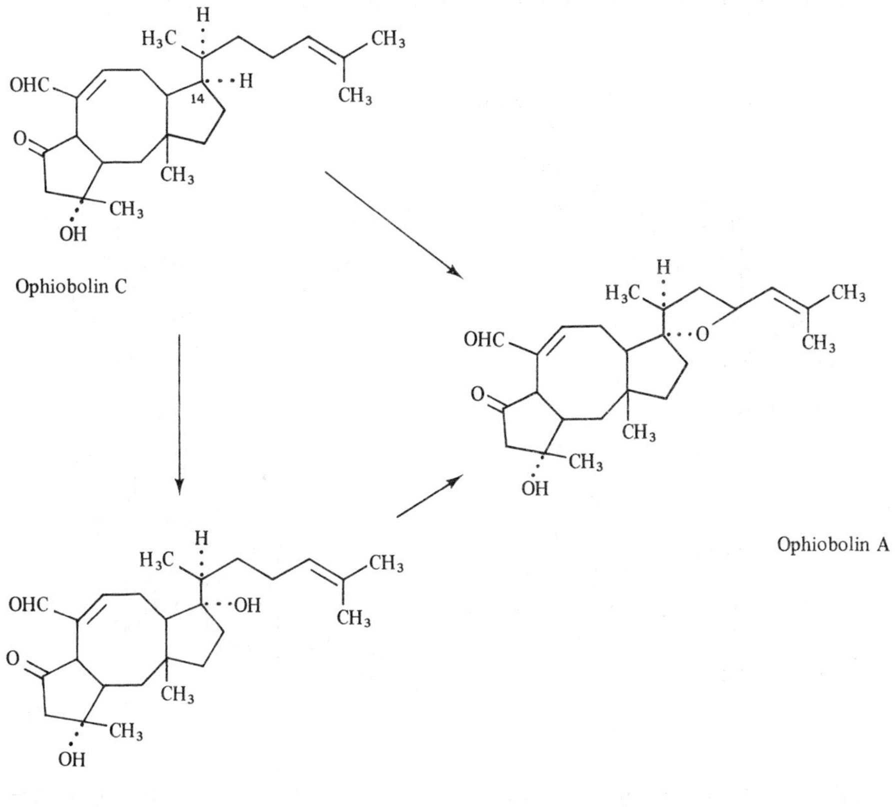

Ophiobolin C

Ophiobolin A

Ophiobolin B

occurs in the presence of an aldehyde function that is itself not oxidized further in the process -- a feat that could hardly be matched
by the classical methods of organic chemistry.

Triterpenes (C_{30}).   In spite of the abundance of triterpenes,
these compounds have yielded very few results in the form of microbial oxidations. Whether this is because few experiments have been done
or because many have failed is, of course, unknown.   In view of the
nature of the successful reports, however, it may be speculated that
little work has been done in this area.

Investigators at the University of Milan used a washed culture
of *Trichothecium roseum* to convert glycyrrhetic acid to 7β-hydroxy-
glycyrrhetic acid (1% yield), 15α-hydroxyglycyrrhetic acid (5% yield)

and 7β,15α-dihydroxyglycyrrhetic acid (30% yield).[88] With *Curvularia*

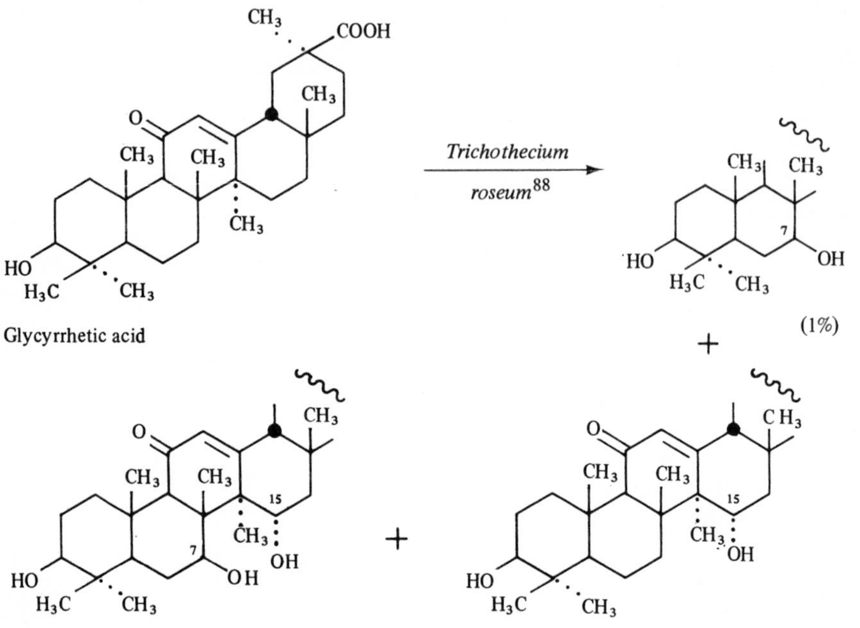

Glycyrrhetic acid

(1%)

+

(30%)

(5%)

lunata, 7β-hydroxyglycyrrhetic acid is obtained in 5% yield from the same substrate, but oxidation of the 3-hydroxyl to ketone constitutes the preponderant reaction (25% yield).[89]

The oxidation of oleanolic acid by *Cunninghamella blakesleeana*[90] gives a multiplicity of products, all in very low yields, including some of the type under discussion here.

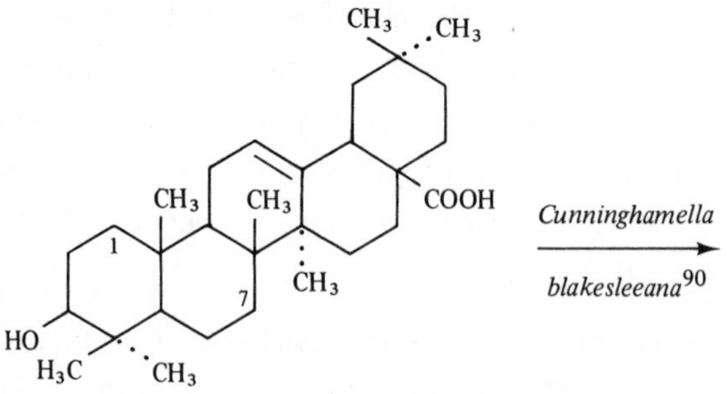

Oleanolic acid

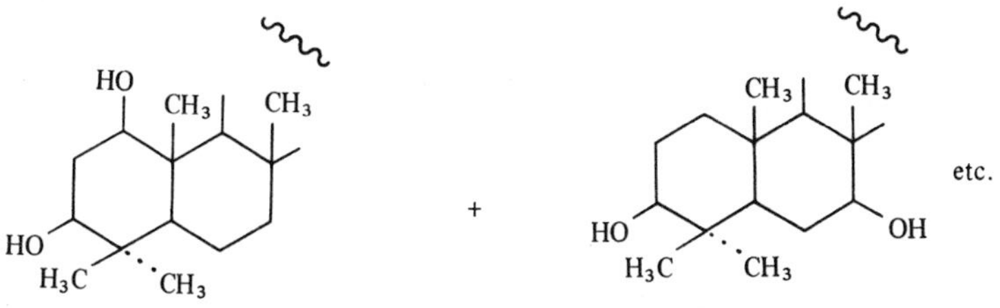

ALKALOIDS

The microbiological oxidation of alkaloids has not had the bene-
fit of the same measure of commercial impetus as have the steroids,
but a few reactions of this kind are known.

Yohimbe Alkaloids. Yohimbine and α-yohimbine, both containing
the indole moiety, whose susceptibility to many chemical oxidative
reactions limits oxidative opportunities at other portions of the
molecule, undergo 18α-hydroxylation with *Streptomyces aureofaciens*
(5% yield), *S. rimosus*,[91,92] and *Calonectria decora*[93] (13% yield).

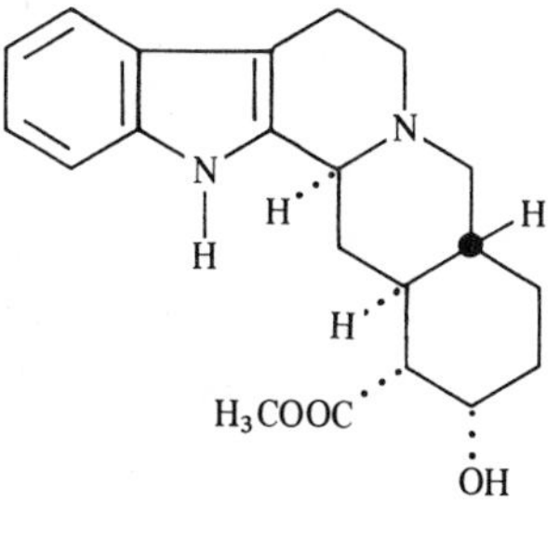

Yohimbine 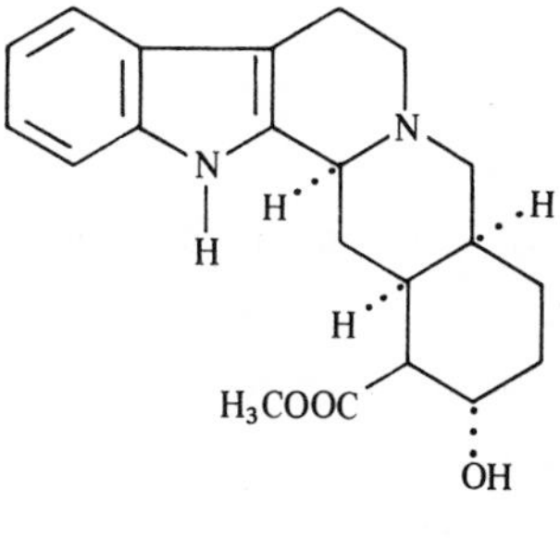   α-Yohimbine

Although the compound is not an alkaloid, its close resemblance
prompts the citation here of the microbiological oxygenation of 1-
acetyl-2,3-cyclooctanoindoline by *Calonectria decora* to the isomeric
ketones A (37% yield) and B (6% yield).[94]

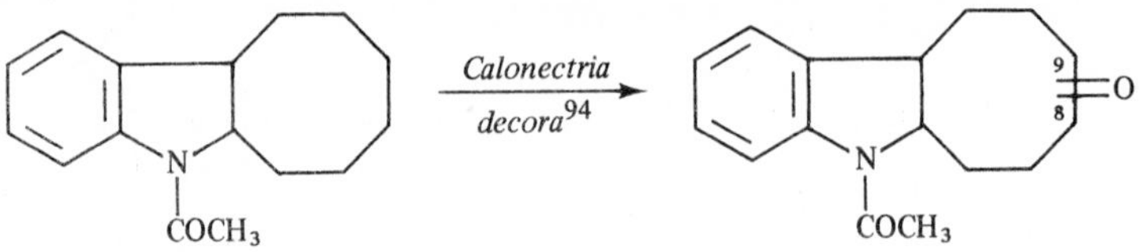

A, 9-Keto
B, 8-Keto

1-Acetyl-2,3-cyclooctanoindoline

Opium Alkaloids. Microbiological oxygenations of opium alkaloids have all involved an allylic (or potentially allylic) position, and are discussed in Chapter 2.

Lupine Alkaloids. Both lupanine and sparteine are attacked by microorganisms. The oxidation of lupanine by *Pseudomonas lupanini* produces 17-hydroxylupanine,[95] probably by way of an initial dehydrogenation, followed by the addition of water to the double bond.

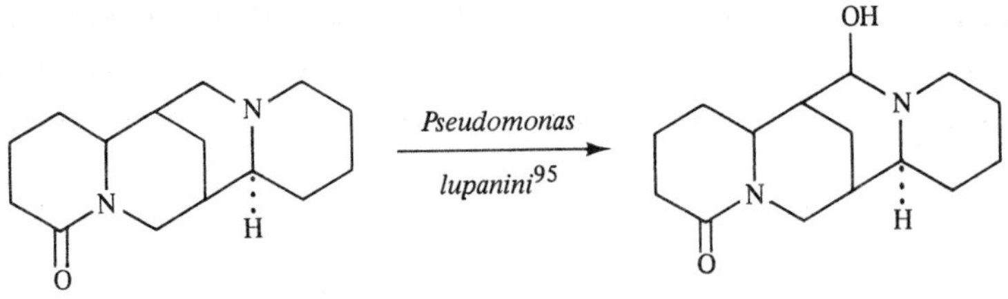

Lupanine

17-Hydroxylupanine

17-Hydroxylupanine is degraded further by the same microorganism. Thus, the assay yield reaches a peak value (55% residual lupanine, 43% 17-hydroxylupanine) after two or three days of fermentation, and then drops off.[96] Since the same net oxidation of lupanine can be carried out chemically,[97] the microbiological method would seem to be of academic interest only, although it suggests the possible application of the method to other alkaloids.

MISCELLANEOUS COMPOUNDS

A variety of other functionally substituted molecules, not readily classified in the categories discussed earlier, have been successfully subjected to oxidation by microorganisms. The following discussion is organized rather loosely on a structural basis, with most acyclic substrates considered first, followed by monocyclic, bicyclic, and finally polycyclic or complex compounds. This mode of organization is not intended to reflect either the historical or the biochemical relationships of these processes, but is designed solely to simplify its perusal and useful evaluation by an organic chemist in search of the right reagent.

Acyclic. Interest in the microbial oxygenations of alkanes has followed a different line of development than in the case of the oxygenation of steroids. Early work on alkane oxygenations was carried out by microbiologists concerned only with whether alkanes (as in petroleum) would support microbial growth. After it was found that various alkanes supported the growth of many microorganisms, interest in the metabolism of the alkane substrates followed. To date, several pathways have been discovered for the metabolism of alkanes, and some of these investigations suggest a great potential for the microbial preparation of chemical intermediates from alkanes. In this section the microbial oxygenation of acyclic alkanes and of the alkane portions of other molecules is discussed.

There are two major pathways for the microbiological metabolism of alkanes. One of these (Fig. 2) involves initial oxygenation at the terminal (or α-) carbon of the alkane chain. The first product that can be isolated in nearly all cases is the terminal alcohol. Any intermediates between the hydrocarbon substrate and hydroxylated product have yet to be firmly characterized, although olefins and hydroperoxides have been suggested. Oxidation of the alcohol may continue, presumably *via* the aldehyde, to the carboxylic acid level of oxidation. At this point further metabolism of the alkyl carboxyl acid may occur *via* β-oxidation (Chapter 7).

A second major pathway in alkane metabolism is oxygenation at the β-carbon of the alkane chain (Fig. 3). The initial products of this pathway are alcohols or ketones.[98] The ketones presumably arise from oxidation of the alcohol by alcohol dehydrogenase enzymes, known to be present throughout cellular systems. Further degradation of the

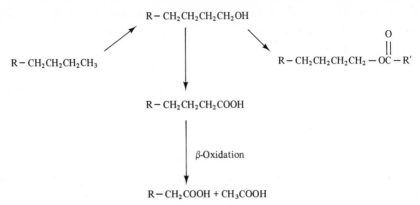

Fig. 2. Oxidative metabolism of alkanes via initial terminal oxygenation.

ketone leads to the formation of a primary alcohol, which has a chain length two carbons shorter than the original substrate.[99] This primary alcohol may then undergo dehydrogenation, as illustrated in Fig. 2, and enter into the fatty acid β-oxidation pathway.

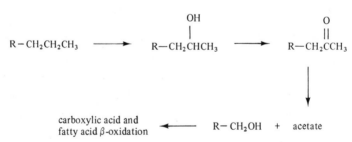

Fig. 3. Oxidative metabolism of alkanes via initial β-oxygenation.

The alcohol formed as the initial product in the hydrocarbon pathway shown in Fig. 2 may combine with a fatty acid to form an ester (often described as a wax), which can be saponified to provide a convenient source of the alcohol. Hexadecyl palmitate, for example, is formed in 30% yield from the oxidation of hexadecane by *Micrococcus cerifans*,[100] and similar oxidations occur with decane, dodecane, tetradecane, and octadecane[101] (Fig. 4). The yield of ester increases

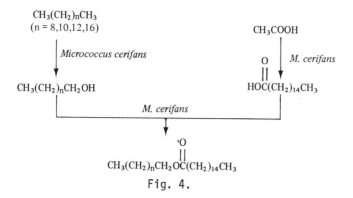

Fig. 4.

with increasing chain length of the substrate, and the appropriate alkyl palmitate is obtained in each case, with the exception of octadecane, which gives an equal amount of octadecyl stearate and of octadecyl palmitate. (The acyl moiety is built up by condensation of acetate units, some of which are undoubtedly derived from the hydrocarbon substrate.[102])

Both terminal and subterminal hydroxylation are seen in the oxygenation of a series of alkanes (C_{14}-C_{18}) by *Candida lipolytica*.[103] The ratio of primary to secondary alcohol is about 2:1. The same organism also oxygenates the analogous 1-alkenes, giving diols and oxidation products thereof from attack at the double bond, as well as alcohols from attack at the saturated end.[103,104]

The inherent compatibility, and possibly even identity, of science and recreation is illustrated by a pseudomonad "isolated from oily mud on the underside of a motorcycle gear box" that can utilize paraffinic hydrocarbons.[105] Cells of this organism grown, for example, on octadecane, are almost one-third lipid; this lipid is rich in octadecanol, which is obtained in up to 15% yield.

Terminal oxidation may occur at both ends of an alkane chain. *Corynebacterium* cultures[106] bring about the oxidations shown in Table 4, and a pseudomonad[107] is known to oxidize octane to suberic acid (isol.) and a lesser amount of adipic acid (isol.).

Good yields of monocarboxylic acids are obtained from the oxidation of heptane and 2-methylhexane by *Pseudomonas aeruginosa*.[108] The

Table 4. Diterminal Oxidation of C_{10}-C_{14} Hydrocarbons by a *Coryne-bacterium*[106]

Substrate	Acid product(s)
Decane	10-Hydroxydecanoic
	1,10-Decanedioic
Dodecane	3-Hydroxydodecanoic
	12-Hydroxydodecanoic
	12-Oxododecanoic
	1,12-Dodecanedioic
Tridecane	1,13-Tridecanedioic
Tetradecane	1,14-Tetradecanedioic

former substrate gives propionic acid (60% yield), while isovaleric acid (30-40%) and 2-methylhexanoic acid (3-7%) are obtained from the latter.   The method used to obtain these yields is more notable than the actual products obtained.   The cultures are initially grown on hexane (when heptane is used as substrate) or on heptane (when 2-methylhexane is used).   Then, a small amount of chloramphenicol is added along with the substrate and the bioconversion allowed to proceed.   This procedure is thought to cause induction of enzymes that are relatively specific for the first substrate but that are also capable of oxygenating the closely related substrates added subsequently.   The chloramphenicol presumably inhibits the ability of the organism's genetic material to make adaptations in the enzymes necessary to the continuation of the metabolism of the intermediate products derived from the substrate molecules.   The potential of techniques such as this to block the further metabolism of useful chemical intermediates remains to be explored.

It is both interesting and important that the microorganism itself may contrive to remove an oxidized substrate from the jeopardy of further metabolism by its own mechanisms. Consider *Torulopsis apicola*.[109,110] Acidic methanolysis of the oily extracellular glycolipid, produced by the normal growth of this yeast, gives predomini-

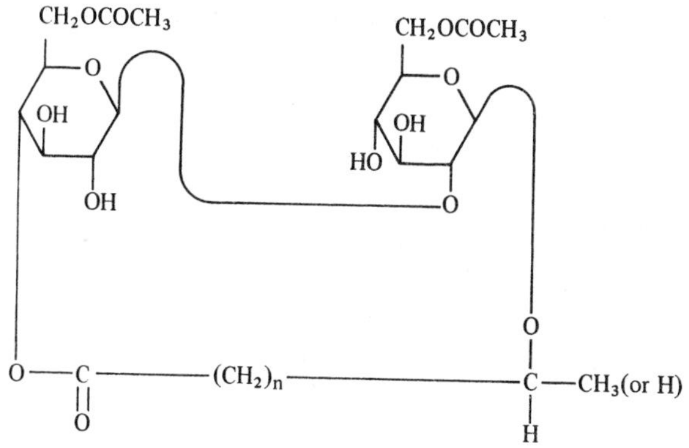

Torulopsis Glycolipid

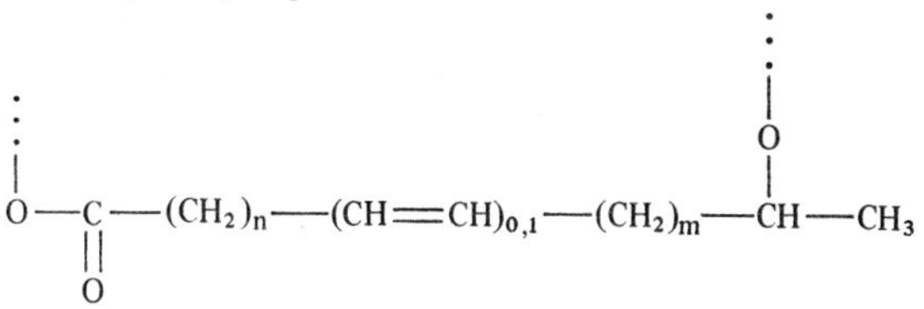

nantly the esters of L-17-hydroxyoctadecanoic acid and L-17-hydroxy-9-octadecenoic acid. When long-chain fatty acids (or esters, including glycerides) or long-chain hydrocarbons are added to the fermentation as substrates, the production of the glycolipid is greatly increased, and its composition bears a direct structural relationship to the added substrate, as illustrated in Table 5. Some of the longer substrates undergo chain-shortening during the fermentation. Since eicosenoic acid and erucic acid both give 17-hydroxyoleic acid, the missing atoms must come from the carboxyl end of the substrate, and it is safe to assume that this is also the case for the other shortened products.

In *Torulopsis apicola*, and also in *T. gropengiesseri*, a pure hydrocarbon substrate evidently first undergoes oxygenation of a terminal methyl group to an alcohol, which is then dehydrogenated to a carboxylic acid. This fatty acid then undergoes further oxidation at the far end of the molecule to give the ω-hydroxy acid, and/or (ω-1)-hydroxyacid. Fatty acid ester substrates are first hydrolyzed and enter into the metabolic scheme at this point, as do fatty acid substrates. An optimum distance between the carboxyl group and the site

Table 5. Fatty Acid and Hydrocarbon Oxygenations by *Torulopsis api-cola*

Substrate	Major Products (>10% assay yields)
$CH_3-(CH_2)_{14}-COOCH_3$	$CH_3-\underset{\underset{OH}{\vert}}{CH}-(CH_2)_{13}-COOCH_3$ (20%) + $HOCH_2-(CH_2)_{14}-COOCH_3$ (29%)
$CH_3-(CH_2)_{16}-COOCH_3$	$CH_3-\underset{\underset{OH}{\vert}}{CH}-(CH_2)_{15}-COOCH_3$ (55%)
$CH_3-(CH_2)_{15}-COOCH_3$	$CH_3-\underset{\underset{OH}{\vert}}{CH}-(CH_2)_{14}-COOCH_3$ (56%)
$CH_3-(CH_2)_{17}-COOCH_3$	$CH_3-\underset{\underset{OH}{\vert}}{CH}-(CH_2)_{16}-COOCH_3$ (33%) + $CH_3-\underset{\underset{OH}{\vert}}{CH}-(CH_2)_{14}-COOCH_3$ (15%)
$CH_3-(CH_2)_{19}-COOCH_3$	$CH_3-\underset{\underset{OH}{\vert}}{CH}-(CH_2)_{14}-COOCH_3$ (32%) + $HOCH_2-(CH_2)_{15}-COOCH_3$ (10%)
$CH_3-(CH_2)_7-CH=CH-(CH_2)_7-$ $COOCH_3$ Methyl oleate	$CH_3-\underset{\underset{OH}{\vert}}{CH}-(CH_2)_6-CH=CH-(CH_2)_7-COOCH_3$ (65%) + $HOCH_2-(CH_2)_7-CH=CH-(CH_2)_7-COOCH_3$ (11%)
$CH_3-(CH_2)_4-CH=CH-CH_2-CH=$ $CH-(CH_2)_7-COOCH_3$ Methyl linoleate	$HOCH_2-(CH_2)_4-CH=CH-CH_2-CH=CH-(CH_2)_7-COOCH_3$ (33%) + $CH_3-\underset{\underset{OH}{\vert}}{CH}-(CH_2)_3-CH=CH-CH_2-CH=CH-$ $(CH_2)_7-COOCH_3$ (22%)
$CH_3-(CH_2)_7-CH=CH-(CH_2)_9-$ $COOCH_3$ Methyl 11-eicosenoate	$CH_3-\underset{\underset{OH}{\vert}}{CH}-(CH_2)_6-CH=CH-(CH_2)_7-COOCH_3$ (29%)
$CH_3-(CH_2)_7-CH=CH-(CH_2)_{11}-$ $COOCH_3$ Methyl erucate	$CH_3-\underset{\underset{OH}{\vert}}{CH}-(CH_2)_6-CH=CH-(CH_2)_7-COOCH_3$ (33%)
$CH_3-(CH_2)_{14}-CH_3$	$CH_3-\underset{\underset{OH}{\vert}}{CH}-(CH_2)_{13}-COOCH_3$ (20%) + $HOCH_2-(CH_2)_{14}-COOCH_3$ (24%)
$CH_3-(CH_2)_{16}-CH_3$	$CH_3-\underset{\underset{OH}{\vert}}{CH}-(CH_2)_{15}-COOCH_3$ (47%)

$CH_3-(CH_2)_{18}-CH_3$ $CH_3-\underset{\underset{OH}{|}}{CH}-(CH_2)_{15}-COOCH_3$ (31%)

$CH_3-(CH_2)_{20}-CH_3$ $CH_3-\underset{\underset{OH}{|}}{CH}-(CH_2)_{15}-COOCH_3$ (43%)

$CH_3-(CH_2)_{22}-CH_3$ $CH_3-\underset{\underset{OH}{|}}{CH}-(CH_2)_{15}-COOCH_3$ (32%)

of ω-oxidation is apparent and can be expressed in terms either of over-all fatty acid chain length (23.3 Å, total length of fatty acid molecule[109]) or of an optimum number (14) of methylene groups[111] between the carboxyl group and the site of hydroxylation. Fatty acids of lengths greater than this optimum undergo β-oxidation, thereby reducing the chain length by two carbons per each β-oxidation, until the optimum length is reached. Fatty acid β-oxidation, of course, represents a competing route for the metabolism of these compounds that is quite effectively defeated by the formation of the glycolipid molecules.

Introduction of methyl groups near the end of a hydrocarbon chain, *i.e.*, branch points in the chain, has a marked effect on the oxygenation of the hydrocarbon by microorganisms. The preferential oxidation of 2-methylhexane by *Pseudomonas aeruginosa*[112] *via* the unsubstituted terminus is analogous to that seen with *Torulopsis gropengiesseri*, which shows a similar preference for initial oxygenation at the least branched end of a variety of substrate molecules (Table 6).[113] This may be followed, however, by further oxygenation (hy-

Table 6. Hydrocarbon Oxidations by *Torulopsis gropengiesseri*

Substrate	Products (yields^a)
$CH_3-\underset{\overset{|}{CH_3}}{CH}-(CH_2)_{13}-CH_3$	$HO-CH_2-\underset{\overset{|}{CH_3}}{CH}-(CH_2)_{13}-COOCH_3$ (27%)

Substrates	Products (yields[a])
CH_3 $CH_3-\overset{\displaystyle CH_3}{\underset{\displaystyle CH_3}{C}}-(CH_2)_{13}-CH_3$	$HO-CH_2-\overset{\displaystyle CH_3}{\underset{\displaystyle CH_3}{C}}-(CH_2)_{13}-COOCH_3$ (35%)
$CH_3-\overset{\displaystyle CH_3}{CH}-(CH_2)_{12}-\overset{\displaystyle CH_3}{CH}-CH_3$	$\Big[HO-CH_2-\overset{\displaystyle CH_3}{CH}-(CH_2)_{12}-\overset{\displaystyle CH_3}{CH}-CH_3$ (12%) $\Big\{HO-CH_2-\overset{\displaystyle CH_3}{CH}-(CH_2)_{12}-\overset{\displaystyle CH_3}{CH}-CH_2OH$ (20%) $\Big[CH_3OOC-\overset{\displaystyle CH_3}{CH}-(CH_2)_{12}-\overset{\displaystyle CH_3}{CH}-CH_2OH$ (20%)
$CH_3-(\overset{\displaystyle CH_3}{CH}-CH_2CH_2CH_2)_3-\overset{\displaystyle CH_3}{CH}-CH_3$	$\Big[HO-CH_2(\overset{\displaystyle CH_3}{CH}-CH_2CH_2CH_2)_3-\overset{\displaystyle CH_3}{CH}-CH_3$ $\Big\{HO-CH_2(\overset{\displaystyle CH_3}{CH}-CH_2CH_2CH_2)_3-\overset{\displaystyle CH_3}{CH}-CH_2OH$ (16%)
$CH_3-(\overset{\displaystyle CH_3}{CH}-CH_2CH_2CH_2)_3-\overset{\displaystyle CH_3}{CH}-CH_2CH_3$	$\Big[HOCH_2(\overset{\displaystyle CH_3}{CH}-CH_2CH_2CH_2)_3-\overset{\displaystyle CH_3}{CH}CH_2CH_3$ $\Big\{CH_3-(\overset{\displaystyle CH_3}{CH}-CH_2CH_2CH_2)_3-\overset{\displaystyle CH_3}{CH}-CH_2CH_2OH$ (above two products yield 20%) $\Big[HOCH_2-(\overset{\displaystyle CH_3}{CH}-CH_2CH_2CH_2)_3-\overset{\displaystyle CH_3}{CH}-CH_2-$ CH_2OH (16%)

a Assay yields, corrected for recovered substrate.

droxylation) at the more highly branched end. When initial oxygenation does occur at the more highly branched end, the resultant alcohol is fairly resistant to further oxidation to the carboxylic acid level (Tables 7 and 8).

Table 7. Fatty Alcohol Oxygenations by *Torulopsis gropengiesseri*

Substrate	Products (yields[a])
$CH_3-(CH_2)_{13}-\overset{\overset{\displaystyle CH_3}{\mid}}{CH}-CH_2OH$	$CH_3OOC-(CH_2)_{13}-\overset{\overset{\displaystyle CH_3}{\mid}}{CH}-CH_2OH$ (10%)
$CH_3-(CH_2)_{13}-\overset{\overset{\displaystyle CH_3}{\mid}}{\underset{\underset{\displaystyle CH_3}{\mid}}{C}}-CH_2OH$	$CH_3OOC-(CH_2)_{13}-\overset{\overset{\displaystyle CH_3}{\mid}}{\underset{\underset{\displaystyle CH_3}{\mid}}{C}}-CH_2OH$ (12%) $CH_3\underset{\underset{\displaystyle OH}{\mid}}{CH}-(CH_2)_{12}-\overset{\overset{\displaystyle CH_3}{\mid}}{\underset{\underset{\displaystyle CH_3}{\mid}}{C}}-CH_2OH$ (10%)
$CH_3-(CH_2)_{13}-\overset{\overset{\displaystyle CH_3}{\mid}}{\underset{\underset{\displaystyle CH_3}{\mid}}{C}}-CH_2CH_2OH$	$CH_3OOC-(CH_2)_{13}-\overset{\overset{\displaystyle CH_3}{\mid}}{\underset{\underset{\displaystyle CH_3}{\mid}}{C}}-CH_2CH_2OH$ (12%)
$CH_3-(\overset{\overset{\displaystyle CH_3}{\mid}}{CH}-CH_2CH_2CH_2)_3-\overset{\overset{\displaystyle CH_3}{\mid}}{CH}-CH_2CH_2OH$	$HOCH_2-(\overset{\overset{\displaystyle CH_3}{\mid}}{CH}-CH_2CH_2CH_2)_3-\overset{\overset{\displaystyle CH_3}{\mid}}{CH}-$ CH_2CH_2OH (16%)

[a] Assay yields, corrected for recovered substrate.

Table 8. Fatty Acid Oxygenations by *Torulopsis gropengiesseri*

Substrate	Products (yields[a])
$CH_3(CH_2)_{13}-\overset{\overset{\displaystyle CH_3}{\mid}}{CH}-COOCH_3$	$HOCH_2(CH_2)_{13}-\overset{\overset{\displaystyle CH_3}{\mid}}{CH}-COOCH_3$ (15%) $CH_3\underset{\underset{\displaystyle OH}{\mid}}{CH}(CH_2)_{12}-\overset{\overset{\displaystyle CH_3}{\mid}}{CH}-COOCH_3$ (12%)
$CH_3\overset{\overset{\displaystyle CH_3}{\mid}}{CH}-(CH_2)_{13}-COOCH_3$	$HOCH_2\overset{\overset{\displaystyle CH_3}{\mid}}{CH}-(CH_2)_{13}-COOCH_3$ (40%)
$CH_3(CH_2)_{13}-\overset{\overset{\displaystyle CH_3}{\mid}}{\underset{\underset{\displaystyle CH_3}{\mid}}{C}}-COOCH_3$	$CH_3\underset{\underset{\displaystyle OH}{\mid}}{CH}-(CH_2)_{12}-\overset{\overset{\displaystyle CH_3}{\mid}}{\underset{\underset{\displaystyle CH_3}{\mid}}{C}}-COOCH_3$ (20%)

Substrate	Products (yields^a)
CH_3 $\mid$ $CH_3-(CH_2)_{13}-C-CH_2-COOCH_3$ $\mid$ CH_3	CH_3 $\mid$ $CH_3-CH-(CH_2)_{12}-C-CH_2COOCH_3$ (30%) $\mid$ $\quad\quad\quad\quad$ $\mid$ OH $\quad\quad\quad\quad\quad$ CH_3
CH_3 $\quad\quad\quad$ CH_3 $\mid$ $\quad\quad\quad\quad$ $\mid$ $CH_3-(CH-CH_2CH_2CH_2)_3-CHCH_2COOCH_3$	CH_3 $\quad\quad\quad\quad$ CH_3 $\mid$ $\quad\quad\quad\quad\quad$ $\mid$ $HOCH_2-(CH-CH_2CH_2CH_2)_3-CH-$ CH_2COOCH_3 (35%)

a Assay yields, corrected for recovered substrate.

Hydrocarbons having terminal halogen substituents (with the exception of fluoroalkanes) serve as excellent substrates for oxidation by *T. gropengiesseri*, giving exclusively α,ω-dicarboxylic acids in assay yields of 15-50% (after methanolysis of the glycolipid)[114] (Table 9). Initial oxidation appears to occur at the terminal methyl

Table 9. Alkyl Halide Oxidations by *Torulopsis gropengiesseri*

Alkyl Halide Substrate	α,ω-Dicarboxylic Methyl Ester Product(s) (yield^a)
C_{15} Pentadecyl bromide	C_{15} (trace)
C_{16} Hexadecyl iodide	C_{16} (43%)
C_{16} Hexadecyl bromide	C_{16} (50%)
C_{16} Hexadecyl chloride	C_{16} (33%)
C_{16} Hexadecyl fluoride	C_{16} (9%)
C_{17} Heptadecyl bromide	C_{17} (45%)
C_{17} 1-Bromo-2-methylhexadecane	C_{17} (50%)
C_{18} Octadecyl iodide	C_{18} (15%) + C_{16} (30%)
C_{18} Octadecyl bromide	C_{18} (17%) + C_{16} (29%)
C_{18} Octadecyl chloride	C_{18} (18%) + C_{16} (29%)
C_{18} Octadecyl fluoride	C_{18} (9%) + C_{16} (4.5%)
C_{18} Oleyl bromide	Δ^9-C_{18} (26%) + Δ^7-C_{16} (9%)
C_{20} Eicosyl bromide	C_{16} (34%)
C_{22} Docosyl bromide	C_{16} (36%)

a Assay yields, but the C_{16} and C_{17} products were also isolated.

group, leading to ω-halofatty acids which then undergo a second oxidation at the halogen substituted terminal carbon. The ω-halo-ω-hydroxy fatty acid is then readily oxidized to the final dicarboxylic acid level. This sequence is supported by several observations: (a) ω-haloacids are metabolized to α,ω-diacids when used as substrates; and (b) no ω-hydroxyacids (which would be expected if an initial hydrolysis or oxidative attack occurred at the halogen bearing carbon) are obtained from the glycolipid. Fermentation of 2-bromohexadecane gives only a low yield of 15-bromohexadecanoic acid, suggesting inhibition to oxidation by halogen at position C-15; however, fermentation of 1-bromo-2-methylhexadecane gives a 50% yield of 2-methylhexadecane-1,16-dioic acid. Haloalkanes of chain length greater than C_{17} are shortened before being incorporated into the glycolipids. This presumably occurs by fatty acid β-oxidation of the intermediate monoacid, as was suggested for shortening of the chains of unsubstituted hydrocarbons.

The oxidation of 1-cyanohexadecane by *Torulopsis gropengiesseri*[114,115] gives 16-oxohexadecanoic acid (8%), 16-hydroxyheptadecanoic acid (10% assay), 16-cyano-2-hexadecanol (8% assay), and additional trace products. Similar formation of an aldehyde acid -- 16-oxohexadec-9-enoic acid -- is observed when ethyl 16-cyanohexadec-9-enoate is the substrate.[114] Detection of hydrogen cyanide during the

$$CH_3(CH_2)_{15}CN$$
1-Cyanohexadecane $\xrightarrow{\text{\textit{Torulopsis gropengiesseri}}^{114, 115}}$

$$
\begin{cases}
O{=}\overset{\overset{\textstyle H}{|}}{C}{-}(CH_2)_{14}COOCH_3 \quad (8\%) \\
\\
CH_3CH{-}(CH_2)_{14}{-}COOCH_3 \quad (6\%\ assay) \\
\quad\ \ \underset{OH}{|} \\
\\
CH_3{-}CH{-}(CH_2)_{14}{-}CN \quad (8\%\ assay) \\
\qquad\ \underset{OH}{|}
\end{cases}
$$

acidic methanolysis of the glycolipids containing these products supports the suggestion that an intermediate cyanohydrin is the species incorporated in the glycolipid.

$$CH_3\!-\!R\!-\!CH_2CN \longrightarrow HOOC\!-\!R\!-\!CH_2CN \longrightarrow HOOC\!-\!R\!-\!\underset{\underset{OH}{\mid}}{CH}\!-\!CN$$

$$HOOC\!-\!R\!-\!CHO + HCN \xleftarrow{\text{Methanolysis}} \text{Glycolipid}$$

Fermentations of 1-alkoxyalkanes with *T. gropengiesseri*[114] also give good yields (10-60%) of oxygenated products (Table 10). 1-Methoxyalkanes give mainly ω-hydroxyalkanoic acids, presumably from hydrolysis or cleavage of the ω-methoxyalkanoic acid. Both ω-hydroxy-acids and (ω-1)-hydroxyalkoxyacetic acids are obtained from 1-ethoxyalkanes. Finally, 1-propoxyalkanes give products that arise from initial oxidation at either end of the molecule, with subsequent oxidation occurring at the opposite end of the molecule.

Several other alkane derivatives undergo oxidative incorporation into *T. gropengiesseri* glycolipids. The predominant product from 1-nitrohexadecane and 1-methylsulfonyloxyhexadecane is the (ω-1)-hydroxyl derivative.[115] 2-Hexadecanone and 2-heptadecanone are reduced

$CH_3(CH_2)_{15}NO_2$ *Torulopsis gropengiesseri* $CH_3\underset{\underset{OH}{\mid}}{CH}\!-\!(CH_2)_{14}NO_2$ (10% assay)

1-Nitrohexadecane $\longrightarrow$

$+ HO(CH_2)_{16}NO_2$ (5% assay)

$CH_3(CH_2)_{15}\!-\!OSO_2CH_3$ *Torulopsis gropengiesseri* $CH_3\underset{\underset{OH}{\mid}}{CH}(CH_2)_{14}OSO_2CH_3$ (15% assay)

1-Methylsulfonyloxyhexadecane $\longrightarrow$

to the alcohols by *T. gropengiesseri*, and the alcohol is the major (25-28% assay) product, but some further oxidation leads to low yields of ω- and of (ω-1)-oxidized products.[115]

It should be noted with care that the discussion of "optimal" substrate characteristics is strictly applicable only to the reactions with the two *Torulopsis* species. Other organisms may give si-

Table 10. Alkoxyalkane Oxidations by *Torulopsis gropengiesseri*

Substrate	Products (yields[a])
$CH_3O-(CH_2)_{13}CH_3$	$HO-(CH_2)_{13}-COOCH_3$ (16%)
$C_2H_5O-(CH_2)_{13}CH_3$	$HO-(CH_2)_{13}-COOCH_3$ (8%)
$n-C_3H_7O-(CH_2)_{13}CH_3$	$CH_3\underset{\underset{OH}{\mid}}{CH}CH_2O-(CH_2)_{13}-COOCH_3$ (30%)
	(isol.) + $CH_3OOC-(CH_2)_2-O-$
	$(CH_2)_{12}-\underset{\underset{OH}{\mid}}{CH}-CH_3$ (30%)(isol.)
$CH_3O-(CH_2)_{14}CH_3$	$HO-(CH_2)_{14}COOCH_3$ (20%)(isol.)
$CH_3-O-(CH_2)_{15}CH_3$	$HO-(CH_2)_{15}COOCH_3$ (30%)(isol.)
$C_2H_5-O-(CH_2)_{15}CH_3$	$HO(CH_2)_{15}COOCH_3$ (8%)
$n-C_3H_7-O-(CH_2)_{15}CH_3$	$CH_3\underset{\underset{OH}{\mid}}{CH}-CH_2-O-(CH_2)_{15}COOCH_3$ (17%)
	(isol.) + $CH_3OOC(CH_2)_2-O-$
	$(CH_2)_{14}-\underset{\underset{OH}{\mid}}{CH}-CH_3$ (35%)(isol.)
$CH_3-O-(CH_2)_{17}CH_3$	$HO(CH_2)_{15}CH_3$ (15%) + CH_3OOC-
	$(CH_2)_7-\underset{\underset{OH}{\mid}}{CH}CH_3$ (14%)
$C_2H_5-O-(CH_2)_{17}CH_3$	$HO(CH_2)_{13}COOCH_3$ (6%)
$CH_3-O-(CH_2)_8-CH=CH(CH_2)_7CH_3$	$HO-(CH_2)_8-CH=CH-(CH_2)_7COOCH_3$
	(16%) + $CH_3OOC-(CH_2)_7-CH=CH-$
	$(CH_2)_6\underset{\underset{OH}{\mid}}{CH}CH_3$ (24%)
	+ $HO(CH_2)_8-CH=CH-(CH_2)_5COOCH_3$
	(9%)
$CH_3OOC-CH_2-O-(CH_2)_{14}CH_3$	$CH_3OOC-CH_2O-(CH_2)_{14}CH_2OH$
	+ $CH_3OOC-CH_2-O-(CH_2)_{13}\underset{\underset{OH}{\mid}}{CH}CH_3$
	(isol.)

[a] Assay yields (some products were also isolated, but in unspecified yields). Products formed in trace amounts are not shown.

milar reactions, but with different optima. For example, the stereospecific synthesis of L(+)-β-hydroxyisobutyric acid from isobutyric
acid with *Pseudomonas putida* in 53% yield[116] represents ω-hydroxyla-
tion, but clearly not at the end of a 23 Å chain length!

Isobutyric acid   L (+)-β-Hydroxyisobutyric acid

   Similarities and differences are also apparent in the oxygenations of the amides of alkylamines. With *Torulopsis gropengiesseri*[115]
as well as with *Sporotrichum sulfurescens*,[117] the preponderant reactions are ω- and (ω-1)-oxygenation, but the latter operates on substrates of much shorter length, as shown in Tables 11 and 12.   (Note
that with *T. gropengiesseri*, the best yields are obtained with long-
chain amine amides, not with long-chain acid amides.)   Furthermore,
the *S. sulfurescens* products are not incorporated into a glycolipid,
but are isolated by direct solvent extraction of the fermentation
beer.

   The wealth of data obtained from the examination of the glycolipids of *Torulopsis* demonstrates the power of these microorganisms
to oxygenate hydrocarbons and hydrocarbon derivatives.   Other microorganisms are equally capable of oxygenating such substrates, but the
special feature of *Torulopsis* that provides the intact metabolites in
the form of extracellular, water-insoluble glycolipids offers the
chemist a particularly good opportunity to examine this capability.
From these studies, trends concerning several ·generalizations about
the hydrocarbon oxygenation by *Torulopsis* may be noted. As mentioned
earlier, initial attack on a hydrocarbon appears to occur preferenti-
ally at a terminal methyl group having the fewest substituents. This
initial attack probably occurs as a hydroxylation, followed by further dehydrogenation of the alcohol to the carboxylic acid oxidation
level. The carboxylic acid group, or a similar group such as amide,

Table 11. Amide Hydroxylations by *Torulopsis gropengiesseri*

Substrate	Products (yields[a])
C_{12} $CH_3(CH_2)_{11}-NH-COOC_2H_5$	$CH_3\underset{\underset{OH}{\vert}}{CH}-(CH_2)_{10}NHCOOC_2H_5$ (27%)
$CH_3(CH_2)_{11}-NH-\underset{\underset{O}{\Vert}}{C}-CH_2CH_3$	$CH_3-\underset{\underset{OH}{\vert}}{CH}-(CH_2)_{10}NH-\underset{\underset{O}{\Vert}}{C}CH_2CH_3$ (40%)
C_{14} $CH_3(CH_2)_{13}-NH\underset{\underset{O}{\Vert}}{C}-CH_3$	$CH_3-\underset{\underset{OH}{\vert}}{CH}-(CH_2)_{12}-NH-\underset{\underset{O}{\Vert}}{C}-CH_3$ (41%)
C_{16} $CH_3(CH_2)_{14}-CONHCH_3$	$CH_3\underset{\underset{OH}{\vert}}{CH}-(CH_2)_{13}-CONHCH_3$ (9%)
$CH_3(CH_2)_{14}-CON(CH_3)_2$	$CH_3\underset{\underset{OH}{\vert}}{CH}-(CH_2)_{13}-CON(CH_3)_2$ (3%) + $-CONHCH_3$ (8%)
$CH_3(CH_2)_{15}NHSO_2CH_3$	$CH_3\underset{\underset{OH}{\vert}}{CH}-(CH_2)_{14}NHSO_2CH_3$ (50%)
$-NH-\underset{\underset{O}{\Vert}}{C}-R$ (R = CH_3, C_2H_5, C_3H_7, C_4H_9)	$CH_3\underset{\underset{OH}{\vert}}{CH}-(CH_2)_{14}\underset{\underset{O}{\Vert}}{NHC}-R$ (40-50%)
C_{18} $CH_3(CH_2)_{16}-CONHCH_3$	$CH_3\underset{\underset{OH}{\vert}}{CH}-(CH_2)_{15}CONHCH_3$ (8%)

[a] Assay yields, corrected for recovered substrate. The ω-hydroxy-
lated product was formed in about 4% yield in each case.

Table 12. Amide Hydroxylations by *Sporotrichum sulfurescens*[a]

Substrate	Products
$CH_3CH_2\underset{\underset{CH_3-CH_2}{\vert}}{CH}-NH\underset{\underset{O}{\Vert}}{C}-C_6H_5$	$(-)-CH_3\overset{\overset{OH}{\vert}}{CH}-CH-NH\underset{\underset{O}{\Vert}}{C}-C_6H_5$(10.5%) $\underset{CH_3-CH_2}{}$

Substrate	Products
$CH_3-CH-CH_2CH_2-NH\overset{O}{\overset{\|}{C}}-C_6H_5$ CH_3	$CH_3-\overset{OH}{\overset{\|}{C}}-CH_2CH_2-NH\overset{O}{\overset{\|}{C}}-C_6H_5$ (13%) CH_3
$(\pm)-CH_3CH_2CH_2-CH-NH\overset{O}{\overset{\|}{C}}-C_6H_5$ CH_3	$HOCH_2CH_2CH_2-CH-NH\overset{O}{\overset{\|}{C}}-C_6H_5$ CH_3 $+ CH_3\overset{O}{\overset{\|}{C}}-CH_2-CH-NH\overset{O}{\overset{\|}{C}}-C_6H_5$ $\phantom{+ CH_3\overset{O}{\overset{\|}{C}}-CH_2-CH}CH_3$ $+ (-)-CH_3\overset{OH}{\overset{\|}{C}}H-CH_2-CH-NH\overset{O}{\overset{\|}{C}}-C_6H_5{}^{b}$ CH_3
$(\pm)-CH_3\overset{}{C}HCH_2-CH-NH\overset{O}{\overset{\|}{C}}-C_6H_5$ $(1R)-(-)-CH_3CH_3$ $(1S)-(+)-$	$CH_3\overset{OH}{\overset{\|}{C}}-CH_2-CH-NH\overset{O}{\overset{\|}{C}}-C_6H_5$ (22%) CH_3CH_3 $(1R)-(-)(41\%)$ $(1S)-(+)(12.5\%)$ $+ HOCH_2-CH-CH_2-CH-NH\overset{O}{\overset{\|}{C}}-C_6H_5(29\%)$ $CH_3CH_3(1R)-(-)(14\%)$ $(1S)-(+)(36\%)$
$CH_3CH_2CH_2CH_2CH-NH\overset{O}{\overset{\|}{C}}-C_6H_5$ CH_3	$CH_3\overset{OH}{\overset{\|}{C}}HCH_2CH_2CH-NH\overset{O}{\overset{\|}{C}}-C_6H_5(ca.\ 70\%)$ CH_3 $+ CH_3\overset{O}{\overset{\|}{C}}-CH_2CH_2-CH-NH\overset{O}{\overset{\|}{C}}-C_6H_5(11\%)$ CH_3
$\overset{CH_3}{\overset{\|}{C}}$ $CH_3-C-CH_2-CH-NH\overset{O}{\overset{\|}{C}}-C_6H_5$ CH_3CH_3	$\overset{CH_3}{\overset{\|}{C}}$ $HOCH_2-C-CH_2-CH-NH\overset{O}{\overset{\|}{C}}-C_6H_5(4\%)$ CH_3CH_3
$CH_3CH_2CH_2CH_2CH-NH\overset{O}{\overset{\|}{C}}-C_6H_5$ CH_3-CH_2	$CH_3\overset{OH}{\overset{\|}{C}}HCH_2CH_2CH-NH\overset{O}{\overset{\|}{C}}-C_6H_5(54\%)$ CH_3-CH_2 $+ CH_3\overset{O}{\overset{\|}{C}}-CH_2CH_2-CH-NH\overset{O}{\overset{\|}{C}}-C_6H_5(15\%)$ CH_3-CH_2

Substrate	Product
$CH_3CH_2\underset{\underset{CH_3}{\mid}}{CH}-CH_2-\underset{\underset{CH_3}{\mid}}{CH}-NH\overset{\overset{O}{\parallel}}{C}-C_6H_5$	$CH_3\underset{\underset{OH}{\mid}}{CH}-CH-CH_2-\underset{\underset{CH_3}{\mid}}{CH}-NH\overset{\overset{O}{\parallel}}{C}-C_6H_5$ (21%)
	$+\ (+)-CH_3\overset{\overset{O}{\parallel}}{C}-\underset{\underset{CH_3}{\mid}}{CH}CH_2\underset{\underset{CH_3}{\mid}}{CH}-NH\overset{\overset{O}{\parallel}}{C}-C_6H_5$ (5%)
$CH_3\underset{\underset{CH_3}{\mid}}{CH}-CH_2CH_2-\underset{\underset{CH_3}{\mid}}{CH}-NH\overset{\overset{O}{\parallel}}{C}-C_6H_5$	$CH_3\underset{\underset{CH_3}{\mid}}{\overset{\overset{OH}{\mid}}{C}}-CH_2CH_2\underset{\underset{CH_3}{\mid}}{CH}-NH\overset{\overset{O}{\parallel}}{C}-C_6H_5$ (59%)
$CH_3CH_2CH_2CH_2\underset{\underset{CH_3}{\mid}}{CH}-NH\overset{\overset{O}{\parallel}}{C}-C_6H_5$	$CH_3\overset{\overset{OH}{\mid}}{CH}CH_2CH_2CH_2\underset{\underset{CH_3}{\mid}}{CH}-NH\overset{\overset{O}{\parallel}}{C}-C_6H_5$ (5%)
	$+\ CH_3\overset{\overset{O}{\parallel}}{C}-CH_2CH_2CH_2\underset{\underset{CH_3}{\mid}}{CH}-NH\overset{\overset{O}{\parallel}}{C}-C_6H_5$ (6%)
	$+\ HOCH_2CH_2CH_2\underset{\underset{CH_3}{\mid}}{CH}-NH\overset{\overset{O}{\parallel}}{C}-C_6H_5$ (2%)
$CH_3\underset{\underset{CH_3}{\mid}}{\overset{\overset{CH_3}{\mid}}{C}}-CH_2-\underset{\underset{CH_3}{\mid}}{\overset{\overset{CH_3}{\mid}}{C}}-NH\overset{\overset{O}{\parallel}}{C}-C_6H_5$	$HOCH_2\underset{\underset{CH_3}{\mid}}{\overset{\overset{CH_3}{\mid}}{C}}-CH_2-\underset{\underset{CH_3}{\mid}}{\overset{\overset{CH_3}{\mid}}{C}}-NH\overset{\overset{O}{\parallel}}{C}-C_6H_5$ (11%)
$CH_3-\underset{\underset{CH_3}{\mid}}{CH}-CH_2CH_2CH_2\underset{\underset{CH_3}{\mid}}{CH}NH\overset{\overset{O}{\parallel}}{C}-C_6H_5$	$CH_3\underset{\underset{CH_3}{\mid}}{\overset{\overset{OH}{\mid}}{C}}-CH_2CH_2CH_2\underset{\underset{CH_3}{\mid}}{CH}-NH\overset{\overset{O}{\parallel}}{C}-C_6H_5$ (50%)

nitro, or sulfonamide but not cyano, halogen, or alkoxyl, then may act as a point of attachment for a second enzyme system, which oxidizes the remaining terminus of the hydrocarbon chain at a preferential distance (see earlier) from the polar group. This second enzyme, if indeed a single enzyme, is a remarkably versatile entity, being able to effect oxygenation at positions carrying groups such as halogen, nitrile, alkoxy, as well as in the presence of additional methyl substituents (*i.e.*, branching). This latter enzyme has also been found to be stereospecific in its hydroxylation in at least one case.[118,119] Oxygenation of octadecanoic acid at the C-17 position gives L-17-hydroxyoctadecanoic acid, in which the hydroxyl oxygen is derived from molecular oxygen and is introduced without inversion of configuration at the 17-carbon atom. Retention of configuration is consistent with similar observations in cases of steroid hydroxylations.

Finally, it may be noted that a cell-free extract having the ability to convert oleic acid to L-17-hydroxyoleic acid has been prepared from *Torulopsis*.[120]

Although the oxygenation of long-chain hydrocarbons generally is seen to occur at terminal or subterminal positions, this is not exclusively so, as demonstrated by two interesting exceptions. The first is the oxygenation of hexadecane by an *Arthrobacter* species, giving 3-hexadecanone and 4-hexadecanone in addition to 2-hexadecanone in a ratio of 26:6:68[121] and in an over-all yield of about 3%. Other alkanes are also oxygenated, but in even lower yields.

A second example is the hydroxylation of the long chain, polyunsaturated acid, 5,8,11,14-eicosatetraenoic acid by *Ophiobolus graminis*. This reaction gives two products, the 18-alcohol (19%) and the 19-alcohol (16%),[122] which serve as intermediates in a biochemical synthesis of oxygenated prostaglandins.

5,8,11,14-Eicosatetraenoic acid

Monocyclic. Few examples are known of the oxygenation of the simplest cycloalkanes, but since the products of such reactions are usually readily available, this does not represent real poverty. A bacterium (identified as JOB5) isolated from soil is capable of oxygenating cyclopropane (to propionaldehyde), cyclopentane (cyclopentanone, 77%), methylcyclopentane (3-methylcyclopentanone), cyclohexane (cyclohexanone), methylcyclohexane (4-methylcyclohexanone), cycloheptane (cycloheptanone), and cyclooctane (cyclooctanone), but not cyclodecane.[123] It is likely that the corresponding monoalcohols are intermediate precursors to the ketones in these oxygenations.

Ethylcyclohexane is oxidized to *trans*-4-ethylcyclohexanol(isol.) by *Alcaligenes faecalis*,[124] and phenylcyclohexane and p-methoxyphenylcyclohexane are oxidized to 4-phenylcyclohexanol (10% yield) and 4-(p-methoxyphenyl)cyclohexanol (5% yield), respectively, by *Sporotrichum sulfurescens* (and a variety of other organisms).[125] Several microorganisms oxygenate bicyclohexyl to mixtures of diols, from which *trans,trans*-4,4'-dihydroxybicyclohexyl, the preponderant product, is obtained in 10-30% yields. The yield of this same dihydroxybicyclohexyl is increased by the use of *trans*-4-hydroxybicyclohexyl as the substrate. Other hydroxybicyclohexyls are also oxygenated by many microorganisms, but mixtures of diols (or hydroxyketones) are obtained that are difficult to separate cleanly.[125]

Cyclic ketals of 4-cyclohexylcyclohexanone are hydroxylated by *S. sulfurescens* to the corresponding (4-hydroxycyclohexyl)cyclohexanone ketals in 5-20% yield,[125] and the same organism also hydroxylates amino substituted bicyclohexyls, such as 4-morpholinobicyclohexyl, principally at the 4'-position.

Under the conditions used for the above-described hydroxylations with *S. sulfurescens*, cyclododecanol is readily oxygenated (see below), but cyclododecane is not. However, by the addition of wetting agent (*e.g.*, Ultrawet 30DS) to the fermentation, a modest yield of dioxygenated products can be obtained from cyclododecane itself.[126]

These results initiated a broad study of the microbiological oxygenation of cyclic compounds with the single organism, *S. sulfurescens*. This organism transforms cyclododecanol to a mixture of dioxy products (diols, hydroxyketones, and diketones).[127] Chemical oxidation of this mixture to diones simplifies the identification of the products, from which cyclododecane-1,6-dione and cyclododecan-1,7-dione are isolated (11% each). A trace amount of the 1,5-dione can also be isolated. These isolated yields are much lower than the quantities actually present, as shown by chromatographic analysis. Cyclotridecanol and cyclotetradecanol, oxygenated by the same microorganism, give about 5% yields of cyclotridecane-1,7-dione and of cyclotetradecane-1,6-dione, respectively.[127]

Microbiological oxidation of cyclohexanol, cycloheptanol, or cyclooctanol to tangible products by *S. sulfurescens* could not be demonstrated, but when the volatility of the substrate, and perhaps the biodegradability of a dioxy product, is reduced by formation of, for example, the N-phenylcarbamate of cyclohexanol, a 35% yield of 4-hydroxycyclohexyl N-phenylcarbamate is obtained.[127]

It has been proposed, on the basis of these findings, that the oxidative reaction occurs by binding of a functional (electron-rich) group of the substrate to a site on the oxidative enzyme system about 5.5 Å away from the enzyme site where oxidation occurs. (It has also

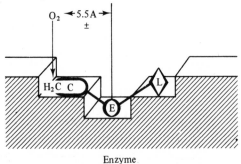

Enzyme

been proposed that there may be some over-all requirement for lipo-
philicity on the part of the substrate, to ensure adequate binding to
the enzyme system.[127]) Similar proposals have been made for the sub-
strate-enzyme relationships of other types of substrates that are oxygenated by microorganisms. The rate of sludge microorganism oxidative degradation of dialkylbenzenesulfonates, for example, may be
correlated with attachment of the sulfonate group to enzyme, followed
by alkane chain attack at some preferred distance from that attachment.[128] Certain aromatic ring openings (Chapter 5) may also be con-
trolled by a binding-distance relationship.[129]

This hypothesis suggests, as a logical extension, the use of cycloalkylamines as substrates in which the amino group can serve as an
electron-rich attachment site. Surprisingly, cyclododecylamine is
converted not to an oxygenated cyclododecylamine but rather to a mixture of monooxygenated cyclododecylacetamides by *S. sulfurescens*. The
same products are obtained when cyclododecylacetamide was used as a
substrate. When the fermentation is carried out with other microorganisms or with *S. sulfurescens* in a medium (Czapek-Dox) in which the
N-acetylation does not occur, no oxygenation of cyclododecylamine is
detected.[130] It can be concluded that N-acylation is a necessary pre-
requisite to oxygenation of cyclic amines, an observation which is
borne out by the numerous successful oxygenations of various N-acyl-
cycloalkylamines, as opposed to the relatively few oxygenations of
free amines.

Numerous primary and secondary cycloalkylamines have been acyla-
ted chemically and successfully oxygenated by *S. sulfurescens* in a
manner consistent with the 5.5 Å hypothesis, as outlined in Table 13.
Various N-acyl groups that can be used successfully are acetamides,
benzamides and substituted benzamides, p-toluenesulfonamides, ure-
thans, ureas, and sulfonylureas. In all of these amide categories,
the carbonyl (or sulfonyl) oxygen atom is considered the most proba-
ble site of attachment to the enzyme in terms of the above hypothesis.[131] The oxygenation products of some of the sulfonylureas are of
medicinal interest since the fungal metabolites are among those seen
when the parent compound is administered to animals and since the metabolites also have hypoglycemic activity.[132]

Table 13. Oxygenation of Alicyclic Amides by *Sporotrichum sulfurescens*

Substrate	Product(s)	Yield, %	Ref.
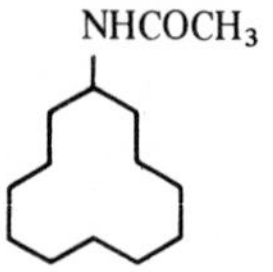	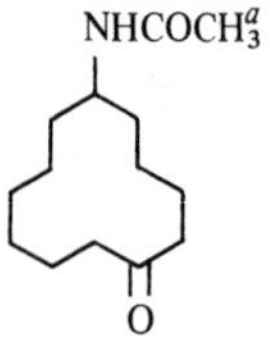	30	130
	+		
	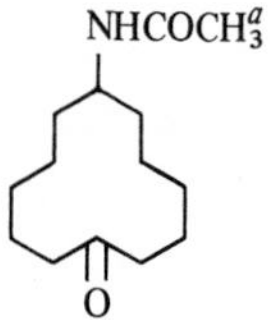	8	130
	+		
	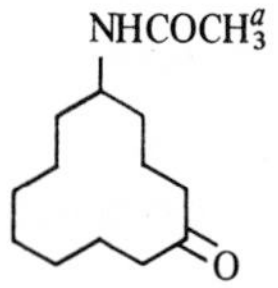	trace	130
		30	130
	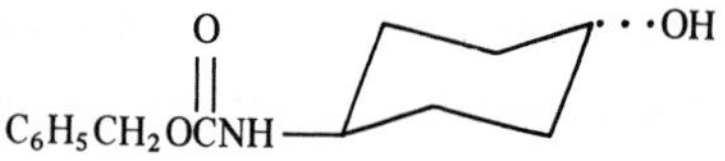	24	130
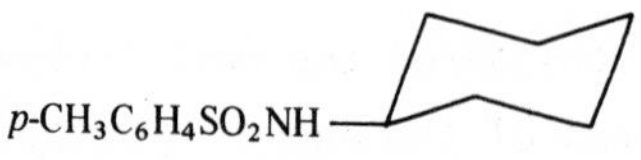	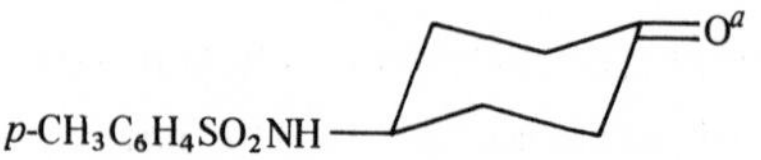	30	130
		30	133

Substrate	Product(s)	Yield, %	Ref.
 $m\text{-ClC}_6\text{H}_4\text{CONH}$ (cyclohexane)	 $m\text{-ClC}_6\text{H}_4\text{CONH}$ (cyclohexane)$\cdots\text{OH}$	38	133
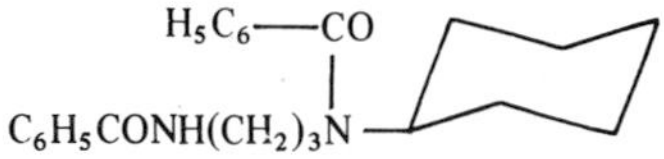 $\text{H}_5\text{C}_6\text{—CO}$, $\text{C}_6\text{H}_5\text{CONH(CH}_2)_3\text{N}$ (cyclohexane)	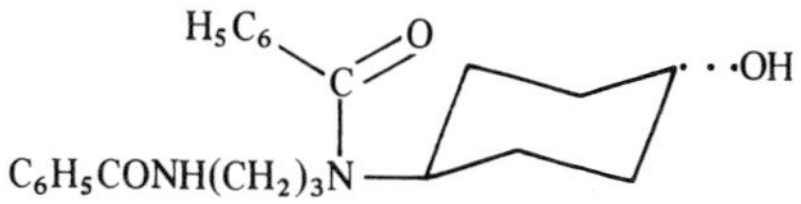 $\text{H}_5\text{C}_6\text{—C=O}$, $\text{C}_6\text{H}_5\text{CONH(CH}_2)_3\text{N}$ (cyclohexane)$\cdots\text{OH}$	52	133
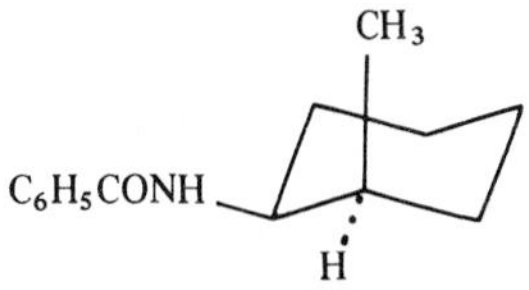 CH_3 , $\text{C}_6\text{H}_5\text{CONH}$, H	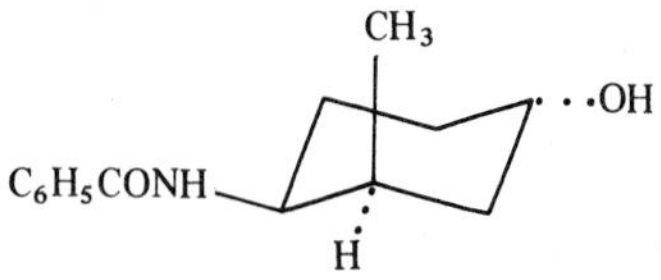 CH_3 , $\text{C}_6\text{H}_5\text{CONH}$, H $\cdots\text{OH}$	25	133
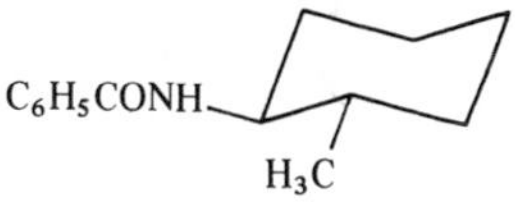 $\text{C}_6\text{H}_5\text{CONH}$, H_3C	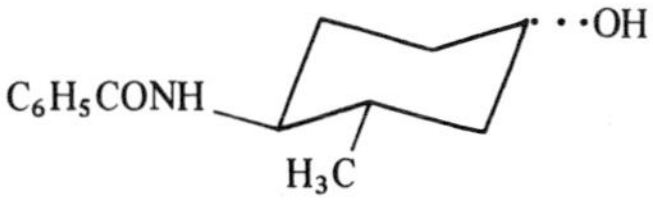 $\text{C}_6\text{H}_5\text{CONH}$, H_3C $\cdots\text{OH}$	9,21	133
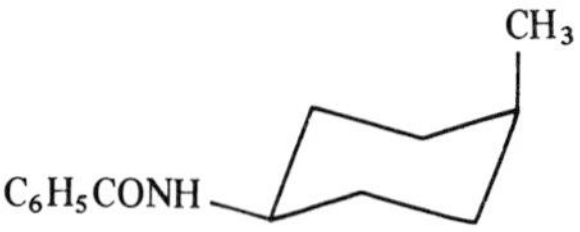 CH_3 , $\text{C}_6\text{H}_5\text{CONH}$	 CH_3 , $\text{C}_6\text{H}_5\text{CONH}$ $\cdots\text{OH}$	30	134
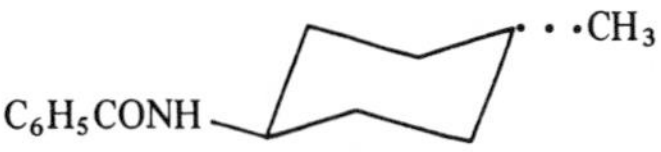 $\text{C}_6\text{H}_5\text{CONH}$ $\cdots\text{CH}_3$	 OH , $\text{C}_6\text{H}_5\text{CONH}$ $\cdots\text{CH}_3$	1	134
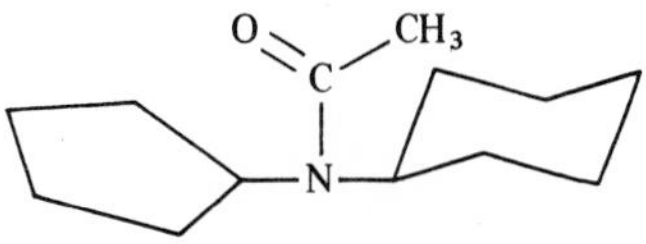 O=C—CH_3 , N (cyclopentane, cyclohexane)	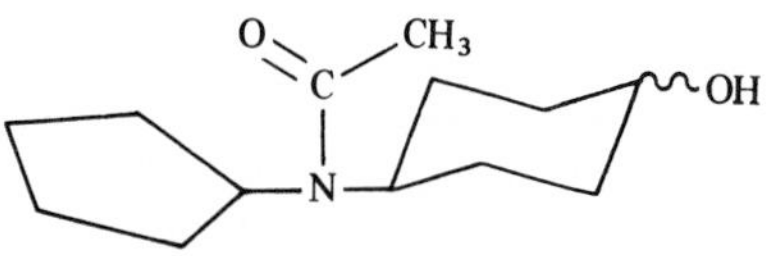 O=C—CH_3 , N $\sim\text{OH}$	36	130
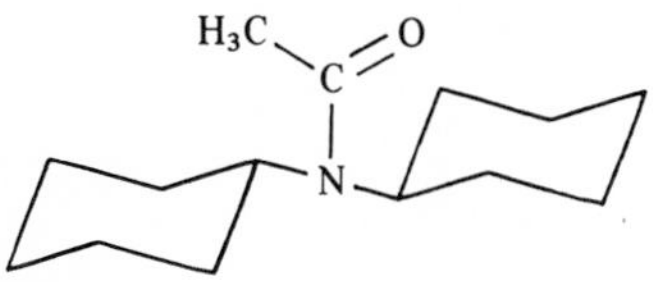 $\text{H}_3\text{C—C=O}$, N	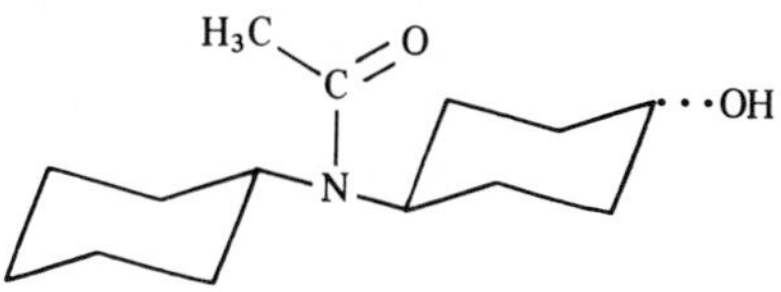 $\text{H}_3\text{C—C=O}$, N $\cdots\text{OH}$	52	130

Substrate	Product(s)	Yield, %	Ref.

| 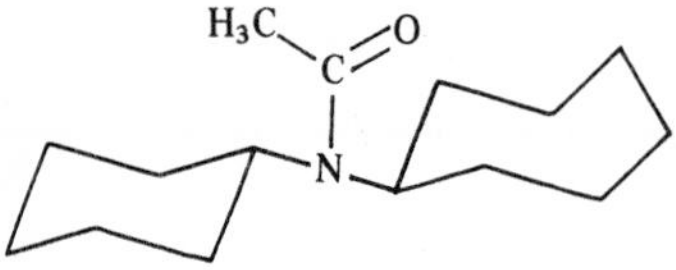 | 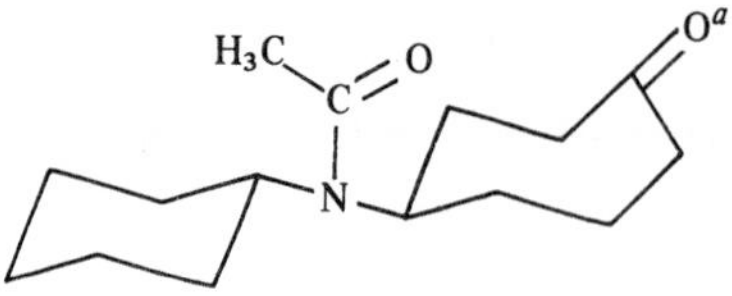 | 36 | 130 |

| 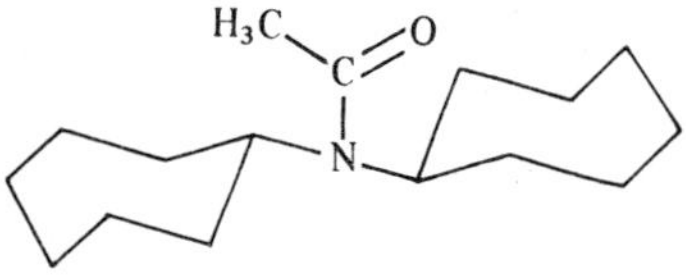 | 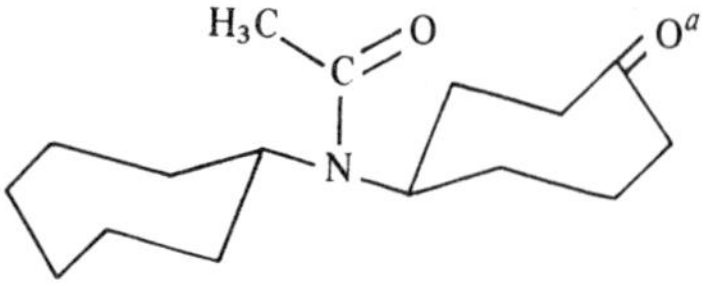 | 16 | 130 |

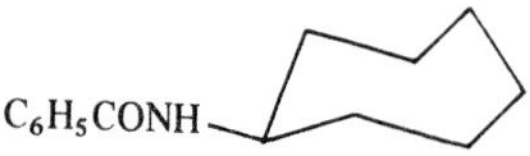

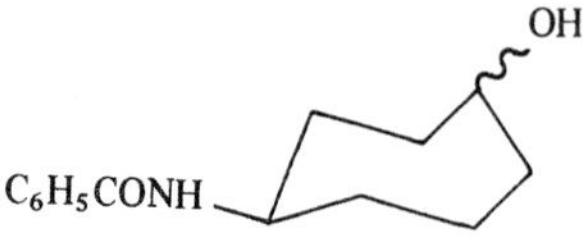

$[\alpha]_D + 25°$

+

$[\alpha]_D + 73°$ [b]

| | 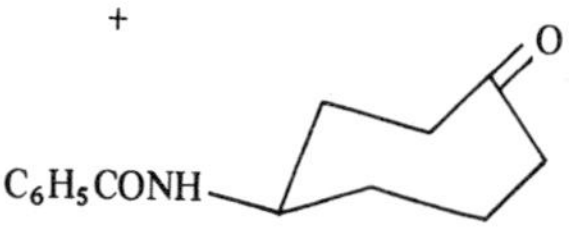 | 14 | 126 |
| | | 40 [c] | 130 |

$[\alpha]_D + 9°$

| 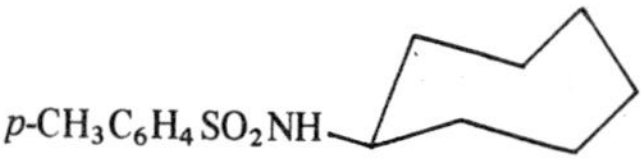 | 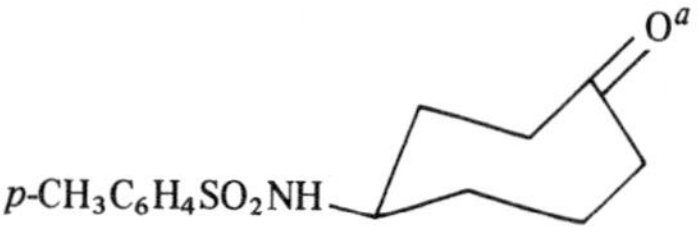 | 40 | 130 |

$[\alpha]_D \ 0°$

| | | 40 | 130 |

+

| 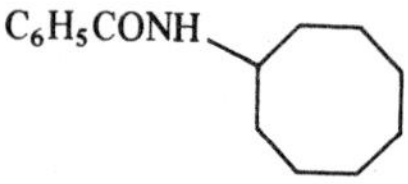 | 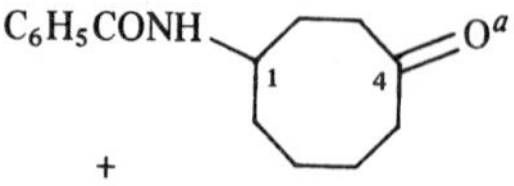 | 15 | 130 |
| | 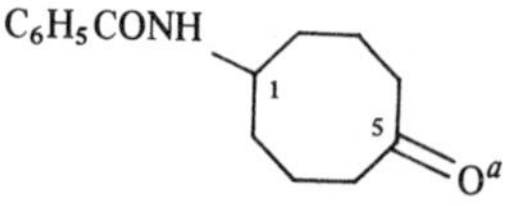 | 45 | 130 |

Substrate	Products	Yield, %	Ref.
p-CH$_3$C$_6$H$_4$SO$_2$NH— (cyclooctane ring)	p-CH$_3$C$_6$H$_4$SO$_2$NH— (ring with =O^a)	30	130
	+		
	p-CH$_3$C$_6$H$_4$SO$_2$NH— (ring with =O^a)	10	130
p-CH$_3$C$_6$H$_4$SO$_2$NHĊNH (O) + cyclopentane	p-CH$_3$C$_6$H$_4$SO$_2$NHĊNH (O) + ring—OH, cis + trans: 17%		132
p-CH$_3$C$_6$H$_4$SO$_2$NHĊNH (O) + cyclohexane	p-CH$_3$C$_6$H$_4$SO$_2$NHĊNH (O) + ring—OH, cis: 7% trans: 14%		132
p-CH$_3$ĊC$_6$H$_4$SO$_2$NHĊNH (O)(O) + cyclohexane	p-CH$_3$ĊC$_6$H$_4$SO$_2$NHĊNH (O)(O) + ring—OH		132
p-CH$_3$C$_6$H$_4$SO$_2$NHĊNH (O) + cycloheptane	p-CH$_3$C$_6$H$_4$SO$_2$NHĊNH (O) + ring—OH	15	132

^{a}These ketones represent the product following chemical oxidation of the bioconversion alcohol.

The configurations of the hydroxyl groups of several cyclohexyl-amides (Table 13) have been determined and found to be equatorial, which is *trans* to the amide group, in every case.[133]

The results obtained with the dicycloalkylacetamides (Table 13) suggest that in this series the oxidizing microorganism has a preference for attacking the cycloheptyl ring over a cyclohexyl ring over a cyclopentyl ring.[130]

The generation of optically active products in the instance of N-cycloheptylbenzamide is noteworthy. On the other hand, the products from oxygenation of the p-toluenesulfonyl and benzyloxycarbonyl de-rivatives of cycloheptylamine show little or no optical activity. No explanation has been advanced for this unusual behavior.[130]

61

Just as microbial oxygenation is seen with alicyclic amides, so oxygenation of heteroalicyclic amides is effected by *S. sulfurescens*. Included in this class are acyl derivatives of various piperidines and of hexamethyleneimine, heptamethyleneimine, and octamethyleneimine.[135,136] The results are summarized in Table 14.

Oxygenation of the piperidine series of substrates produces several noteworthy results.[136] Optically active products are obtained in several cases, generally as a result of the oxygenation of substrate enantiomers at different positions. The reader is reminded that alkyl groups at the 2-position of the N-benzoyl piperidines are forced to assume an axial configuration.[137] This condition presumably exists during the oxygenation of the molecule. Also of interest is the oxygenation, in several cases, of the alkyl side chains of some of the piperidine substrates (see the earlier discussion of alkyl amide oxygenations). In this way, a microbial synthesis, albeit in low yield, of the N-benzoyl derivatives of the alkaloids sedridine and isopelletierine has been achieved.

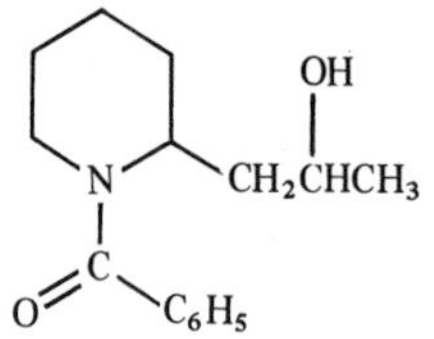

N-Benzoylsedridine 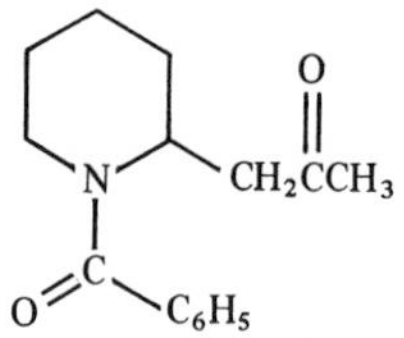   N-Benzoylisopelletierine

Several of the oxygenated products of intermediate ring size have served as useful chemical intermediates, demonstrating the potential of microbial oxygenations as a synthetic method. The seven-membered ring ketone, 1-benzoylhexahydro-4H-azepin-4-one, has found its way into a series of biologically active indole derivatives.[138]

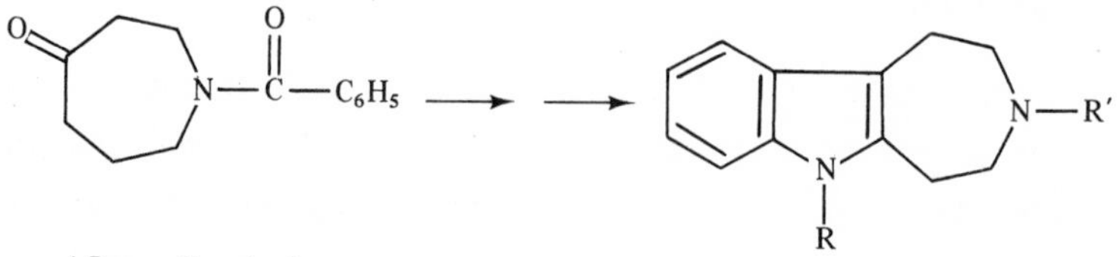

1-Benzoylhexahydro-
4H-azepin-4-one

Table 14. Oxygenation of Heteroalicyclic Amides by *Sporotrichum sulfurescens*

Substrate	Products	Yield, %	Ref.
C_6H_5CO—N (piperidine)	C_6H_5CO—N—OH	18	135
$(\pm)C_6H_5CO$—N (H_3C-substituted)	C_6H_5CO—N (2S,3S) H_3C, OH	10	136
	+ C_6H_5CO—N $\cdots$OH (2R,4S) H_3C	21	136
(2S)C_6H_5CO—N (H_3C-substituted)	C_6H_5CO—N (2S,3S) H_3C, OH	36	136
$(\pm)C_6H_5CO$—N (CH_3-substituted)	C_6H_5CO—N CH_3—OH	7	136
	+ $(-)C_6H_5CO$—N $\cdots$OH CH_3	6	136
C_6H_5CO—N—CH_3	C_6H_5CO—N CH_3 / OH	13	136
	+ C_6H_5CO—N—CH_2OH	23	136
$(\pm)C_6H_5CO$—N (CH_3CH_2-substituted)	$(+)C_6H_5CO$—N CH_3CH_2 OH	20	136
	+ C_6H_5CO—N $\cdots$OH CH_3CH_2	8	136

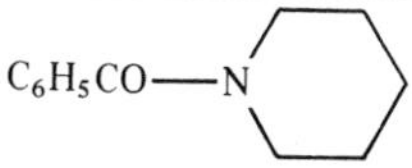

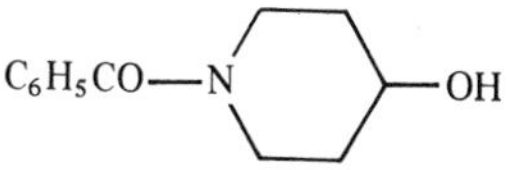

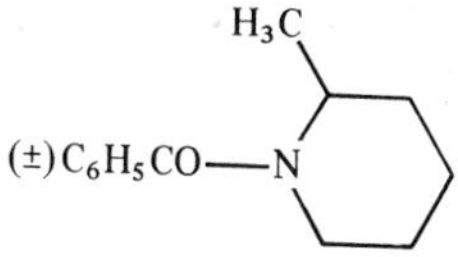

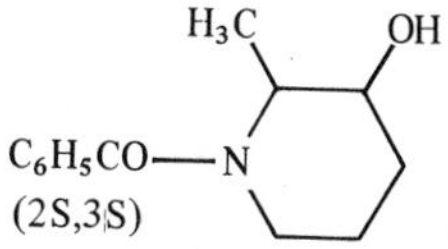

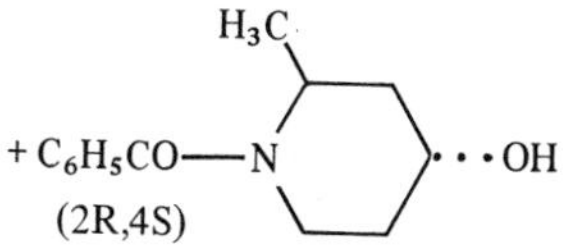

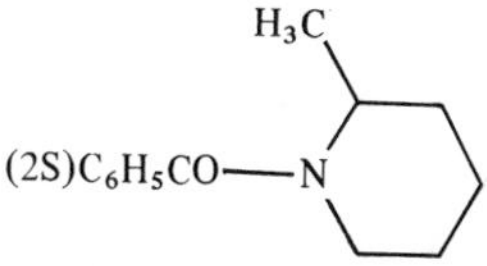

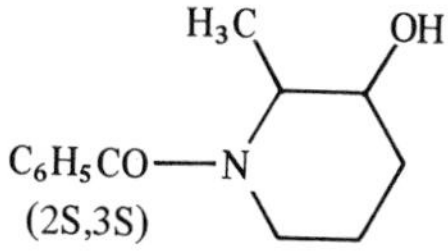

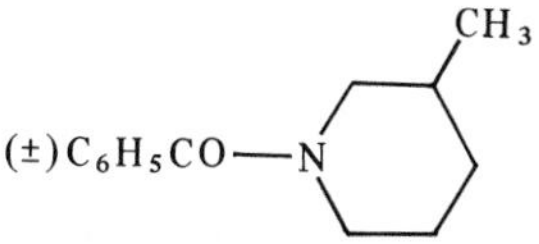

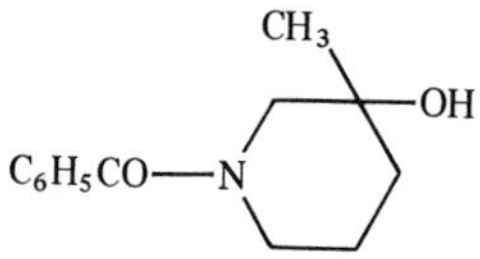

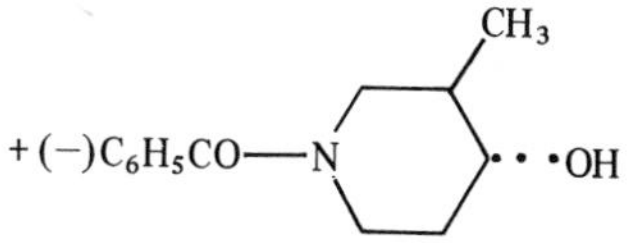

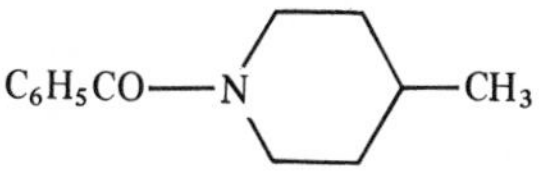

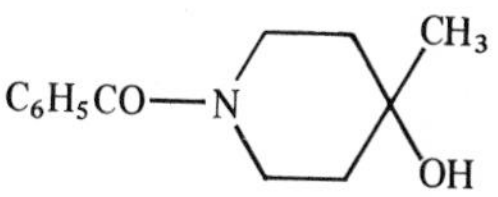

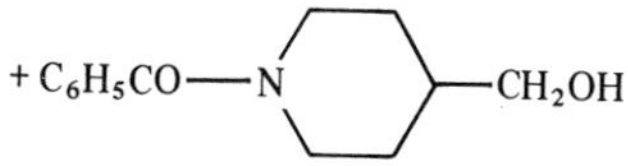

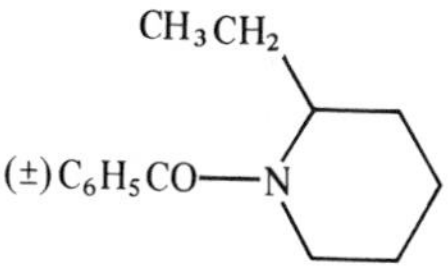

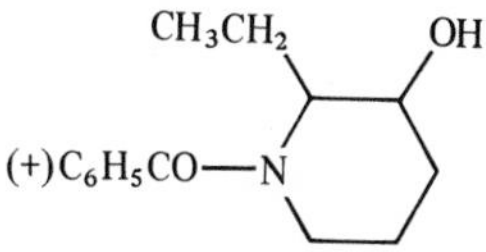

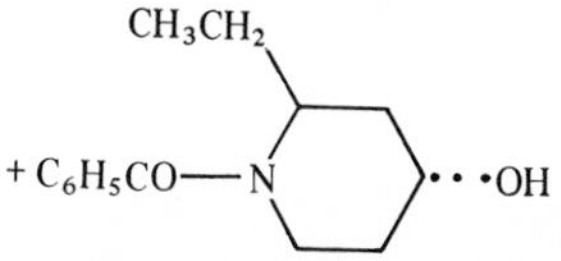

Substrate	Products	Yield, %	Ref.
CH₃CH₂CH₂ (±) C₆H₅CO—N (piperidine ring)	CH₃CH₂CH₂ OH (+)C₆H₅CO—N (piperidine ring)	14	136
	CH₃CH₂CH₂ + C₆H₅CO—N ⋯OH (piperidine ring)	8	
	CH₃CH(OH)CH₂ + C₆H₅CO—N (piperidine ring)	2	
CH₂CH₃ | C₆H₅CO—N—CH₂ (piperidine ring)	CH₃ C₆H₅CO—N—CH₂—C (piperidine ring) ║O	30	136
H₃C C₆H₅CO—N (piperidine ring) H₃C	CH₃ OH C₆H₅CO—N (piperidine ring) H₃C	49	136
C₆H₅CO—N (azepane ring)	O [a] C₆H₅CO—N (azepane ring)	55	135
	O [a] + C₆H₅CO—N (azepane ring)	10	135
CH₃ C₆H₅CO—N (azepane ring)	CH₃ OH C₆H₅CO—N (azepane ring)	29	135
	CH₃ [a] + C₆H₅CO—N (azepane ring) ═O	11	135
p-CH₃C₆H₄SO₂—N (azepane ring)	O [a] p-CH₃C₆H₄SO₂—N (azepane ring)	10	135

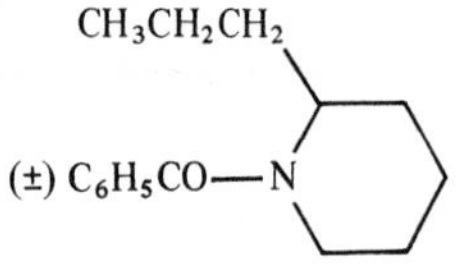

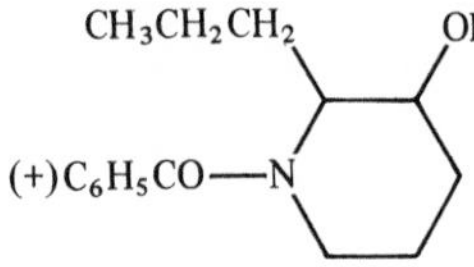

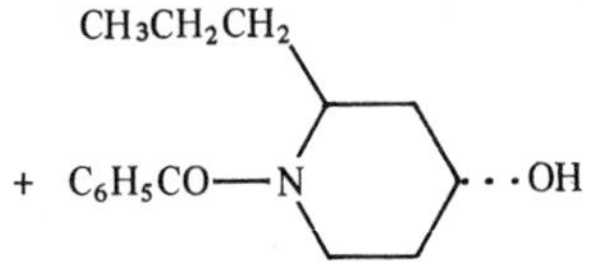

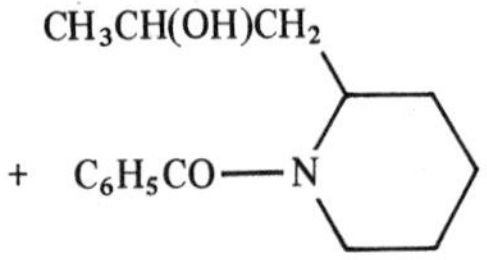

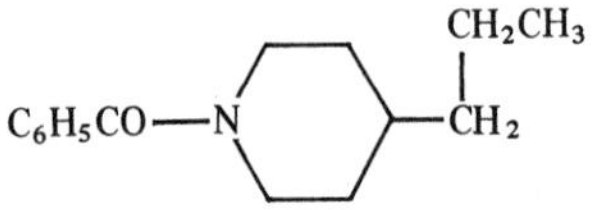

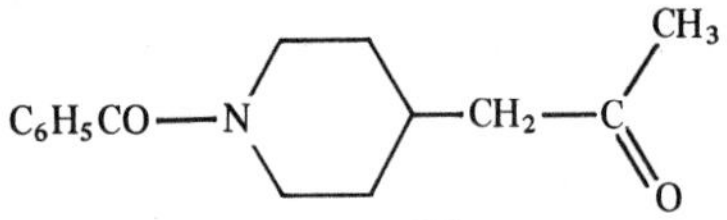

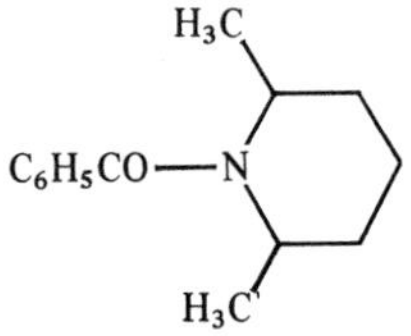

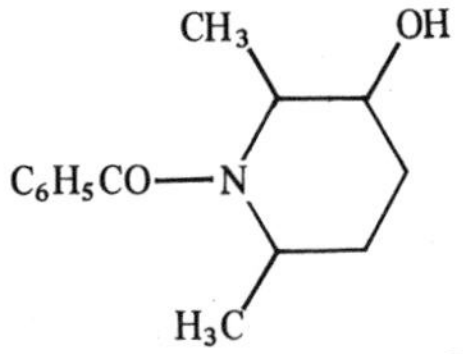

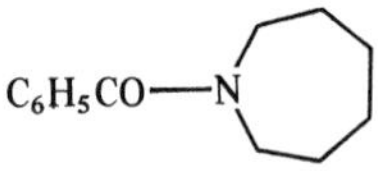

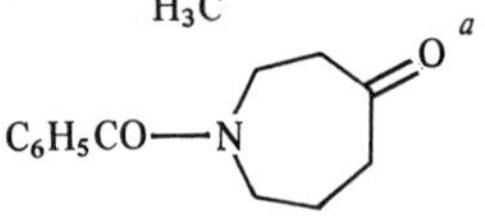

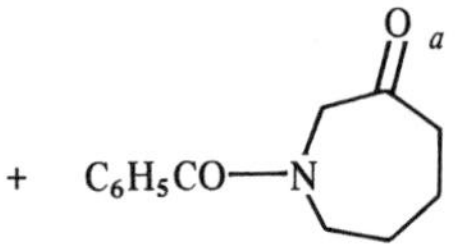

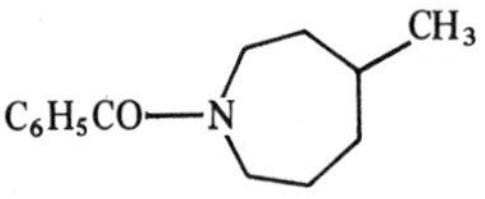

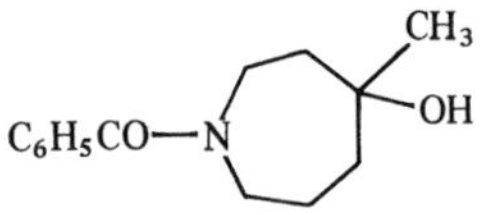

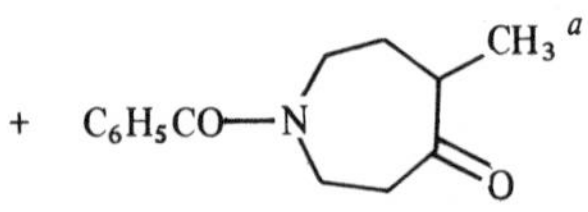

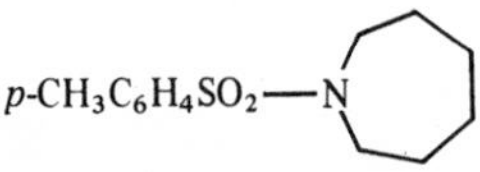

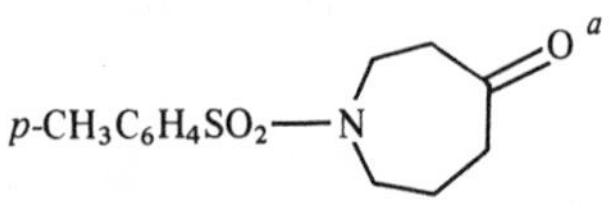

Substrate	Products	Yield, %	Ref.

C_6H_5CO—N (azacyclo ring) | C_6H_5CO—N (ring with =O) [a] | 59 | 135 |

+ C_6H_5CO—N (ring with =O) [a] | | 19 | |

C_6H_5CO—N (larger ring) | C_6H_5CO—N (ring with =O) [a] | 42 | 135 |

+ C_6H_5CO—N (ring with =O) [a] | | 17 | |

[a]After chemical oxidation of the bioconversion alcohol.

Microbiological oxygenation of the medium size heteroalicyclic rings provides an alternate method for synthesis of intermediates useful in studies of transannular reactions.[135]

Oxygenation of the 4-methyl derivative of 1-benzoylhexahydro-4H-azepine gives 4- and 5-oxygenated products in ratio of 3:1, indicative of a preference for oxygenation at the tertiary methyl substituted carbon.[134]

Cycloalkanes bearing an alkylsulfonyl substituent, i.e., sulfones, also are oxygenated by S. sulfurescens (and other organisms).[139] Cyclohexyl sulfone, for example, affords a 43% yield of a mixture of 3-hydroxylated and 4-hydroxylated products, in the approximate ratio 2:1. When given a choice of ring size, as in cycloheptyl cyclohexyl sulfone, the organism produces 4-hydroxycycloheptyl cyclohexyl sulfone (17%) and 4-oxocycloheptyl cyclohexyl sulfone (16%).[139] This is the same ring-size preference seen with dicycloalkylacetamide oxygenations (Table 13).

Bicyclic.   Turning to microbial oxygenations of bicyclic struc- tures, the pioneering work of Prelog directed primarily toward the reduction of decalones by *Curvularia falcata* also has yielded hydrox- ylation products.[140]   (±)-2-Oxo-*trans*-decalin affords (2S,9R)-2,10- dihydroxy-*trans*-decalin in 3% yield, while (±)-2-oxo-*cis*-decalin gives (2S,9S)-2,10-dihydroxy-*cis*-decalin in 4% yield as well as race- mic 2,9-dihydroxy-*cis*-decalin in less than 1% yield. When racemic 2- hydroxy-*cis*-decalin is used as substrate, an 11% yield of (2S,9S)- 2,10-dihydroxy-*cis*-decalin is obtained.   As in the steroid area, the microbial introduction of the hydroxyl group occurs with retention of configuration.   There is also selectivity with regard to the stereo- chemistry of the substrate since the racemic decalones each give rise to optically active products.   The postulate that a decalone under- goes microbial reduction to the decalol prior to oxygenation seems reasonable.   Judging from the somewhat higher yield of diol obtained when 2-hydroxydecalin is used as substrate instead of 2-decalone, it might be expected that the use of the alcohol in general would im- prove the efficiency of the microbial oxidation.   It would also be interesting to know why certain stereoisomers apparently fail to un- dergo hydroxylation.

The stereochemical selectivity of *Calonectria decora* for one or the other enantiomer of ethyl (±)-1,3,4,5,6,7-hexahydro-7-oxo-4a(2H)- naphthalenecarboxylate is apparent from the levorotatory character of the 2-hydroxylated product, obtained in 12% yield.[141]

dl   (−)

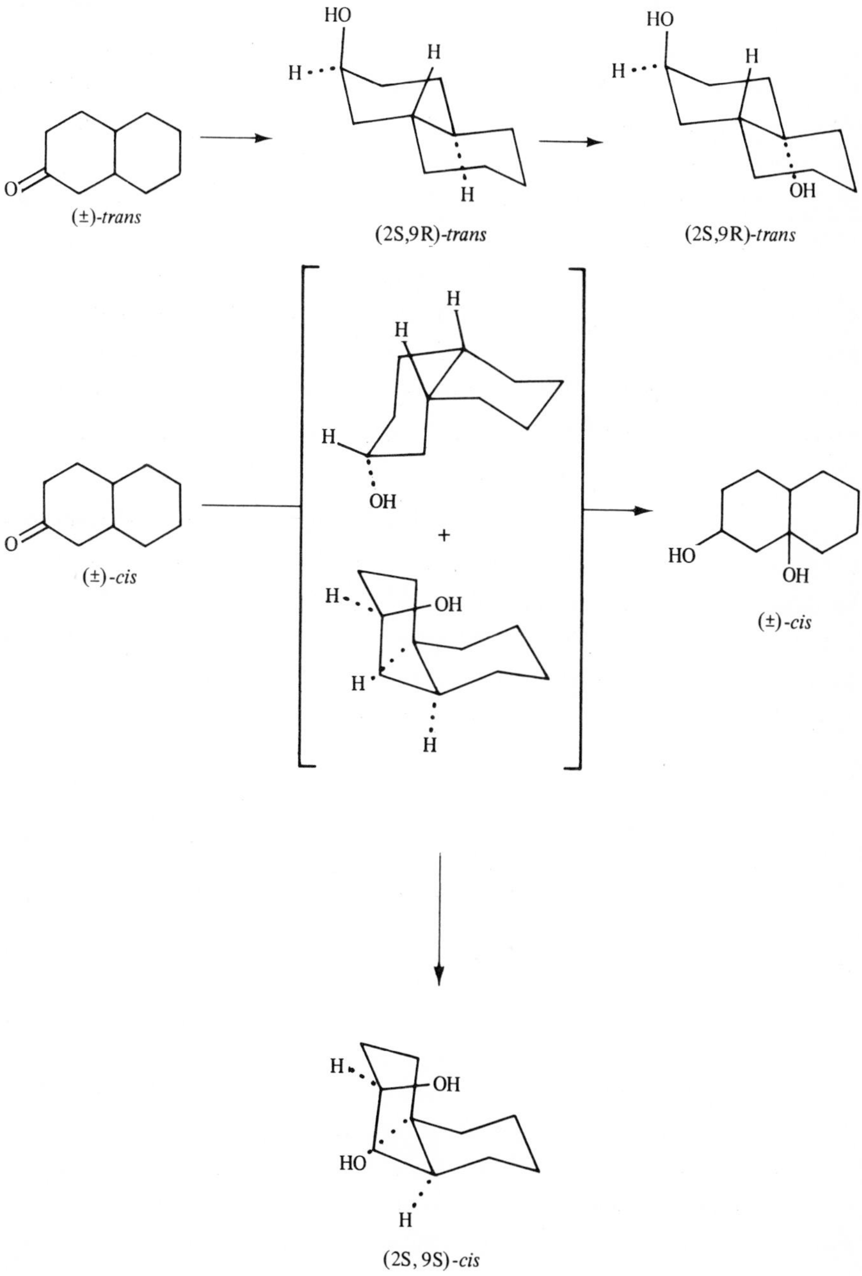

Fig. 5. Oxygenation of 2-oxo-*trans*- and *cis*-decalins by *Curvularia falcata*.

A variety of bicyclic compounds, both alicyclic and heteroalicyclic, are oxygenated by *Sporotrichum sulfurescens*. In the alicyclic
area, the N-benzoyl derivative of the potent hypotensive (ganglionic
blocking) agent mecamylamine (N,2,3,3-tetramethyl-2-norbornanamine)
is hydroxylated by *S. sulfurescens* to give a 21% yield of *exo*-6-hy-
droxy-N-benzoyl-N,2,3,3-tetramethyl-2-norbornanamine by direct isolation.[142] Other microbial oxidation products can be recovered by first
oxidizing the crude fermentation product (hydroxyamides) and then
isolating two ketoamides: 6-oxo-N-benzoyl-N,2,3,3-tetramethyl-2-nor-
bornanamine (24% yield) and 7-oxo-N-benzoyl-N,2,3,3-tetramethyl-2-
norbornanamine (29% yield). This is another example of the use of
microbiological methods to prepare compounds that could arise from
animal metabolism of a drug.

Both (+) and (-)-N-benzoylisopinocampheylamine give the corresponding enantiomer of the same hydroxymethyl product when oxygenated
by *S. sulfurescens*, the former in 62% yield and the latter in 28%
yield.[133] Since the introduced oxygen is not actually on the ring,
this reaction should also be considered in the framework of the earlier discussion of alkylamide oxygenations.

N-Benzoylisopinocampheylamine

Among the examples from the heteroalicyclic area, oxygenation of
(±)-1-benzoyl-*trans*-decahydroquinoline with *S. sulfurescens* gives a
mixture of three alcohols in a total yield of 80-90%.[143] Repeated
column chromatography results in isolation of the optically active 5-
hydroxy and 7-hydroxy products and of a racemic 6-hydroxy product.
Under the same conditions, hydroxylation of the (-)-enantiomer of the
substrate gives optically pure (4aS,5S,8aR)-1-benzoyl-*trans*-decahy-
droquinolin-5-ol and (4aR,6R,8aR)-1-benzoyl-*trans*-decahydroquinolin-

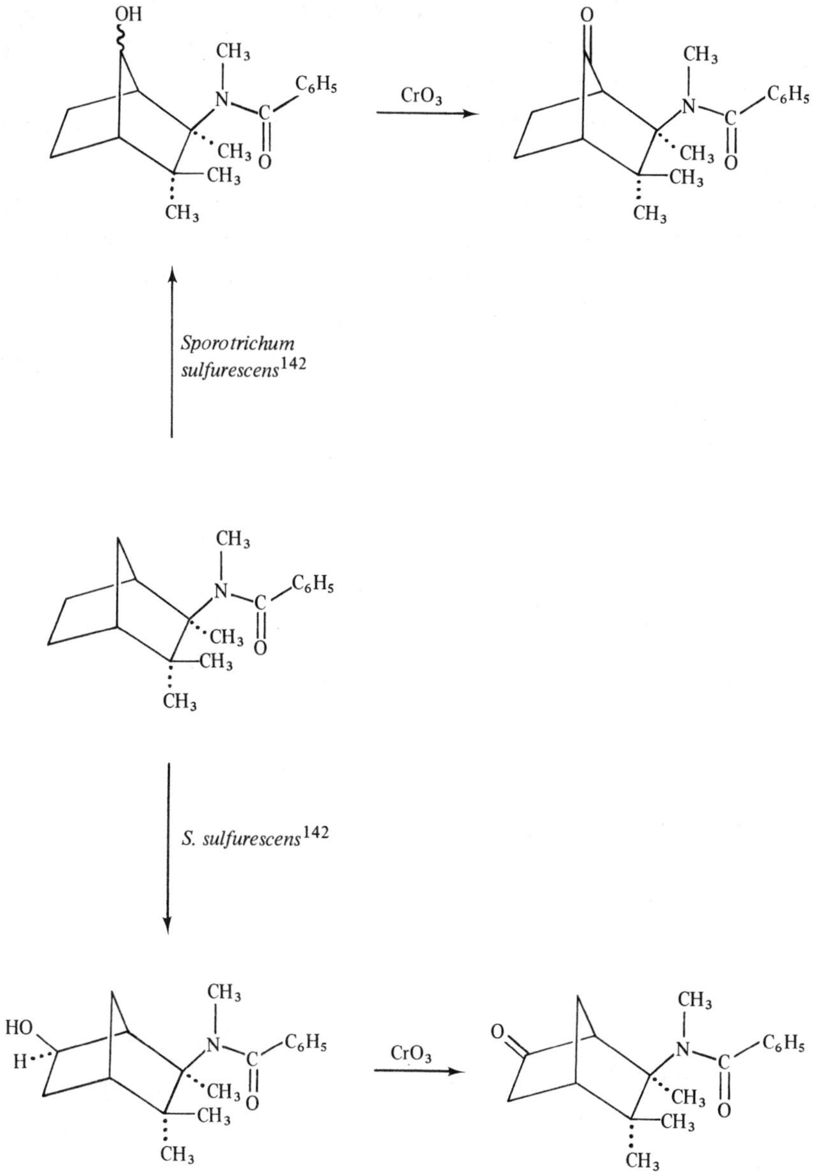

Fig. 6. Oxygenation of N-benzoyl-N,2,3,3-tetramethyl-2-norbornanamine (N-benzoylmecamylamine) by *Sporotrichum sulfurescens*.

6-ol in a ratio of 87:13.   Similarly, hydroxylation of the (+)-enan-
tiomer of the substrate gives optically pure (4aS,7S,8aR)-1-benzoyl-
trans-decahydroquinolin-7-ol and (4aS,6S,8aS)-1-benzoyl-*trans*-decahy-
droquinolin-6-ol in a ratio of 35:65.   (The absolute configurations
were established by modification of the products to the decahydro-
quinolinones and application of the octant rule to the optical rota-
tory dispersion curves of these products.) It is clear from these experiments that the (-)-enantiomer is hydroxylated predominantly at
the 5-position and to a lesser extent at the 6-position;   the (+)-en-
antiomer is hydroxylated predominantly at the 6-position and, to a
lesser amount, at the 7-position.[143]

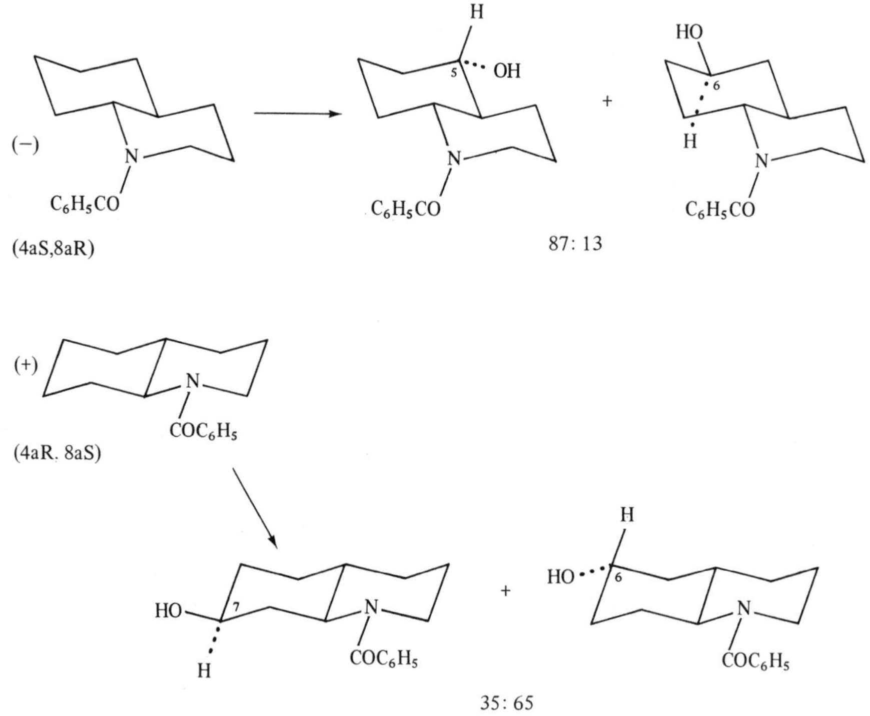

Fig. 7. Oxygenation of 1-benzoyl-*trans*-decahydroquinoline by
         Sporotrichum sulfurescens.

The spiroheterocycle, 1-benzoyl-1-azaspiro[4.5]decane, also is oxygenated by *S. sulfurescens*, giving a single product in 54% yield.

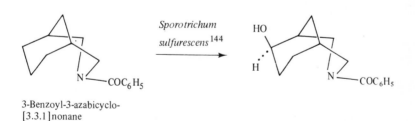

1-Benzoyl-1-
azaspiro[4.5]decane

Sporotrichum
sulfurescens[133]

Other examples of heteroalicyclic (bicyclic) microbial oxidation are the reactions of 3-benzoyl-3-azabicyclo[3.3.1]nonane and 3-ben-zoyl-3-azabicyclo[3.2.2]nonane. The former compound with *S. sulfur-escens* gives 3-benzoyl-*endo*-3-azabicyclo[3.3.1]nonan-6-ol (71% yield) while the latter gives a mixture of 3-benzoyl-*endo*-3-azabicyclo-[3.2.2]nonan-6-ol (50% yield) and 3-benzoyl-3-azabicyclo[3.2.2]nonan-6-one (22% yield).[144] When the microbial oxidation of 3-benzoyl-3-

3-Benzoyl-3-azabicyclo-
[3.3.1]nonane

Sporotrichum
sulfurescens[144]

3-Benzoyl-3-azabicyclo-
[3.2.2]nonane

Sporotrichum
sulfurescens[144]

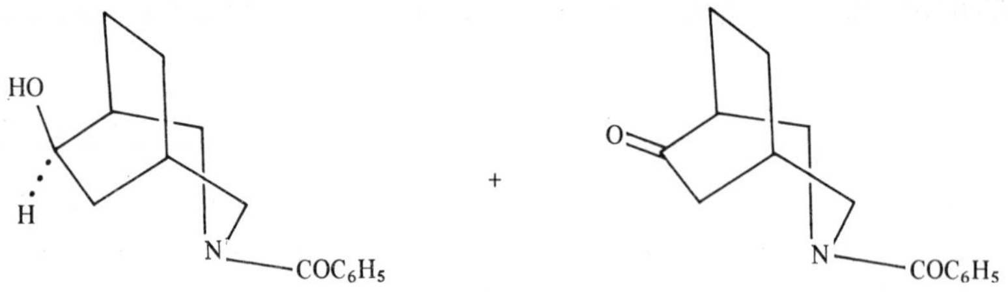

azabicyclo[3.3.1]nonane is carried out using the organism *Rhizopus arrhizus* the products are 3-benzoyl-*endo*-3-azabicyclo[3.3.1]nonan-6-ol (22% yield) and optically active (1R,5R)-3-benzoyl-3-azabicyclo-[3.3.1]nonan-1-ol (25% yield).[145]

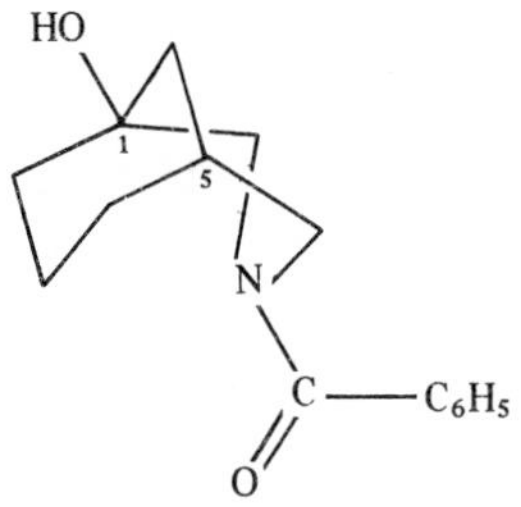

(1R, 5R)-3-Benzoyl-3-aza-
bicyclo[3.3.1]nonan-1-ol

Polycyclic. Microbial oxidations of steroids have been discussed earlier. These represent the bulk of the information on polycyclic systems, with relatively minor contributions from the areas of alkaloids, terpenes, etc.

It remains only to note that the microbial oxidation of N-acylated adamantanamines has been studied and represents the only practical source of certain polyfunctional adamantanes. N-Acetyl-1-adamantanamine with *Sporotrichum sulfurescens* gives N-acetyl-1-adamantanamin-4α-ol (50% yield) and N-acetyl-1-adamantanamin-3-ol (10% yield).[146]

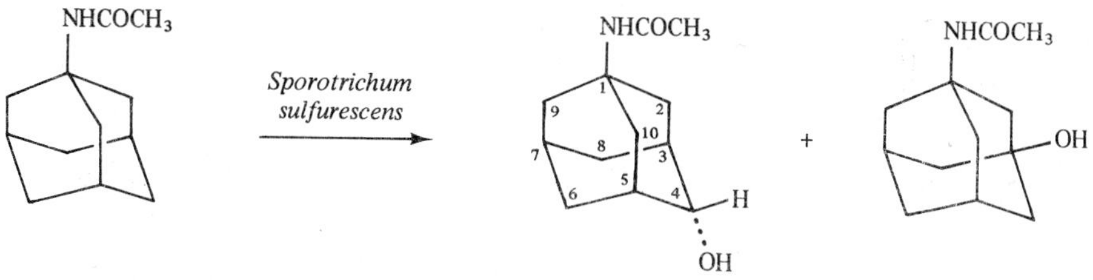

N-Acetyl-1-adamantanamine

Interestingly, alteration of the substitution of the nitrogen in the substrate leads to a somewhat different oxidation pattern. In the case of N-benzoyl-N-methyl-1-adamantanamine, for example, the major product (55% yield) of *S. sulfurescens* biooxidation is N-benzoyl-N-methyl-1-adamantanamine-4α,6α-diol, while N-benzoyl-1-adamantanamin-4α-ol is formed in 14% yield. (The monohydroxyamide is oxidized by *S. sulfurescens* giving the same 4α,6α-diol in 35% yield.) Several other bioconversions of substituted adamantanes with varied nitrogen-substitutions are shown in Fig. 8.

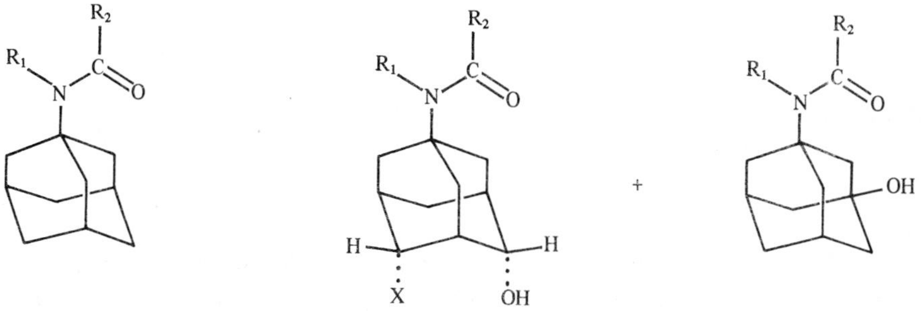

Substrate	*Product(s)*	
$R_1 = R_2 = CH_3$	X = H (35%)	(6%)
$R_1 = H; R_2 = CH_2C_6H_5$	X = OH (65%)	—
$R_1 = CH_3; R_2 = C_6H_{11}$	X = OH (10%)	—
$R_1, R_2 = —C—C_6H_4—$ $\quad\quad\quad ‖$ $\quad\quad\quad O$	X = OH (isol.)	—

Fig. 8. Nitrogen-substituent effects on adamantane oxygenation with *Sporotrichum sulfurescens*.

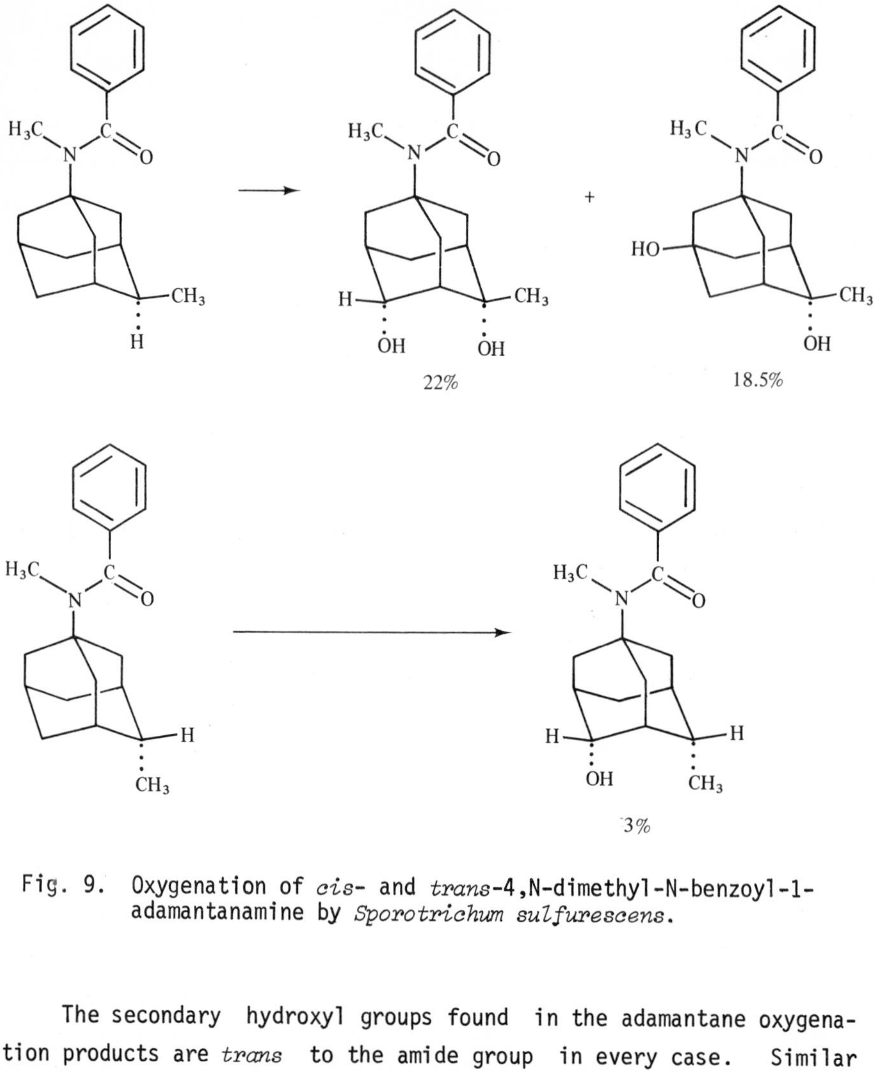

Fig. 9. Oxygenation of *cis*- and *trans*-4,N-dimethyl-N-benzoyl-1-adamantanamine by *Sporotrichum sulfurescens*.

The secondary hydroxyl groups found in the adamantane oxygenation products are *trans* to the amide group in every case. Similar observations of a *trans* relationship between hydroxyl group and amide group have been noted earlier in the discussion of this section. Consequently, it can be stated that there exists some gross stereochemical feature of the oxygenation enzyme that results in the introduction of oxygen *trans* to the amide substituent in these cyclic substrates.[131] Conversely, it is a necessary condition for oxygenation that the substrate have a carbon-hydrogen bond in a *trans* relationship to the amide group.

Confirmation of the above proposition is found in consideration
of the oxygenation of the *cis* and *trans*-4-methyl-N-methyl-N-benzoyl-
1-adamantanamines (Fig. 9)[147] and of several substrates discussed
earlier in this chapter. Consistent with the results shown in Fig. 9
are those obtained with the N-benzoyl derivatives of *cis*- and *trans*-
4-methylcyclohexylamine.   The *cis* isomer is oxygenated in good yield
at the 4-position, whereas the *trans* isomer is oxygenated in very
poor yield at this position (see Table 13).[134]   When taken with the
preferential oxygenation at the tertiary carbon of 4-methyl-1-ben-
zoylhexahydro-4H-azepine (Table 14), these results suggest that a
methyl substituent on a cyclic substrate facilitates oxygenation of
the ring carbon to which it is attached unless the methyl group is
oriented *trans* to the amide function.[134]

REFERENCES

1. E. C. Gottschalk, Jr., *The Wall Street Journal*, July 9, 1968,
    p. 1.
2. D. H. Peterson and H. C. Murray, *J. Amer. Chem. Soc. 74*, 1871
    (1952).
3. E. A. Weaver, *U. S. Patent* 3,019,170 (January 30, 1962).
4. E. A. Weaver, H. E. Kenney, and M. E. Wall, *Appl. Microbiol. 8*,
    345 (1960).
5. D. H. Peterson, H. C. Murray, S. H. Eppstein, L. M. Reineke, A.
    Weintraub, P. D. Meister, and H. M. Leigh, *J. Amer. Chem. Soc.
    74*, 5933 (1952).
6. H. C. Murray and D. H. Peterson, *U. S. Patent* 2,602,769 (July 8,
    1952).
7. E. Riedl-Tumova, O. Siblikova-Zbudovska, and O. Hanc, *Cesk.
    Farm. 4*, 65 (1955).
8. A. G. Timofeeva, O. S. Madaeva, E. G. Gusakova, N. F. Kovylkina,
    N. I. Men'skova, and V. M. Novikova, *Izvestiya Akad. Nauk. SSSR,
    Ser. Biol.*, 712 (1958); *Chem. Abstr. 53*, 8302g (1959).
9. A. Bowers, J. S. Mills, C. Casas-Campillo, and C. Djerassi, *J.
    Org. Chem. 27*, 301 (1962).
10. C. Casas-Campillo and J. Ruiz-Herrera, *Rev. Latinoamer. Micro-*

biol. 3, 213 (1960).

11. D. H. Peterson, A. H. Nathan, P. D. Meister, S. H. Eppstein,
     H. C. Murray, A. Weintraub, L. M. Reineke, and H. M. Leigh, *J.
     Amer. Chem. Soc. 75*, 419 (1953).

12. D. H. Peterson, P. D. Meister, A. Weintraub, L. M. Reineke,
     S. H. Eppstein, H. C. Murray, and H. M. (Leigh) Osborn, *J. Amer.
     Chem. Soc. 77*, 4428 (1955).

13. H. C. Murray and D. H. Peterson, *U. S. Patent* 2,920,073 (January 5, 1960).

14. A. Ercoli, P. deRuggieri, and D. Della Morte, *Gazz. Chim. Ital.
     85*, 628 (1955).

15. H. E. Kenney, S. Serota, E. A. Weaver, and M. E. Wael, *J. Amer.
     Chem. Soc. 74*, 3689 (1960).

16. W. J. Wechter and H. C. Murray, *Chem. Ind.*, 411 (1962).

17. W. J. Wechter and H. C. Murray, *J. Org. Chem. 28*, 755 (1963).

18. H. Wehrli, M. Cereghetti, K. Schaffner, J. Urech, and E.
     Vischer, *Helv. Chim. Acta 44*, 1927 (1961).

19. L. J. Lerner, A. I. Laskin, and F. L. Weisenborn, *U. S. Patent*
     3,005,017 (October 17, 1961).

20. A. I. Laskin and F. L. Weisenborn, *Bacteriol. Proc.*, 26 (1962).

21. A. Schubert and R. Siebert, *Chem. Ber. 91*, 1856 (1958).

22. A. Schubert, G. Langbein, and R. Siebert, *Chem. Ber. 90*, 2576
     (1957).

23. A. Gubler and C. Tamm, *Helv. Chim. Acta 41*, 301 (1958).

24. R. D. Muir and R. M. Dodson, *U. S. Patent* 2,823,170 (February
     11, 1958).

25. E. O. Karow and D. N. Petsiavas, *Ind. Eng. Chem. 48*, 2213
     (1956).

26. C. Vezina, S. N. Sehgal, and K. Singh, *U. S. Patent* 3,352,760
     (November 14, 1967).

27. P. D. Meister and A. Weintraub, *U. S. Patent* 2,877,162 (March
     10, 1959).

28. D. van der Sijde, J. de Flines, and W. F. van der Waard, *Rec.
     Trav. Chim. 85*, 721 (1966).

29. W. F. van der Waard, D. van der Sijde, and J. de Flines, *Rec.
     Trav. Chim. 85*, 712 (1966).

30. H. Els, G. Englert, A. Furst, P. Reusser, and A. J. Schocher, *Helv. Chim. Acta 52*, 1157 (1969).

31. J. de Flines, D. van der Sijde, and W. F. van der Waard, *Rec. Trav. Chim. 85*, 701 (1966).

32. C. Meystre, E. Vischer, and A. Wettstein, *Helv. Chim. Acta 38*, 381 (1955).

33. W. J. McAleer, M. A. Kozlowski, T. H. Stoudt, and J. M. Chemerda, *J. Org. Chem. 23*, 958 (1958).

34. O. K. Sebek, *U. S. Patent* 3,116,220 (December 31, 1963).

35. J. Fried, R. W. Thoma, J. R. Gerke, J. E. Herz, M. N. Donin, and D. J. Perlman, *J. Amer. Chem. Soc. 74*, 3962 (1952).

36. C. Tamm, A. Gubler, G. Juhasz, E. Weiss-Berg, and W. Zurcher, *Helv. Chim. Acta 46*, 889 (1963).

37. E. Weiss-Berg and C. Tamm, *Helv. Chim. Acta 46*, 1166 (1963).

38. K. M. Mann, F. R. Hanson, P. W. O'Connell, H. V. Anderson, M. P. Brunner, and J. N. Karnemaat, *Appl. Microbiol. 3*, 14 (1955).

39. a. M. H. J. Zuidweg, *Biochim. Biophys. Acta 152*, 144 (1968).
    b. K. Mosbach and P.-O. Larrson, *Biotechnol. Bioeng. 12*, 19 (1970).

40. E. Kondo and E. Masuo, *Ann. Repts. Shionogi Res. Lab. 10*, 103 (1960).

41. H. L. Herzog, M. J. Gentles, W. Charney, D. Sutter, E. Townley, M. Yudis, P. Kabasakalian, and E. B. Hershberg, *J. Org. Chem. 24*, 691 (1959).

42. W. Charney, H. L. Herzog, and D. Sutter, *U. S. Patent* 3,014,051 (December 19, 1961).

43. W. Charney, H. L. Herzog, and D. Sutter, *U. S. Patent* 2,958,631 (November 1, 1960).

44. C. Spalla, A. M. Amici, and M. L. Bianchi, *Giorn. Microbiol. 9*, 249 (1961).

45. K. Singh, S. N. Sehgal, and C. Vezina, *Appl. Microbiol. 16*, 393 (1968).

46. S. D. Levine and P. A. Principe, *U. S. Patent* 3,382,155 (May 7, 1968).

47. A. R. Hanze, O. K. Sebek, and H. C. Murray, *J. Org. Chem. 25*, 1968 (1960).

48. A. R. Hanze, O. K. Sebek, and H. C. Murray, *U. S. Patent* 3,038,913 (June 12, 1962).

49. K. Kieslich, K. Petzoldt, H. Kosmol, and W. Koch, *Justus Liebigs Ann. Chem. 726*, 168 (1969).

50. J. de Flines, W. F. van der Waard, W. J. Mijs, and S. A. Szpilfogel, *Rec. Trav. Chim. 82*, 129 (1963).

51. P. Crabbe and C. Casas-Campillo, *U. S. Patent* 3,375,174 (March 26, 1968).

52. C. J. Sih, *Biochim. Biophys. Acta 62*, 541 (1962).

53. J. E. Bridgeman, P. C. Cherry, E. R. H. Jones, and G. D. Meakins, *Chem. Commun.*, 482 (1967).

54. P. C. Cherry, E. R. H. Jones, and G. D. Meakins, *Chem. Commun.*, 587 (1966).

55. J. E. Bridgeman, J. W. Browne, P. C. Cherry, M. G. Combe, J. M. Evans, E. R. H. Jones, A. Kasal, G. D. Meakins, V. Morisawa, and P. D. Woodgate, *Chem. Commun.*, 463 (1969).

56. G. Wix, K. G. Büki, E. Tömörkeny, and G. Ambrus, *Steroids 11*, 401 (1968).

57. Y. Y. Lin, M. Shibahara, and L. L. Smith, *J. Org. Chem. 34*, 3530 (1969).

58. L. L. Smith, G. Greenspan, R. Rees, and T. Foell, *J. Amer. Chem. Soc. 88*, 3120 (1966).

59. I. I. Zaretskaya, L. M. Kogan, O. B. Tikhomirova, Jh. D. Sis, N. S. Wulfson, V. I. Zaretskii, V. G. Zaikin, G. K. Skryabin, and I. V. Torgov, *Tetrahedron 24*, 1595 (1968).

60. H. C. Murray and P. D. Meister, *U. S. Patent* 2,889,255 (June 2, 1959).

61. P. D. Meister, D. H. Peterson, S. H. Eppstein, H. C. Murray, L. M. Reineke, A. Weintraub, H. M. (Leigh) Osborn, *J. Amer. Chem. Soc. 76*, 5679 (1954).

62. W. von Daehne, H. Lorch, and W. O. Godtfredsen, *Tetrahedron Lett.*, 4843 (1968).

63. S. Okuda, Y. Sato, T. Hattori, and M. Wakabayashi, *Tetrahedron Lett.*, 4847 (1968).

64. L. Gsell and C. Tamm, *Helv. Chim. Acta 52*, 150 (1969).

65. M. Okada, A. Yamada, and M. Ishidate, *Chem. Pharm. Bull. 8*, 530 (1960).

66. Y. Sato, Y. Sato, and K. Tanabe, *Steroids 9*, 553 (1967).

67. Y. Sato and S. Hayakawa, *J. Org. Chem. 28*, 2739 (1963).

68. W. H. Bradshaw, H. E. Conrad, E. J. Corey, I. C. Gunsalus, and D. Lednicer, *J. Amer. Chem. Soc. 81*, 5507 (1959).

69. H. E. Conrad, J. Hedegaard, I. C. Gunsalus, E. J. Corey, and H. Uda, *Tetrahedron Lett.*, 561 (1965).

70. P. J. Chapman, G. Meerman, I. C. Gunsalus, R. Srinivasan, and K. L. Rinehart, *J. Amer. Chem. Soc. 88*, 618 (1966).

71. J. Hedegaard and I. C. Gunsalus, *J. Biol. Chem. 240*, 4038 (1965).

72. I. C. Gunsalus, P. J. Chapman, and J. K. Kuo, *Biochem. Biophys. Res. Commun. 18*, 924 (1965).

73. B. Pfrunder and C. Tamm, *Helv. Chim. Acta 52*, 1644 (1969).

74. C. Tamm, *Abstr. Papers, Amer. Chem. Soc.*, No. 151, I31 (1966).

75. H. Ishii, T. Tozyo, and H. Minato, *Chem. Commun.*, 649 (1968).

76. H. Hikino, Y. Tokuoka, Y. Hikino, and T. Takemoto, *Tetrahedron 24*, 3147 (1968).

77. H. Hikino, T. Kohama, and T. Takemoto, *Chem. Pharm. Bull. 17*, 1659 (1969).

78. H. Hikino, K. Aota, Y. Tokuoka, and T. Takemoto, *Chem. Pharm. Bull. 16*, 1088 (1968).

79. J. F. Biellmann, R. Wennig, P. Daste, and M. Raynaud, *Chem. Commun.*, 168 (1968).

80. M. Raynaud, J. F. Biellmann, and P. Daste, *Compt. rend. Soc. Biol. 160*, 371 (1965).

81. D. R. Brannon, H. Boaz, B. J. Wiley, J. Mabe, and D. Horton, *J. Org. Chem. 33*, 4462 (1968).

82. B. E. Cross and P. L. Myers, *Biochem. J. 108*, 303 (1968).

83. B. E. Cross and K. Norton, *Chem. Commun.*, 535 (1965).

84. B. E. Cross, R. H. B. Galt, and J. R. Hanson, *J. Chem. Soc.*, 295 (1964).

85. J. R. Hanson and A. F. White, *Tetrahedron 24*, 6291 (1968).

86. S. Nozae, M. Morisaki, K. Tsuda, and S. Okuda, *Tetrahedron Lett.*, 3365 (1967).

87. L. Canonica, A. Fiecchi, M. Galli Kienle, B. M. Ranzi, A. Scala,
     T. Salvatori, and E. Pella, *Tetrahedron Lett.*, 3371 (1967).

88. L. Canonica, M. Ferrari, G. Jommi, U. M. Paguoni, F. Pelizzoni,
     B. M. Ranzi, S. Maroni, G. Neucini, and T. Salvatori, *Gazz.
     Chim. Ital. 97*, 1032 (1967).

89. L. Canonica, G. Jommi, U. M. Paguoni, F. Pelizzoni, B. M.
     Ranzi, and C. Scolastico, *Gazz. Chim. Ital. 96*, 820 (1966).

90. H. Hikino, S. Nabetani, and T. Takemoto, *Yakugaku Zasshi 89*,
     809 (1969).

91. S. C. Pan and F. L. Weisenborn, *J. Amer. Chem. Soc. 80*, 4749
     (1958).

92. F. Weisenborn and S. C. Pan, *U. S. Patent* 3,003,926 (October 10,
     1961).

93. E. L. Patterson, W. W. Andres, E. F. Krause, R. E. Hartman, and
     L. A. Mitscher, *Arch. Biochem. Biophys. 103*, 117 (1963).

94. T. L. Lemke, R. A. Johnson, H. C. Murray, D. J. Duchamp, C. G.
     Chidester, J. B. Hester, Jr., and R. V. Heinzelman, *J. Org.
     Chem. 36*, in press.

95. M. Toczko, *Biochim. Biophys. Acta 128*, 570 (1966).

96. M. Mozejko-Toczko, *Acta Microbiol. Polonica 9*, 157 (1960).

97. L. Marion and N. J. Leonard, *Canad. J. Chem. 29*, 355 (1951).

98. H. B. Lukins and J. W. Foster, *J. Bacteriol. 85*, 1074 (1963).

99. F. W. Forney, A. J. Markovetz, and R. E. Kallio, *J. Bacteriol.
     93*, 649 (1967).

100. J. E. Stewart, R. E. Kallio, D. P. Stevenson, A. C. Jones, and
     D. O. Schissler, *J. Bacteriol. 78*, 441 (1959).

101. J. E. Stewart and R. E. Kallio, *J. Bacteriol. 78*, 726 (1959).

102. D. P. Stevenson, W. R. Finnerty, and R. E. Kallio, *Biochem.
     Biophys. Res. Commun. 9*, 426 (1962).

103. M. J. Klug and A. J. Markovetz, *J. Bacteriol. 93*, 1847 (1967).

104. M. J. Klug and A. J. Markovetz, *J. Bacteriol. 96*, 1116 (1968).

105. M. T. Heydeman, *Biochim. Biophys. Acta 42*, 557 (1960).

106. A. S. Kester and J. W. Foster, *J. Bacteriol. 85*, 859 (1963).

107. M. Y. Ali Khan, A. H. Hall, and D. S. Robinson, *Nature 198*, 289
     (1963).

108. G. J. E. Thijsse and A. C. van der Linden, *Antonie van Leeuwen-hoek 29*, 89 (1963).

109. A. P. Tulloch, J. F. T. Spencer, and P. A. J. Gorin, *Canad. J. Chem. 40*, 1326 (1962).

110. A. P. Tulloch, A. Hill, and J. F. T. Spencer, *Chem. Commun.*, 584 (1967).

111. D. F. Jones and R. Howe, *J. Chem. Soc. (C)*, 2801 (1968).

112. G. J. E. Thijsse and A. C. van der Linden, *Antonie van Leeuwen-hoek 27*, 171 (1961).

113. D. F. Jones, *J. Chem. Soc. (C)*, 2809 (1968).

114. D. F. Jones and R. Howe, *J. Chem. Soc. (C)*, 2816 (1968).

115. D. F. Jones and R. Howe, *J. Chem. Soc. (C)*, 2821 (1968).

116. C. T. Goodhue, J. R. Schaeffer, R. E. Stevens, and H. A. Risley, *Abstr. Papers, Amer. Chem. Soc.*, No. 158, MICR 48 (1969).

117. R. A. Johnson, H. C. Murray, and L. M. Reineke, *J. Amer. Chem. Soc. 93*, in press.

118. D. F. Jones, *J. Chem. Soc. (C)*, 2827 (1968).

119. E. Heinz, A. P. Tulloch, and J. F. T. Spencer, *J. Biol. Chem. 244*, 882 (1969).

120. E. Heinz, A. P. Tulloch, and J. F. T. Spencer, *Biochim. Bio-phys. Acta 202*, 49 (1970).

121. D. A. Klein, J. A. Davis, and L. E. Casida, *Antonie van Leeuw-enhoek 34*, 495 (1968).

122. C. J. Sih, G. Ambrus, P. Foss, and C. J. Lai, *J. Amer. Chem. Soc. 91*, 3685 (1969).

123. J. Ooyama and J. W. Foster, *Antonie van Leeuwenhoek 31*, 45 (1965).

124. Y. Arai and K. Yamada, *Agr. Biol. Chem. 33*, 63 (1969).

125. G. S. Fonken, M. E. Herr, and H. C. Murray, *U. S. Patent* 3,281,330 (October 25, 1966).

126. H. C. Murray and R. A. Johnson, unpublished results.

127. G. S. Fonken, M. E. Herr, H. C. Murray, and L. M. Reineke, *J. Amer. Chem. Soc. 89*, 672 (1967).

128. R. D. Swisher, *Develop. Ind. Microbiol. 6*, 39 (1964).

129. M. H. Rogoff, *J. Bacteriol. 83*, 998 (1962).

130. G. S. Fonken, M. E. Herr, H. C. Murray, and L. M. Reineke, *J. Org. Chem. 33*, 3182 (1968).
131. R. A. Johnson, M. E. Herr, H. C. Murray, and G. S. Fonken, *J. Org. Chem. 33*, 3217 (1968).
132. G. S. Fonken, M. E. Herr, and H. C. Murray, *U. S. Patent* 3,352,884 (November 14, 1967).
133. R. A. Johnson, M. E. Herr, H. C. Murray, and G. S. Fonken, *J. Org. Chem. 35*, 622 (1970).
134. R. A. Johnson, M. E. Herr, H. C. Murray, and L. M. Reineke, *J. Amer. Chem. Soc. 93*, in press.
135. R. A. Johnson, M. E. Herr, H. C. Murray, and G. S. Fonken, *J. Org. Chem. 33*, 3187 (1968).
136. R. A. Johnson, H. C. Murray, L. M. Reineke, and G. S. Fonken, *J. Org. Chem. 34*, 2279 (1969).
137. R. A. Johnson, *J. Org. Chem. 33*, 3627 (1968).
138. J. B. Hester, A. H. Tang, H. H. Keasling, and W. Veldkamp, *J. Med. Chem. 11*, 101 (1968).
139. G. S. Fonken, M. E. Herr, and H. C. Murray, *U. S. Patent* 3,304,323 (February 14, 1967).
140. V. Prelog and H. E. Smith, *Helv. Chim. Acta 42*, 2624 (1959).
141. A. Schubert, A. Rieche, G. Hilgetag, R. Siebert, and S. Schwarz, *Naturwiss. 24*, 623 (1958).
142. M. E. Herr, H. C. Murray, and G. S. Fonken, *J. Med. Chem. 14*, in press.
143. R. A. Johnson, H. C. Murray, L. M. Reineke, and G. S. Fonken, *J. Org. Chem. 33*, 3207 (1968).
144. R. A. Johnson, M. E. Herr, H. C. Murray, L. M. Reineke, and G. S. Fonken, *J. Org. Chem. 33*, 3195 (1968).
145. R. A. Johnson, H. C. Murray, and L. M. Reineke, *J. Org. Chem. 34*, 3834 (1969).
146. M. E. Herr, R. A. Johnson, H. C. Murray, L. M. Reineke, and G. S. Fonken, *J. Org. Chem. 33*, 3201 (1968).
147. M. E. Herr, R. A. Johnson, W. C. Krueger, H. C. Murray, and L. M. Pschigoda, *J. Org. Chem. 35*, 3607 (1970).

General References

General - Ch. Tamm, *Angew. Chem. 74*, 225 (1962); *Angew. Chem. Intern. Ed. 1*, 78 (1962).

D. Perlman (ed.), *Fermentation Advances*, Academic, New York, 1969.

K. Kieslich, *Synthesis*, 120 (1969).

Steroids - W. Charney and H. L. Herzog, *Microbial Transformations of Steroids*, Academic, New York, 1967.

A. Capek, O. Hanc, and M. Tadra, *Microbial Transformations of Steroids*, Academia, Prague, 1966.

Terpenoids - M. Raynaud, Ph. Daste, F. Grossin, J. F. Biellmann, and R. Wennig, *Ann. Inst. Pasteur 115*, 731 (1968).

Alkaloids - H. Iizuka and A. Naito, *Microbial Transformation of Steroids and Alkaloids*, University Park Press, State College, Pennsylvania, 1967.

Hydrocarbons - J. B. Davis, *Petroleum Microbiology*, Elsevier, Amsterdam, 1967.

C. Ratledge, *Chem. Ind.*, 843 (1970).

Chapter 2

ALLYLIC OXYGENATIONS

Allylic oxygenations by microorganisms may be considered to be of three types: (a) oxygenation of the classical monounsaturated three-carbon allyl group, either with or without rearrangement (and occasionally followed by reduction of the double bond); (b) oxygenation at the aliphatic carbon of a benzyl group; and (c) oxygenation adjacent to a heteroaliphatic unsaturated group, such as a ketone.

ALLYLIC

Steroids, Terpenoids, and Related Compounds. It is of some historical interest that the earliest report of the microbiological hydroxylation of a steroid involved an allylic oxidation. Kramli and Horvath used *Proactinomyces roseus* to convert cholesterol to 7-hydroxycholesterol.[1]

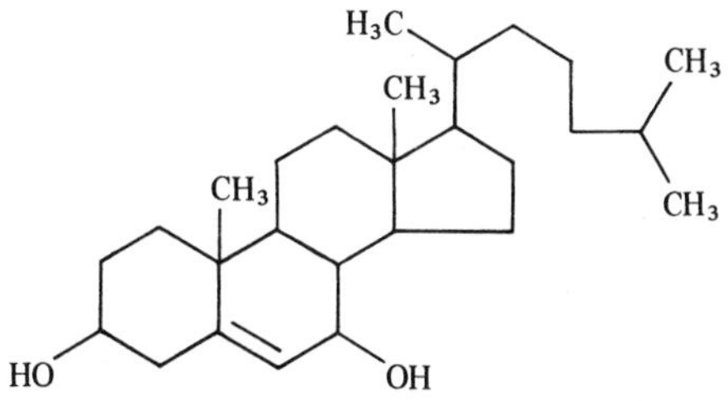

7-Hydroxycholesterol

Since the extent to which allylic oxygenations of steroidal molecules have been examined is well reviewed by Charney and Herzog (see General References), only a few examples are included here to acquaint the reader with some of the possibilities that exist.

85

   The natural abundance of steroids having unsaturation at positions C-4 and C-5 results in the observance of numerous examples of
hydroxylation at the 6-position.   An outstanding example is the hydroxylation of 17-methyltestosterone by *Gibberella saubinetti* in 90%
yield.[2]   As is generally the case, the 6-hydroxyl group has the β-
configuration.

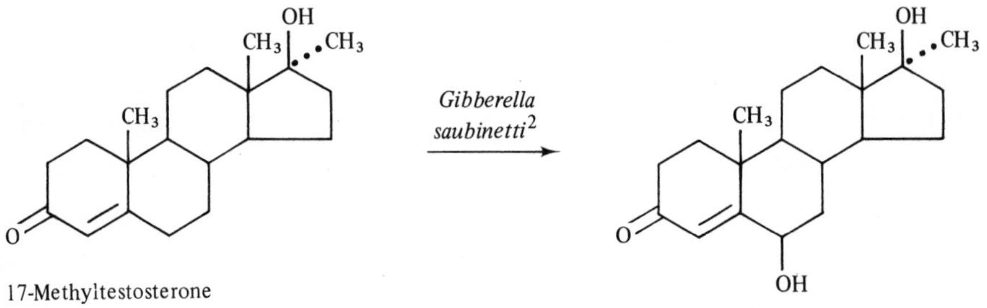

17-Methyltestosterone

   A striking example of two allylic oxidations in the same molecule is the conversion of 9,11-dehydroprogesterone to the 6β,12α-di-
hydroxy derivative by *Colletotrichum phomoides* in about 50% yield.[3]

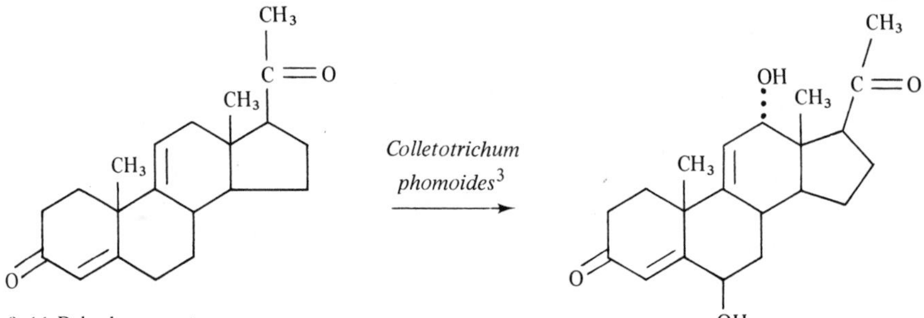

9,11-Dehydroprogesterone

   As is the case for conventional steroid oxidation, *retro* (9β,
10α) steroids also undergo allylic oxidation, usually as one of several reactions. Hydroxylation at the 8-position, rarely observed in
conventional steroids, is observed with the *retro*-pregnadiene derivative shown below.[4]

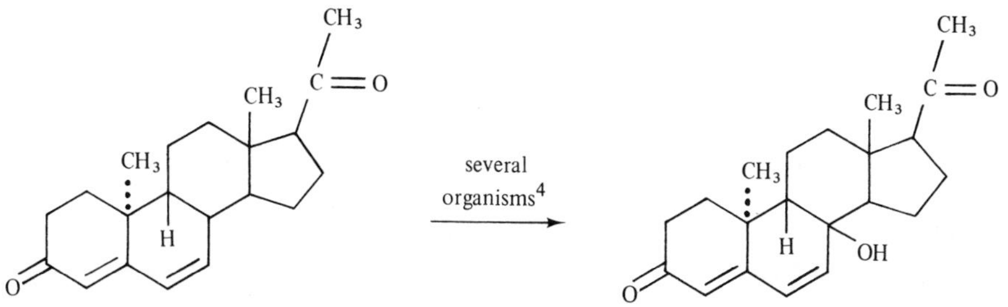

There may be some differences between the microbial oxidation of
d- (natural) and l-steroids, as illustrated by the oxidation of 19-
nortestosterone (either d- or dl-) by *Curvularia lunata*[5]:

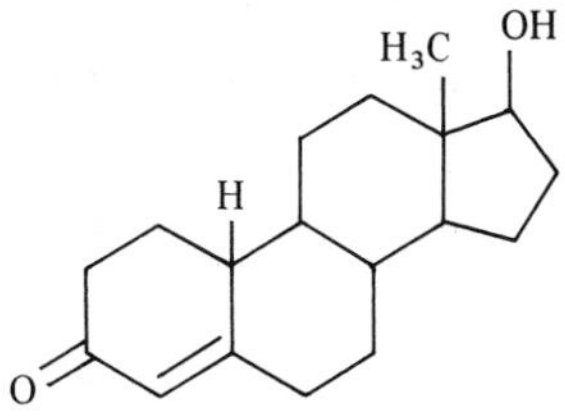

d-19-Nortestosterone

Allylic Products (Yields, %)	
From d	From dl
d-6β-OH (1.4)	dl-6β-OH (4.3)
d-10β-OH (52)	d-10β-OH (18.9)
d-10β,11β-(OH)$_2$	dl-10β,11β-(OH)$_2$
(0.5)	(1.3)

Alteration of the C-13 methyl group in dl-19-nortestosterone to ethyl
or propyl also seems to influence the nature of the hydroxylations,
both allylic and other, seen with *Aspergillus ochraceus*,[6] and with
Curvularia lunata,[5] but definitive conclusions are not yet warranted.

The steroid alkaloids solasodine and conessine undergo microbiological allylic oxidations (although in the case of the solasodine
oxidation, the major product -- 9α-hydroxysolasodine -- is not the
result of allylic oxidation).

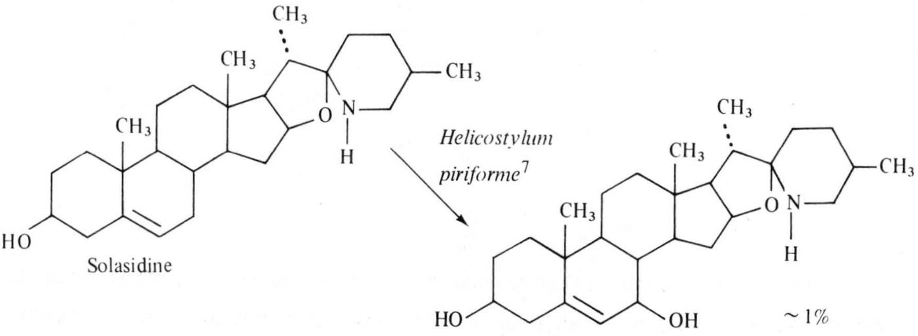

Solasidine

Helicostylum piriforme[7]

~ 1%

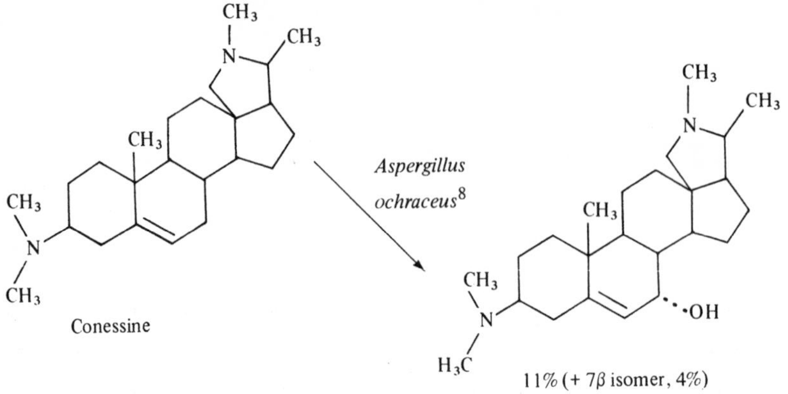

Conessine

Aspergillus ochraceus[8]

11% (+ 7β isomer, 4%)

The tricyclic hydroxyketone (A) resembles the steroid structure sufficiently to warrant its inclusion here. Of the two products obtained by incubation with *Cunninghamella bainieri*, the first (B) has been used as an intermediate toward the synthesis of cassaine.[9]

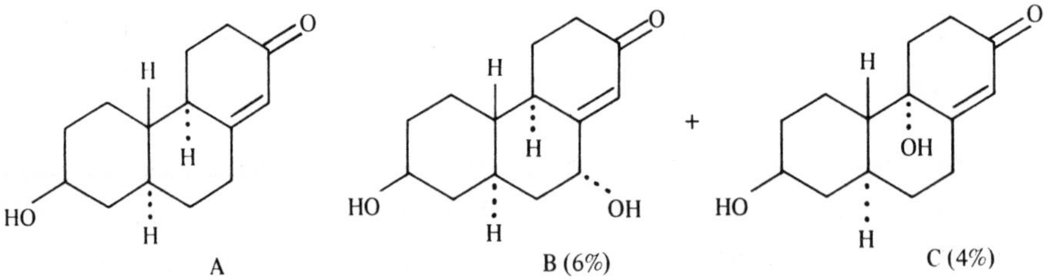

A   B (6%)   C (4%)

An old report describes the oxygenation of dehydronorcholene by a bacterium (*Escherichia coli*).[10]

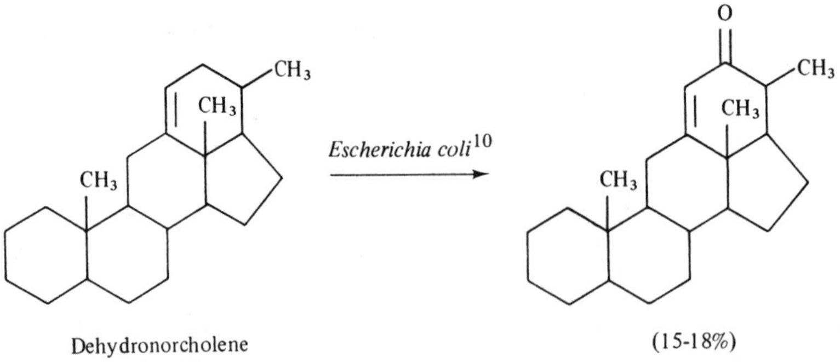

Dehydronorcholene

Escherichia coli[10]

(15-18%)

The hydroxylation of cinerone to cinerolone is important because 2-substituted 3-methyl-4-hydroxycyclopentenones can be used as chemi-

cal intermediates to pyrethrins, a valuable class of naturally occurring insecticides.   Although a variety of microorganisms   carry out
this oxidation, the most effective are *Aspergillus niger* and *Strepto-*
myces aureofaciens.[11]   The conversion   yields (assay) to cinerolone
are 60% and 42%, respectively. A second product of allylic oxidation
(at the terminal methyl group) is also formed (as is some dihydrocin-
erolone). Allethrone is similarly oxygenated at the 4-position, giving allethrolone as the major product (16%).   Other products   result
from oxygenation of the allylic 3-methyl group (4%) and of the doubly
allylic methylene group (1%).[12]

Cinerone, R = CH₃
Allethrone, R = H

Cinerolone, R = CH₃
Allethrolone, R = H

Allylic microbiological oxidations of terpenoids have been limited almost exclusively to the monoterpenes and to a few closely related nonterpenes, with the major contribution coming from Bhattacharyya and co-workers in India.   Their work was originally undertaken to
investigate the possibility that the essential oils of agarwood (*Agu-*
ilaria agallocha) arise from fungal infection of the tree, with conversion of the normally odorless oleoresinous secretions to odorifer-
ous ones.   Technical difficulties, including an inadequate supply of
agarwood hydrocarbon, precluded a satisfactory   resolution of that
problem, but did not prevent an examination of the fungal oxidation
of readily available terpene hydrocarbons.

α-Pinene (70% *d*, 30% *l*) is utilized by a variety of fungi, of
which a strain (NCIM 612) of *Aspergillus niger* is most efficient.[13,14]
A maximum yield of oxygenated products is obtained after about an 8-
hour incubation.   About 75% of the substrate is recovered, and the
reaction products are *d*-verbenone (2% yield), *d-cis*-verbenol (6%),
and *d-trans*-sobrerol (4%). Since verbenol and verbenone are produced

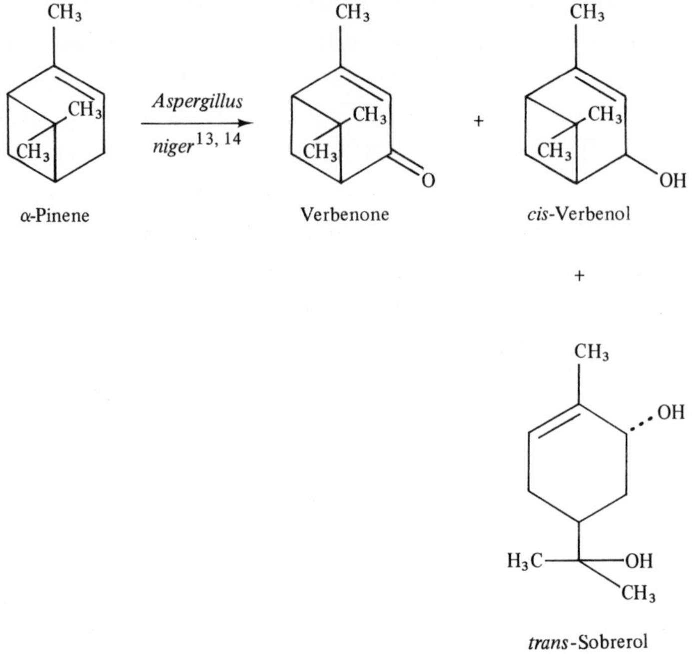

α-Pinene   Verbenone   cis-Verbenol

+

trans-Sobrerol

by autoxidation of α-pinene,[15] it is noteworthy that no appreciable
autoxidation is observed when the microorganism is omitted from the
"fermentation." Furthermore, oxidation of α-pinene by A. *niger* yields
cis-verbenol, while autoxidation affords the *trans*-isomer.   Finally,
the microbial metabolites are obtained optically pure, despite the
use of optically impure α-pinene, while autoxidation gives mixtures
of optical isomers.

When carane, Δ³-carene, α-santalene, or humulene are subjected
to A. *niger*, only the α-santalene gives appreciable amounts of characterizable products.[16] A 24-hour fermentation gives about a 1% yield
of a tertiary alcohol that has not been completely characterized.

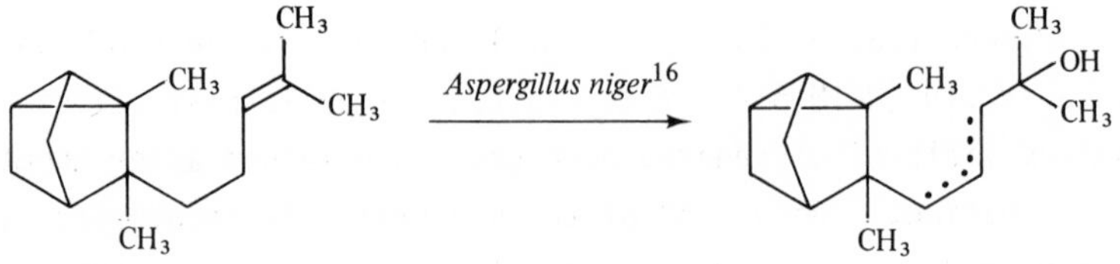

α-Santalene

Other products of the fermentation, representing further degradation, perhaps of the indicated tertiary alcohol, are teresantalic acid (about 10% yield) and teresantalol (0.3%). Here also, as in the case

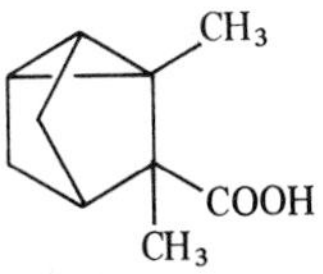

CH$_3$

COOH
CH$_3$

Teresantalic acid

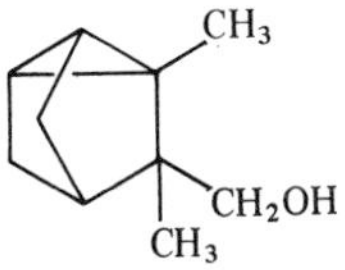

CH$_3$

CH$_2$OH
CH$_3$

Teresantalol

of α-pinene, a striking difference is seen between microbial oxidation and autoxidation. In this instance, however, the three compounds (carane, Δ^3-carene, and humulene) that are readily susceptible to autoxidation are resistant to oxidation by the microorganism.

Some simple model compounds (such as cyclohexene and methylcyclohexenes) that closely resemble the terpenes are also oxidized by A. niger.[17,18] These fermentations present a technical problem: the high volatility of the substrates precludes the use of open shaken flasks, so that closed fermentors must be used. The cyclohexene oxidation gives 2-cyclohexenone, (+)-2-cyclohexenol, and (+)-3-cyclohexene-1,2-cis-diol.[17] Chromatographic assay of samples taken during the

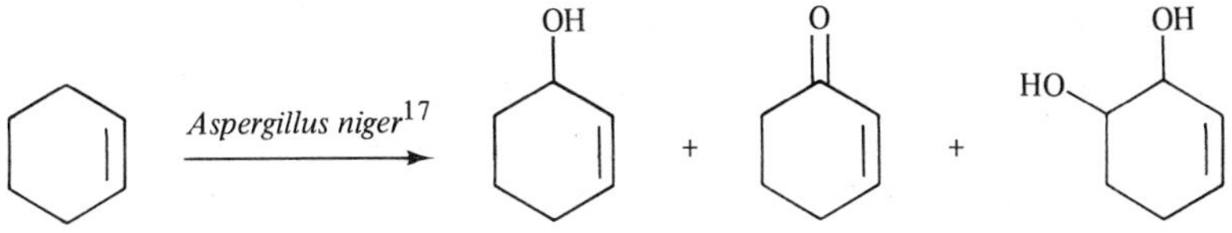

OH

O

OH

HO

Aspergillus niger[17]   +   +

Cyclohexene

course of the fermentation shows that 2-cyclohexenol is the initial oxidation product, and that the other two compounds are formed by its further oxidation. The yields are extremely low, and about 33% of the cyclohexene is recovered.[17]

The microbial oxidations of 1-methylcyclohexene and of 4-methylcyclohexene are carried out essentially as for cyclohexene.[18] The 1-methylcyclohexene gives (+)-1-methylcyclohexen-6-ol (3.5%) and 1-

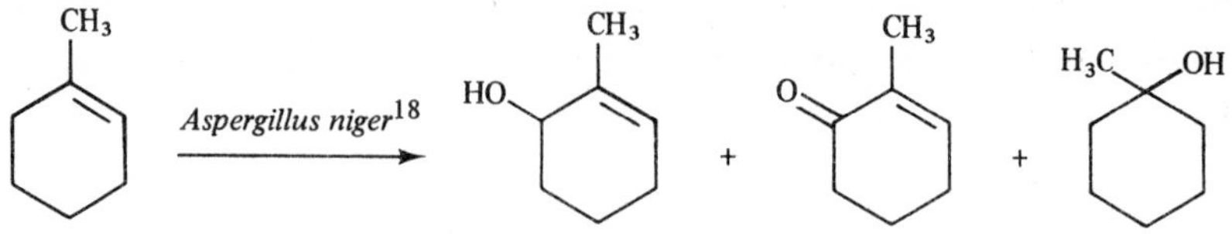

1-Methylcyclohexene

methylcyclohexen-6-one (1%) (together with the hydration product 1-
methylcyclohexanol), while the 4-methylcyclohexene affords (±)-1-
methyl-3-cyclohexen-2-ol (3.6%) and the corresponding ketone (2.4%).
(Substrate is recovered from both fermentations in excess of 50%.[18])

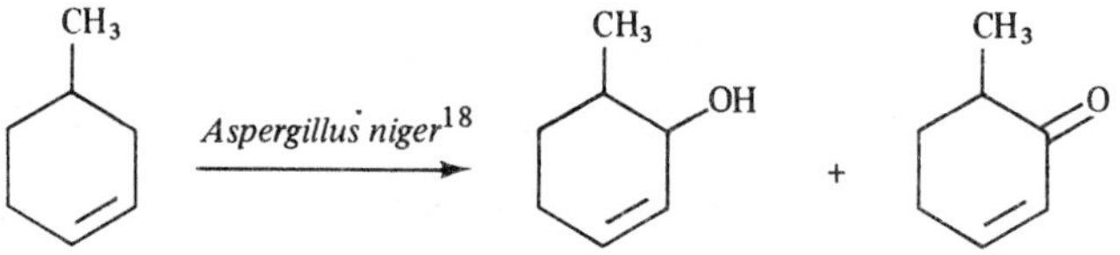

4-Methylcyclohexene

The oxidative transformation of limonene by a soil pseudomonad
gives carvone, cis-carveol, and many other more extensively oxidized
products.[19,20]   These products are thought to arise from the initial
oxidation of limonene to perillyl alcohol, but none of the latter

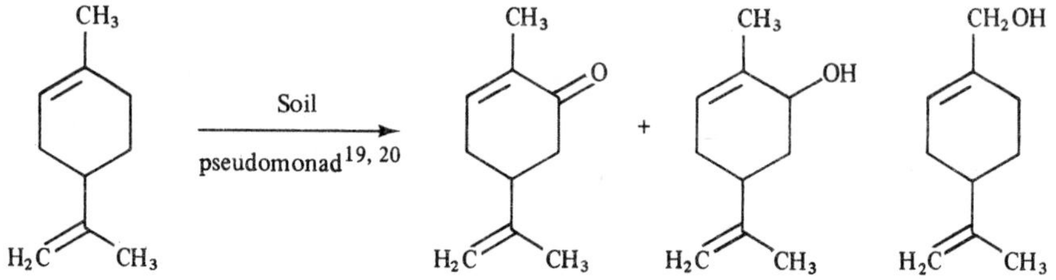

Limonene   Carvone   cis-Carveol   Perillyl
                                                                         alcohol

compound can, in fact, be identified as such in the fermentation extracts.   Since the multiplicity of products is great, this reaction
would not appear to have value as a synthetic process at this stage
of its development.
         The oxygenation of α-pinene is effected by the same pseudomonad
used for the transformation of limonene.   Minor products resulting

from allylic oxygenation are myrtenol and myrtenic acid, but the major products from this transformation result from extensive re-arrangements of the pinane ring.[21,22]

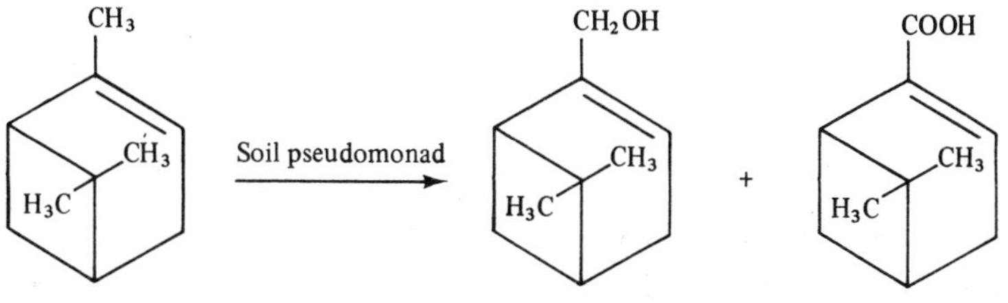

α-Pinene   Myrtenol   Myrtenic acid

The allylic 6α-hydroxylation of the sesquiterpene ketone cypero-tundone by *Corticium sasakii* affords sugeonol.[23] Since sugeonol is also a naturally occurring constituent of Japanese nutgrass, *Cyperus rotundus*, this preparation represents a microbial synthesis of the natural product.

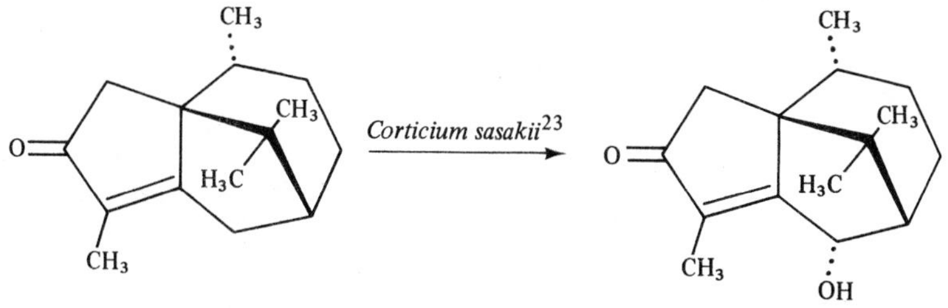

Cyperotundone   Sugeonol

Of the few reported allylic oxidations of larger terpenoids, the ones found in the biosynthetic pathways of the ophiobolins[24,25] and of gibberellic acid[26] probably do not represent useful synthetic re-

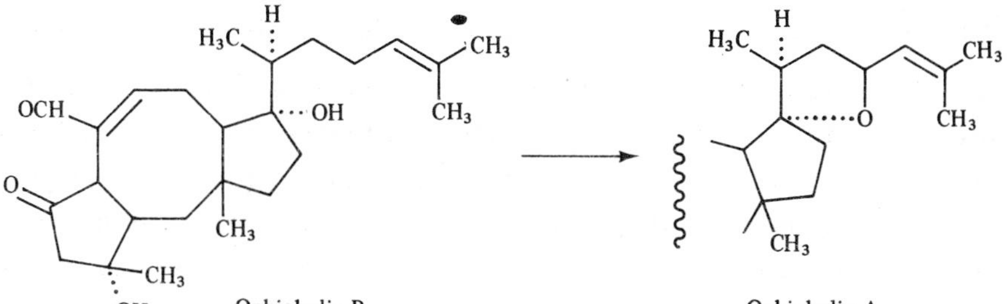

Ophiobolin B   Ophiobolin A

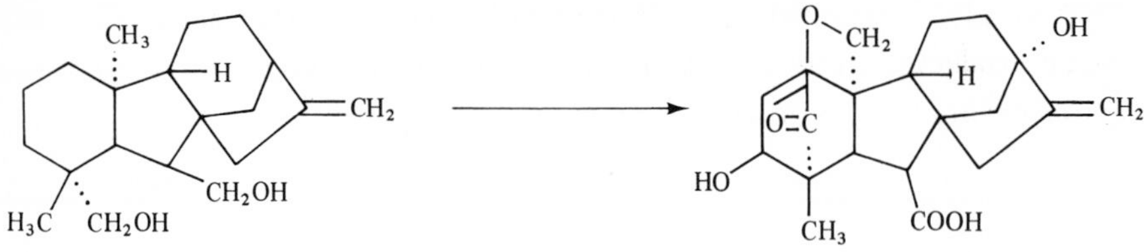

Gibberellic acid

actions since they are accompanied by other reactions. The oxidation of oleanolic acid by *Cunninghamella blakesleeana* to a multiplicity of products, each obtained in less than 1% yield, seems no more promising.[27]

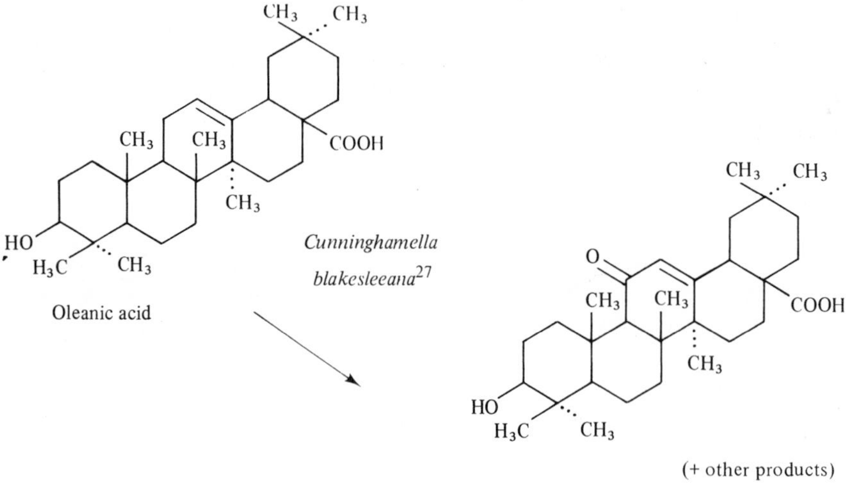

Oleanic acid

Cunninghamella blakesleeana[27]

(+ other products)

Antibiotics. Because of the immense commercial importance of the naturally produced antibiotics tetracycline, hydroxytetracycline, and chlorotetracycline, it is not at all surprising that some effort has been devoted to microbiological oxidation studies in this series.

The oxidation of 5a(11a)-dehydrotetracycline (DHTC) (itself a microbiological oxidation product of anhydrotetracycline[28]) to 5-hydroxytetracycline (HTC) by cell-free sonic extracts of *Streptomyces rimosus* goes by way of the intermediate 5a(11a)-dehydro-5-hydroxytetracycline (DHHTC).[29] The value of this sequence as a synthetic process is unclear. However, essentially the same over-all process

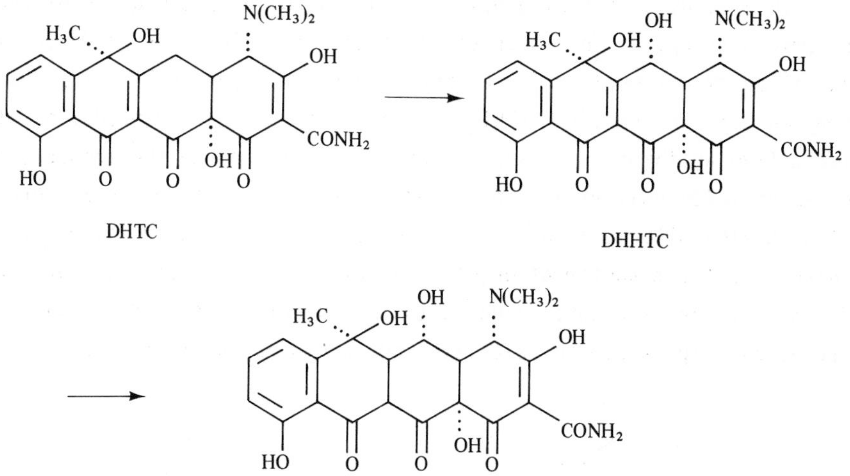

DHTC

DHHTC

HTC

can be used to produce significant amounts of 5-hydroxy-7-chlorotet-
racycline (HCTC), a hybrid antibiotic with the functionality of both
chlorotetracycline and hydroxytetracycline, by incubating 5a(11a)-
dehydro-7-chlorotetracycline (DCTC) with washed cells of *S. rimosus*.[30,31] The biotransformation yield after a 6-hour incubation is
about 40%, but the isolated yield is much lower due to the chemical
instability of the product in neutral or alkaline solutions.

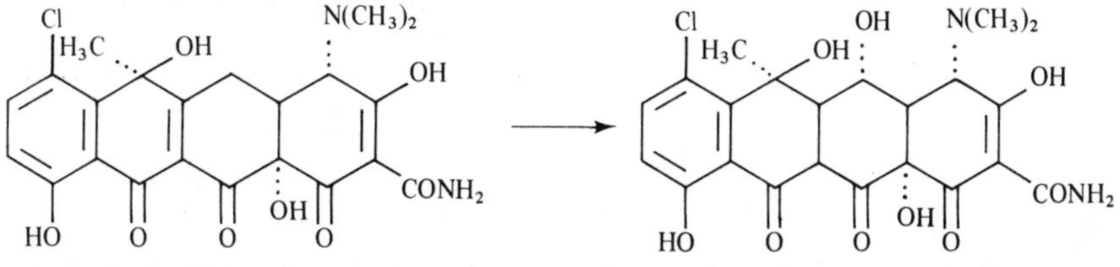

DCTC

HCTC

The introduction of the 5-hydroxyl group into DHTC requires at-
morpheric oxygen and the participation of NADPH (TPNH), which suggests strongly that the introduced oxygen at C-5 is derived from molecular oxygen rather than from water. It is a safe assumption that
the same applies in the case of the chloro compound DCTC.

In a study of the biogenesis of the macrolide antibiotics of the rifamycin group, it has been shown that rifamycin Y is produced by oxidation of rifamycin B by either growing or washed cells of *Streptomyces mediterranei*.[32] (This microorganism ordinarily produces at least five antibiotics *de novo*, but can be limited to the production of essentially only rifamycins B and Y by growing it in the presence of diethylbarbituric acid. The diethylbarbituric acid is not required for the transformation of rifamycin B to rifamycin Y.) Because *de novo* synthesis occurs along with the oxidation of B to Y, yields are almost impossible to calculate, but would appear to be low.

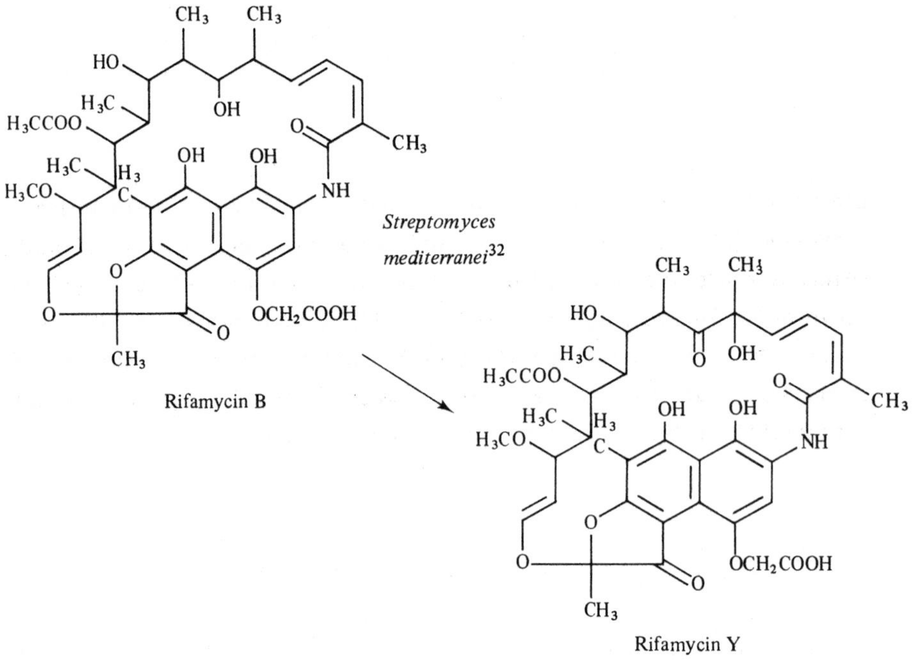

Streptomyces mediterranei[32]

Rifamycin B

Rifamycin Y

Alkaloids. Because of the medicinal value of several of the lysergic acid-related alkaloids produced by the ergot fungus that infects certain grass species, a great deal of effort has been expended on the isolation and elucidation of the biosynthesis of these fungal metabolites. In particular, the clavine alkaloids, substituted ergolines that are biosynthetic intermediates to the lysergic acid alkaloids, have been examined for their microbial oxidative transformations in several laboratories. Only a few selected oxidations will be

considered, since the literature, although extensive, is rarely ex-
plicit in regard to isolated yields of products.

Agroclavine, which is thought to be progenitor to the entire
clavine alkaloid family, is oxidized to elymoclavine, setoclavine,
and isosetoclavine by *Claviceps purpurea*.[33] Elymoclavine then undergoes further microbial oxidation to penniclavine and isopenniclavine.[33] The necessary enzymes for these oxidations appear to be quite

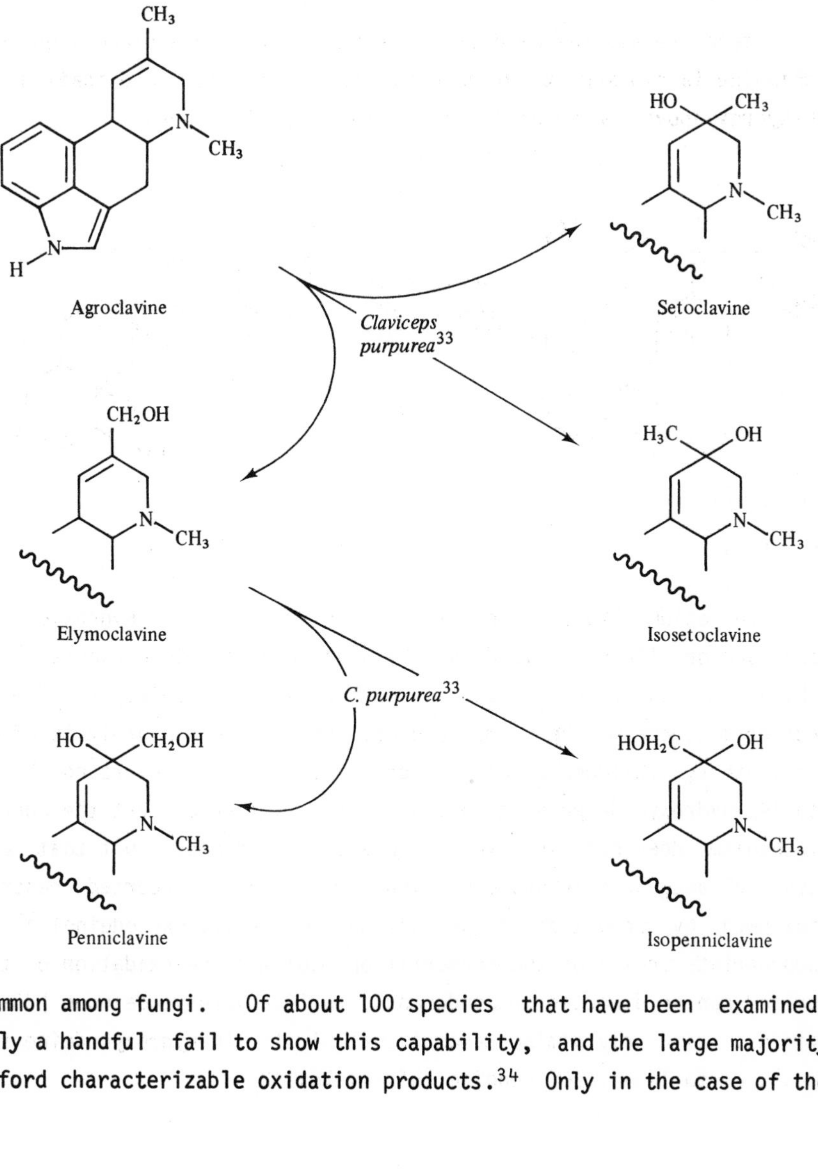

Agroclavine

Claviceps purpurea[33]

Setoclavine

Elymoclavine

Isosetoclavine

C. purpurea[33]

Penniclavine

Isopenniclavine

common among fungi. Of about 100 species that have been examined,
only a handful fail to show this capability, and the large majority
afford characterizable oxidation products.[34] Only in the case of the

biooxidation of agroclavine to setoclavine (and traces of isosetoclavine) by *Psilocybe semperviva* has an isolated yield been reported.
Setoclavine is obtained in 32% yield when the fermentation is carefully monitored and harvested at peak yield.[35] Since agroclavine can
be oxidized chemically (with acidic dichromate) to setoclavine (38%
yield) and isosetoclavine (13% yield), the microbial method may be of
value only as it may apply to other clavine substrates not amenable
to chemical oxidation.

Among the several oxidative reactions that take place when apoyohimbine is exposed to microorganisms is the allylic oxidation to
18-hydroxyapoyohimbine with *Cunninghamella blakesleeana.*[36]

Apoyohimbine

The opium alkaloid, thebaine, is oxidized to 14β-hydroxycodein-
one and/or 14β-hydroxycodeine by a variety of microorganisms, of
which *Trametes sanguinea* has been most carefully studied.[37] The reaction sequence may be formally considered to be a hydrolysis of the
enol ether, followed by allylic oxidation of the unsaturated ketone
to 14β-hydroxycodeinone (although there is evidence that the correct
mechanism does not involve an intermediate dienol, but that enol
ether cleavage and 14-hydroxylation proceed as a concerted reaction
followed by reduction of the ketone to 14β-hydroxycodeine).[38] By
appropriate choice of the fermentation medium, the oxidation of thebaine can be directed preponderantly to 14β-hydroxycodeinone [33-46%
yields, with only a minor amount (0-10%) of 14β-hydroxycodeine for-
med], or to 14β-hydroxycodeine (54% yield).[39]

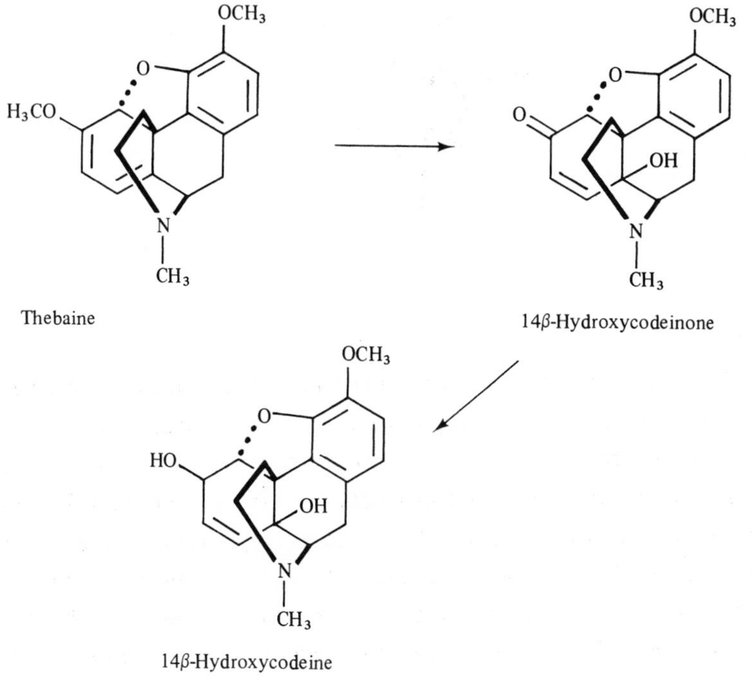

Thebaine

14β-Hydroxycodeinone

14β-Hydroxycodeine

The fermentative oxidation of thebaine to 14β-hydroxycodeinone corresponds to the chemical method that uses hydrogen peroxide,[40] but the microbiological procedure involves the incorporation of atmospheric oxygen into the 14-hydroxy group.[41]

The evidence (alluded to above) for a concerted oxidative-hydrolytic mechanism derives from the fermentative dissimilations of codeinone and neopinone by *T. sanguinea*. By appropriate fermentation medium selection, codeinone yields codeine (24.5%) and 14β-hydroxycodeine (7.6%) or, alternatively, codeine (8.4%), 14β-hydroxycodeine (4.3%), and 14β-hydroxycodeinone (6.6%). Under comparable conditions, neopinone gives 14β-hydroxycodeinone (2.8%) or alternatively, codeine (3.8%) and 14β-hydroxycodeine (8.8%).

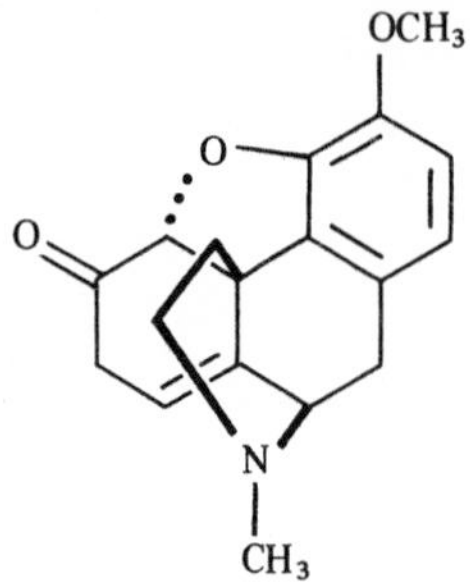

Neopinone

BENZYLIC

Several simple alkyl-substituted aromatic compounds undergo benzylic oxygenation by microorganisms. Further oxidation of the benzylic alcohol to the carboxylic acid is common, and in most instances
represents the only isolated product. *p*-Cymene, for example, is oxidized by a *Pseudomonas* species[42,43] to *p*-isopropylbenzyl alcohol (and
other, nonallylic, oxidation products), but the yield is low, in contrast to the 39% yield of cumic acid that may be obtained.[44-46] Note
that the microorganism oxidizes the methyl group, rather than the
chemically more susceptible isopropyl group.[47]

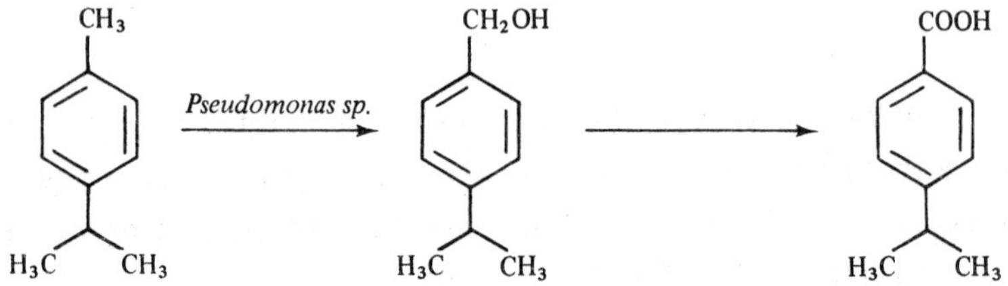

p-Cymene   Cumic acid

A similar oxidation of p-xylene by some *Nocardia* species (not
those that cleave the aromatic ring) leads to p-toluic acid (and
other products),[48] while the oxidation of 2,4-dimethylphenol by a
Pseudomonas species can be stopped at 4-hydroxy-3-methylbenzoic acid
(isol.) by inclusion of 2,2'-bipyridyl in the fermentation, or al-

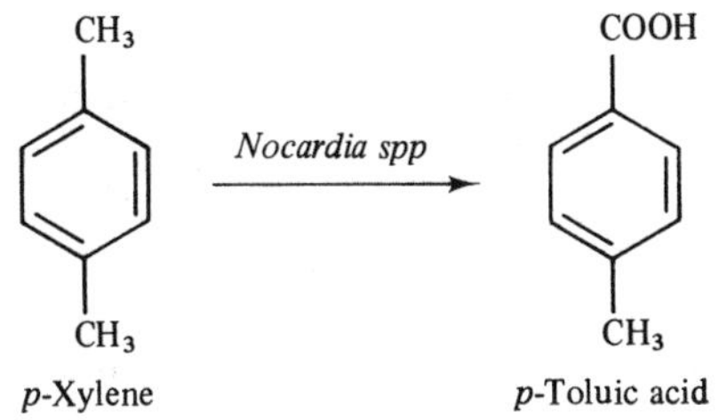

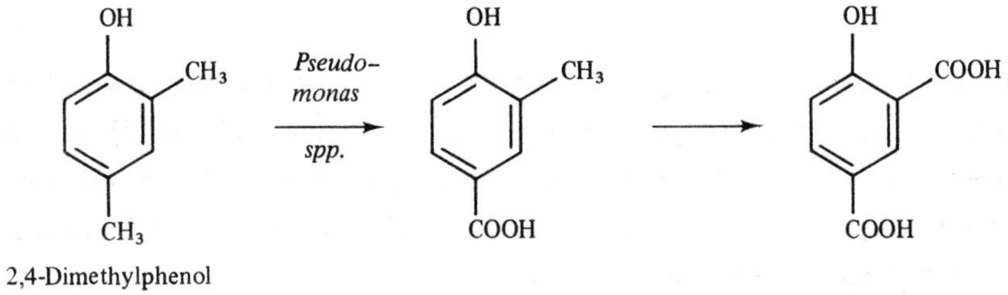

CH₃ → p-Xylene → COOH / p-Toluic acid (*Nocardia spp*)

2,4-Dimethylphenol → (*Pseudomonas spp.*) → →

lowed to proceed to 4-hydroxyisophthalic acid (isol.), which then under-goes oxidative degradation.[49]   2-Methylnaphthalene is oxidized to 2-naphthoic acid (low yield, isol.) by a *Pseudomonas* species *via* the benzyl alcohol, but there are competing metabolic pathways.[50,51]

Tetralin is benzylically oxidized by *Aspergillus niger* to (±)-α-tetralol (1.1%) and α-tetralone (0.8%).[18]

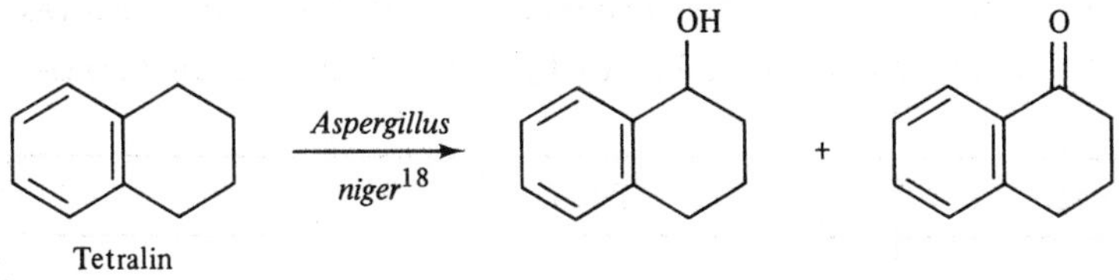

Tetralin → (*Aspergillus niger*[18]) → +

A group of workers at Sterling-Winthrop Research Institute who were interested in schistosomicidal agents found that experimental animals converted lucanthone metabolically to the biologically more active hydroxylated derivative hycanthone (1-[2-(diethylamino)ethyl-amino]-4-hydroxymethylthioxanthen-9-one). The microbiological oxida-tion of lucanthone with *Aspergillus sclerotiorum* gives hycanthone in 60% yield, and affords a convenient source of supply for this medici-nally interesting substance.[52]   By slight alterations of the fermen-

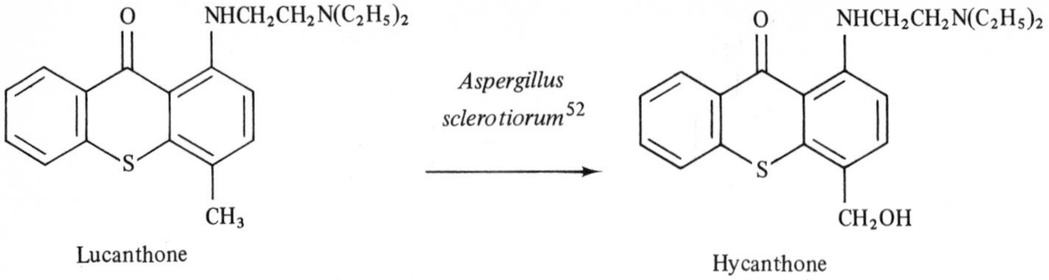

Lucanthone   Hycanthone

tation conditions (notably media changes), several additional minor products (*e.g.* the corresponding aldehyde and carboxylic acid) can be isolated from the lucanthone → hycanthone oxidation. (It is interesting that lucanthone sulfoxide fails to undergo the benzylic oxidation under these conditions.)

Several other schistosomicidal agents that are converted by animal metabolism to the therapeutically active material also are oxidized by microorganisms. 1-(3-Chloro-4-methylphenyl)piperazine hydrochloride is converted to 2-chloro-4-(1-piperazinyl)benzyl alcohol in 73% yield by *A. sclerotiorum*. (A small amount of the corresponding aldehyde is also formed.) This and several other oxidations in this series are shown in Table 1.[53,54]

Table 1. Oxygenation of the Aromatic Methyl Group in a Series of
          Schistosomicidal Agents by *Aspergillus sclerotiorum*

Substrate	Product	Yield, %
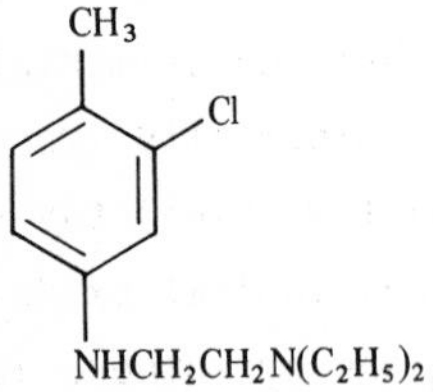	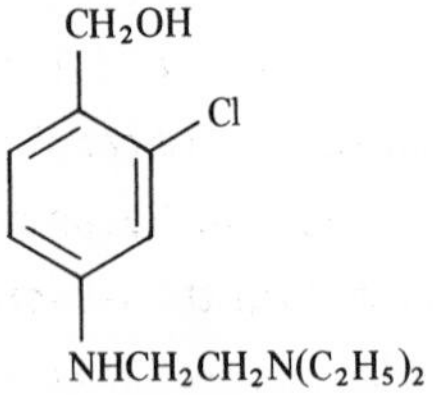	66

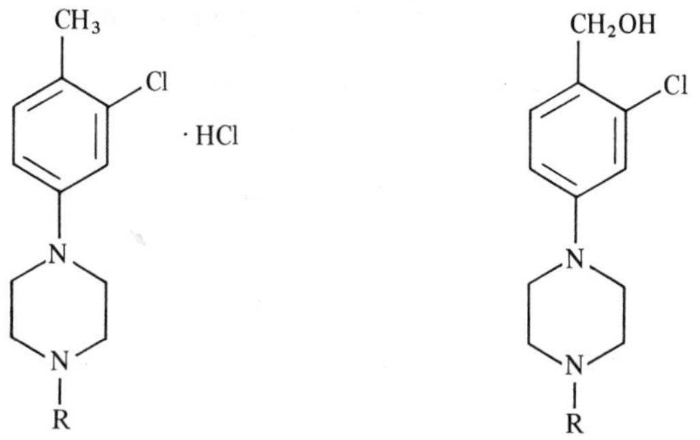

R = —H	R = —H	73
$\overset{S}{\underset{\parallel}{}}$ —CNH$_2$	—H	3
—CH$_2$CH$_2$OH	—CH$_2$CH$_2$OH	52
—CH$_2$CH=CH$_2$	—CH$_2$CH=CH$_2$	54
—CH$_2$C$_6$H$_5$	—CH$_2$C$_6$H$_5$	29

Nalidixic acid (Table 2, R = C$_2$H$_5$), which is metabolized by humans to the corresponding hydroxymethyl compound, is oxidized by a number of microorganisms, perhaps most effectively by *Penicillium adametzi*, to give the same hydroxymethyl derivative in 40-60% yield.[55,56] This fermentation works best by the use of washed cells. Three analogs of nalidixic acid are also oxidized by *P. adametzi*, as shown in Table 2.

Table 2. Oxygenation of Nalidixic Acid and Analogs by *Penicillium adametzi*

Substrate   Product

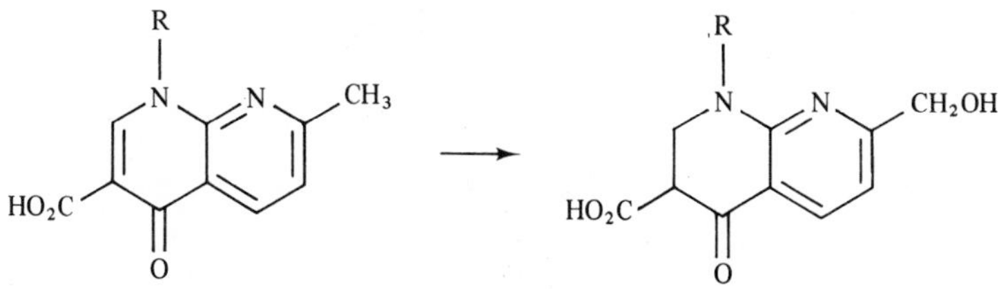

R	Yield (%) of Oxygenated Product
$-C_2H_5$	40-60
$-CH_2CH=CH_2$	62
$-CH_2CH_2CH_3$	-
$-CH_2(CH_2)_4CH_3$	17

The biosynthesis of chloramphenicol by a *Streptomyces* species includes the allylic oxidation of L-*p*-aminophenylalanine to L-*threo*-*p*-aminophenylserine,[57] but since this reaction has not been dissected out of the biosynthesis as a separate entity, its synthetic utility remains to be demonstrated.

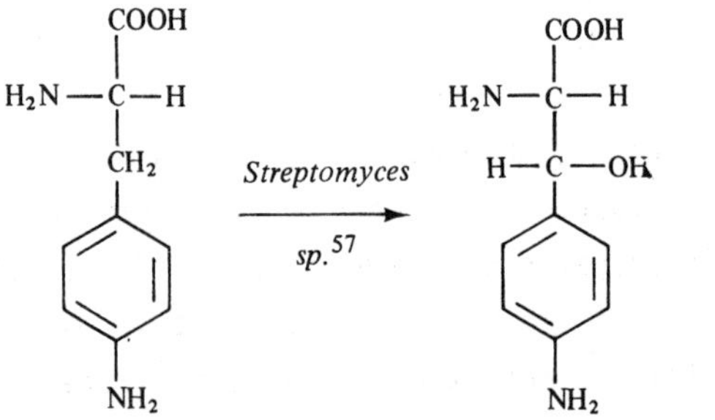

L-*p*-Aminophenylalanine   L-*threo*-*p*-Aminophenylserine

Aromatic steroid substrates have received little attention. The conversion of estrone to 6β-hydroxyestrone by *Mortierella alpina* in less than 1% yield[58] would suggest that purely chemical methods would ordinarily be preferred for benzylic oxidation.

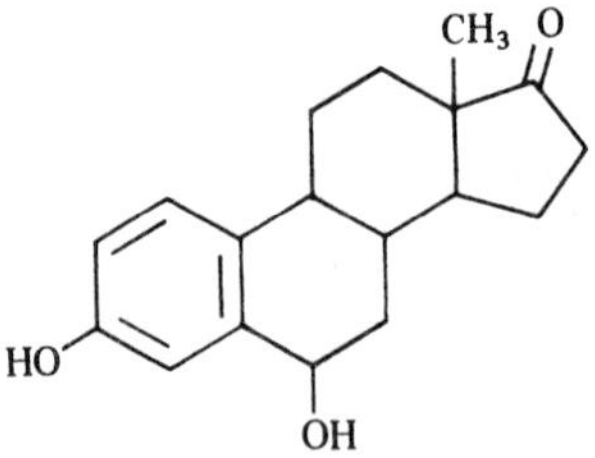

6β-Hydroxyestrone

CARBONYL ACTIVATED

Oxygenation at the carbon next to the carbonyl functional group is most commonly encountered with substrates from the steroid field. Several oxygenations at the C-16 position are illustrated.

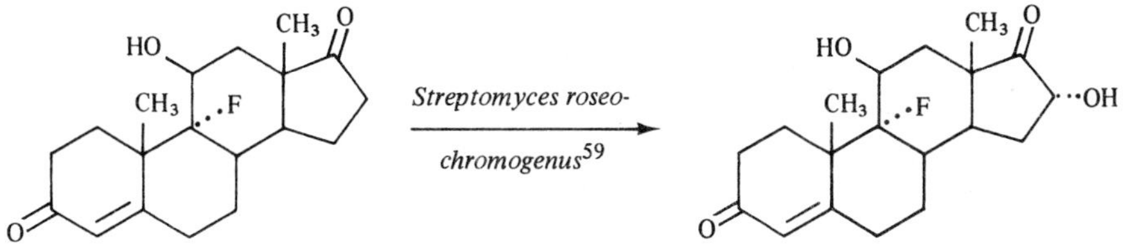

Streptomyces roseo-

chromogenus[59]

(58%)

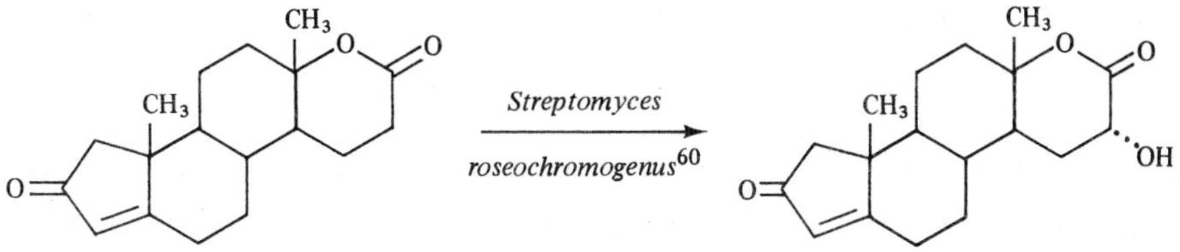

Streptomyces

roseochromogenus[60]

(75%)

Streptomyces spp.[61, 62]

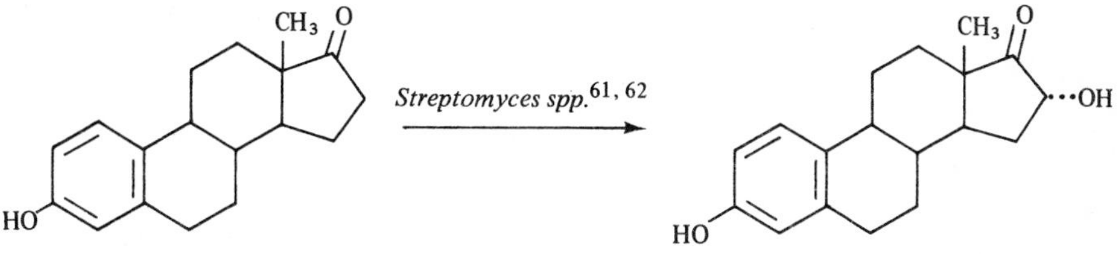

Hydroxylation occurs adjacent to a 20-keto group when several closely related substrates are incubated with *Ophiobolus herpotrichus* or with *Aspergillus niger.*[63-68]

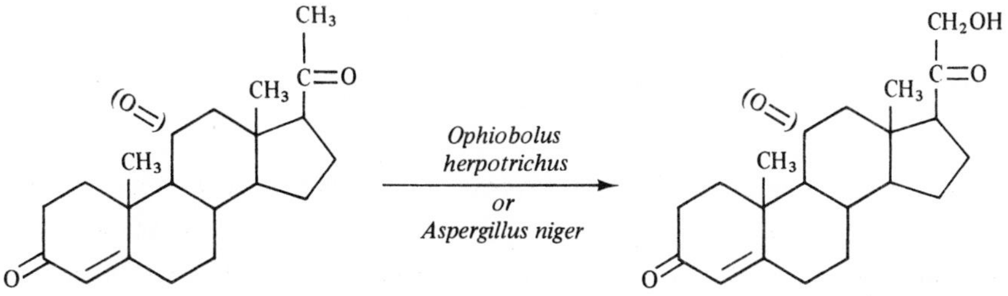

Ophiobolus herpotrichus

or

Aspergillus niger

(11-Keto)progesterone

(75%)

The structurally related DL-steroid, 18-methyl-19-norpregn-4-ene-3,11,20-trione, prepared by total synthesis, is similarly oxygenated on the 21-methyl group by *A. niger*.[69]

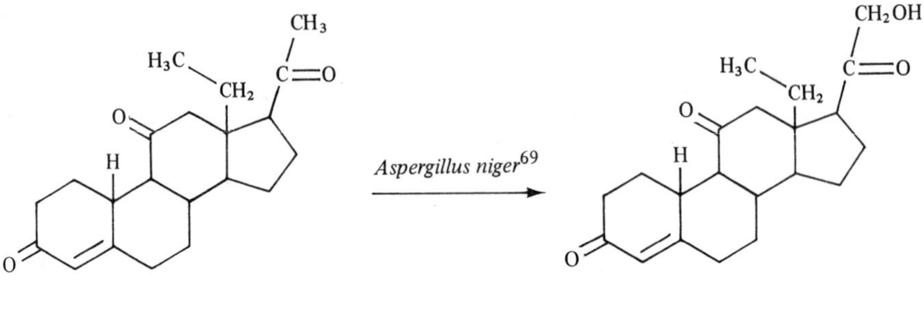

Aspergillus niger[69]

DL-18-Methyl-19-norpregn-
4-ene-3,11,20-trione

(34%)

Hydroxylations at other positions on the steroid nucleus have also been seen and 17α-hydroxylation has received fair attention owing to its potential applicability to corticosteroid synthesis. For example, 11α-hydroxyprogesterone is oxygenated at the 17α-position by *Sepedonium ampullosporum*, giving 11α,17α-dihydroxyprogesterone in 27% yield.[70]

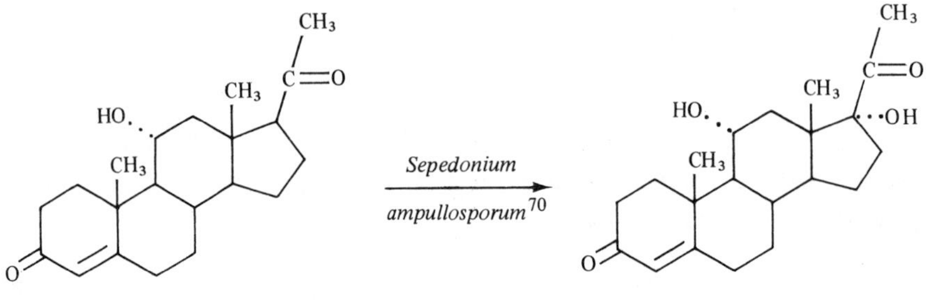

Sepedonium

ampullosporum[70]

11α-Hydroxyprogesterone

11α,17α-Dihydroxyprogesterone

The 2β-hydroxylation of Reichstein's Substance S by *Lactarius quietus* in 8% yield can be carried out with minced fruiting bodies (mushrooms) of this microorganism, a technique that may in certain cases have some practical advantages over the conventional use of vegetative mycelium.[71]

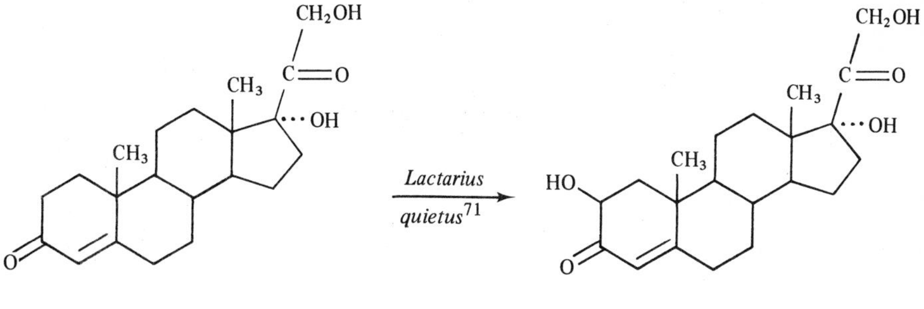

Reichstein's substance S

The 12a-hydroxyl group of tetracycline (TC) can be inserted into the corresponding deoxy substrate, 12a-deoxytetracycline (DOTC), by the action of a variety of chemical oxidizing agents, as well as by a variety of microorganisms. The use of microorganisms for this reaction may thus be rather academic, although at least one patent[72] has been issued on a microbiological process using *Sporormia minima* or *Thielavia terricola*; yields of over 30% are claimed. The same oxidation takes place with *Curvularia lunata* (about 23% yield [bioassay]), as well as with other cultures such as *C. pallescens* and *Botrytis cinerea*.[73] Interestingly, several strains of *Streptomyces aureofaciens* (which produce chlortetracycline) are unable to bring about 12a-hydroxylation.

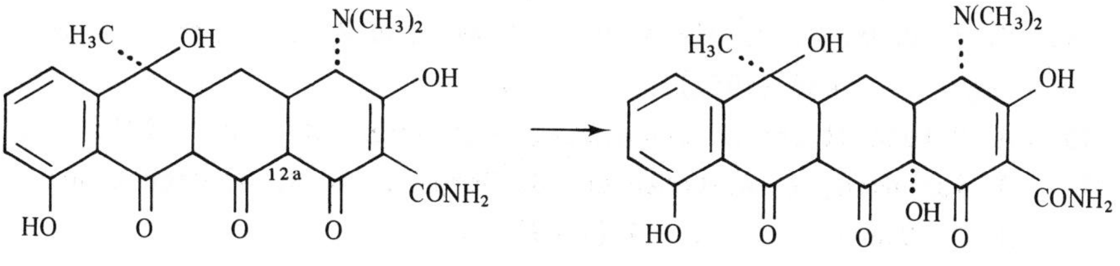

DOTC   TC

Finally, the tranquillizer diazepam (Valium®) and its N-demethyl analog undergo "allylic" oxidation by *Pellicularia filamentosa* to products with a hydroxyl group on the heterocyclic ring.[74]

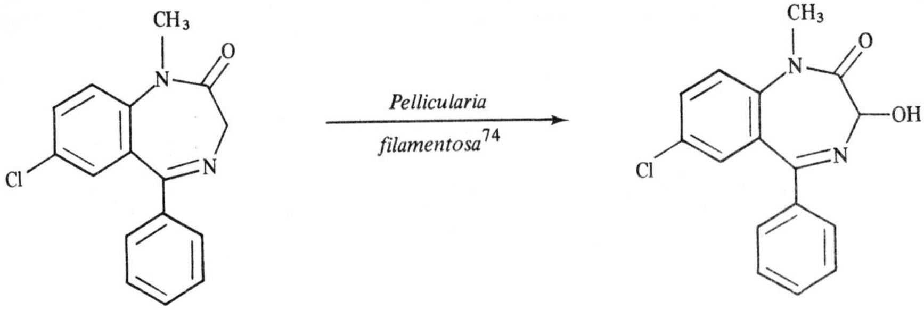

Diazepam

REFERENCES

1. A. Krámli and J. Horvath, *Nature* *162*, 619 (1948).

2. J. Urech, E. Vischer, and A. Wettstein, *Helv. Chim. Acta 43*, 1077 (1960).

3. J. Fried and R. W. Thoma, *U. S. Patent* 2,914,543 (November 24, 1959).

4. H. Els, G. Englert, A. Fürst, P. Reusser, and A. J. Schocher, *Helv. Chim. Acta 52*, 1157 (1969).

5. Y. Y. Lin, M. Shibahara, and L. L. Smith, *J. Org. Chem. 34*, 3530 (1969).

6. L. L. Smith, G. Greenspan, R. Rees, and T. Foell, *J. Amer. Chem. Soc. 88*, 3120 (1966).

7. Y. Sato and S. Hayakawa, *J. Org. Chem. 28*, 2739 (1963).

8. S. M. Kupchan, C. J. Sih, S. Kubata, and A. M. Rahim, *Tetrahedron Lett.*, 1767 (1963).

9. S. J. Daum, M. M. Riano, P. E. Shaw, and R. L. Clarke, *J. Org. Chem. 32*, 1435 (1967).

10. A. Butenandt and H. Dannenberg, *Naturwiss. 30*, 52 (1942).

11. B. Tabenkin, R. A. LeMahieu, J. Berger, and R. W. Kierstead, *Appl. Microbiol. 17*, 714 (1969).

12. R. A. LeMahieu, B. Tabenkin, J. Berger, and R. W. Kierstead, *J. Org. Chem. 35*, 1687 (1970).

13. P. K. Bhattacharyya, B. R. Prema, B. D. Kulkarmi, and S. K. Pradhan, *Nature 187*, 689 (1960).

14. B. R. Prema and P. K. Bhattacharyya, *Appl. Microbiol. 10*, 524 (1962).

15. R. N. Moore, C. Golumbic, and G. S. Fisher, *J. Amer. Chem. Soc.* *78*, 1173 (1956).

16. B. R. Prema and P. K. Bhattacharyya, *Appl. Microbiol. 10*, 529 (1962).

17. P. K. Bhattacharyya and K. Ganapathy, *Indian J. Biochem. 2*, 137 (1965).

18. K. Ganapathy, K. S. Khanchandani, and P. K. Bhattacharyya, *Indian J. Biochem. 3*, 66 (1966).

19. R. S. Dhavalikar and P. K. Bhattacharyya, *Indian J. Biochem. 3*, 144 (1966).

20. R. S. Dhavalikar, P. N. Rangachari, and P. K. Bhattacharyya, *Indian J. Biochem. 3*, 158 (1966).

21. O. P. Shukla, M. N. Moholay, and P. K. Bhattacharyya, *Indian J. Biochem. 5*, 79 (1968).

22. O. P. Shukla and P. K. Bhattacharyya, *Indian J. Biochem. 5*, 92 (1968).

23. H. Hikino, K. Aota, Y. Tokuoka, and T. Takemoto, *Chem. Pharm. Bull. 16*, 1088 (1968).

24. S. Nozoe, M. Morisaki, K. Tsuda, and S. Okuda, *Tetrahedron Lett.*, 3365 (1967).

25. L. Canonica, A. Fiecchi, M. Galli Kienle, B. M. Ranzi, A. Scala, T. Salvatori, and E. Pella, *Tetrahedron Lett.*, 3371 (1967).

26. B. E. Cross and K. Norton, *Chem. Commun.*, 535 (1965).

27. H. Hikino, S. Nabetani, and T. Takemoto, *Yakugaku Zasshi 89*, 809 (1969).

28. P. A. Miller and J. R. D. McCormick, *U. S. Patent* 3,023,148 (February 27, 1962).

29. P. A. Miller, J. H. Hash, M. Lincks, and N. Bohonos, *Biochem. Biophys. Res. Commun. 18*, 325 (1965).

30. L. A. Mitscher, J. H. Martin, P. A. Miller, P. Shu, and N. Bohonos, *J. Amer. Chem. Soc. 88*, 3647 (1966).

31. J. H. Martin, L. A. Mitscher, P. A. Miller, P. Shu, and N. Bohonos, *Antimicrob. Ag. Chemother.*, 563 (1966).

32. G. C. Lancini, J. E. Thiemann, G. Sartori, and P. Sensi, *Experientia 23*, 899 (1967).

33. S. Agurell and E. Ramstad, *Arch. Biochem. Biophys. 98*, 457
     (1962).

34. J. Beliveau and E. Ramstad, *Lloydia 29*, 234 (1966).

35. A. Brack, R. Brunner, and H. Kobel, *Helv. Chim. Acta 45*, 276
     (1962).

36. W. O. Godtfredson, G. Korsby, H. Lorck, and S. Vandegal, *Experi-
     entia 14*, 88 (1958).

37. K. Iizuka, S. Okuda, K. Aida, T. Asai, K. Tsuda, M. Yamada, and
     I. Seki, *Chem. Pharm. Bull. 8*, 1056 (1960).

38. M. Yamada, K. Iizuka, S. Okuda, T. Asai, and K. Tsuda, *Chem.
     Pharm. Bull. 10*, 981 (1962).

39. K. Iizuka, M. Yamada, J. Suzuki, I. Seki, K. Aida, S. Okuda, T.
     Asai, and K. Buda, *Chem. Pharm. Bull. 10*, 67 (1962).

40. I. K. Fel'dman and A. I. Lyutenberg, *J. Appl. Chem. USSR 18*, 715
     (1945); *Chem. Abstr.*, *40*, 6489 (1946).

41. K. Aida, K. Uchida, K. Iizuka, S. Okuda, K. Tsuda, and T.
     Uemura, *Biochem. Biophys. Res. Commun. 22*, 13 (1966).

42. K. M. Madhyastha and P. K. Bhattacharyya, *Indian J. Biochem. 5*,
     102 (1968).

43. K. M. Madhyastha and P. K. Bhattacharyya, *Indian J. Biochem. 5*,
     161 (1968).

44. S. Horiguchi and K. Yamada, *Agr. Biol. Chem. 32*, 555 (1968).

45. A. Kamibayashi, Y. Saeki, T. Higashihara, M. Yamaguchi, and H.
     Ono, *Report of the Fermentation Research Institute*, 43 (Novem-
     ber, 1967).

46. K. Yamada, S. Horiguchi, and J. Takahashi, *Agr. Biol. Chem. 10*,
     943 (1965).

47. J. B. Davis and R. L. Raymond, *Appl. Microbiol. 9*, 383 (1961).

48. R. L. Raymond and V. W. Jamison, *U. S. Patent* 3,383,289 (May 14,
     1968).

49. P. J. Chapman and D. J. Hopper, *Biochem. J. 110*, 491 (1968).

50. M. H. Rogoff and I. Wender, *J. Bacteriol. 77*, 783 (1959).

51. V. Treccani and A. Fiecchi, *Ann. Microbiol. Enzimol. 8*, 36
     (1958).

52. D. Rosi, G. P. Peruzzotti, E. W. Dennis, D. A. Berberian, H.
     Freele, B. F. Tullar, and S. Archer, *J. Med. Chem. 10*, 867
     (1967).

53. D. Rosi, T. R. Lewis, R. Lorenz, H. Freele, D. A. Berberian, and S. Archer, *J. Med. Chem. 10*, 877 (1967).

54. S. Archer and D. Rosi, *U. S. Patent* 3,379,620 (April 23, 1968).

55. P. B. Hamilton, D. Rosi, G. P. Peruzzotti, and E. D. Nielson, *Appl. Microbiol. 17*, 237 (1969).

56. E. D. Nielson, P. B. Hamilton, D. Rosi, and G. P. Peruzzotti, *U. S. Patent* 3,317,401 (May 2, 1967).

57. R. C. McGrath, L. C. Vining, F. Sala, and D. W. S. Westlake, *Can. J. Biochem. 46*, 587 (1968).

58. A. I. Laskin and J. Fried, *U. S. Patent* 3,115,443 (December 24, 1963).

59. G. H. Thomas and R. W. Thoma, *U. S. Patent* 2,853,502 (September 23, 1958).

60. A. I. Laskin and F. L. Weisenborn, *U. S. Patent* 3,098,079 (July 16, 1963).

61. D. A. Kita, J. L. Lardinas, and G. M. Shull, *Nature 190*, 627 (1961).

62. A. I. Laskin, *U. S. Patent* 3,098,796 (July 23, 1963).

63. C. Meystre, E. Vischer, and A. Wettstein, *Helv. Chim. Acta 37*, 1548 (1954).

64. B. A. Rubin, C. Casas-Campillo, G. Hendrichs, F. Cordova, and A. Zaffaroni, *Bacteriol. Proc.*, 33 (1956).

65. E. Weisz, G. Wix, and M. Bodanszky, *Naturwiss. 43*, 39 (1956).

66. A. Wettstein, E. Vischer, and C. Meystre, *U. S. Patent* 2,778,776 (January 22, 1957).

67. A. Zaffaroni, C. Casas-Campillo, F. Cordoba, and G. Rosenkranz, *Experientia 11*, 219 (1955).

68. A. Zaffaroni and B. A. Rubin, *U. S. Patent* 2,812,285 (November 5, 1957).

69. B. Gadsby, M. R. G. Leeming, G. Greenspan, and H. Smith, *J. Chem. Soc. (C)*, 2647 (1968).

70. H. C. Murray and L. M. Reineke, *U. S. Patent* 3,011,951 (December 5, 1961).

71. Z. Prochazka and V. Sasek, *Coll. Czech. Chem. Commun. 32*, 610 (1967).

72. D. Beck and G. M. Shull, *U. S. Patent* 2,970,087 (January 31, 1961).

73. C. E. Holmlund, W. W. Andres, and A. J. Shay, *J. Amer. Chem. Soc. 81*, 4750 (1959).

74. G. Greenspan, H. W. Ruelius, and H. E. Alburn, *U. S. Patent* 3,453,179 (July 1, 1969).

General References

The General References given in Chapter 1 are also applicable to this Chapter.

Chapter 3

OLEFINIC OXYGENATION

Microbially catalyzed oxygenation reactions at olefinic sites may be categorized as (a) those effecting hydration of the double bond to produce a mono-alcohol; (b) those which result in formation of epoxides; and (c) those whose over-all effect is formation of a diol, often *via* an intermediate epoxide. Because all of these reactions may occur in the same substrate-microorganism interaction, no effort has been made to present a rigidly segregated discussion of each category. It is immediately apparent that these reactions can often be done in a highly satisfactory manner by a variety of chemical reagents. Nevertheless, microbial oxygenation of olefins may be of value in cases where other portions of the molecule are susceptible to chemical attack and may be particularly useful when a stereospecific reaction at the olefin is desired.

Hydration of the double bond of oleic acid by a *Pseudomonas* species gives 10-hydroxystearic acid in 14% yield.[1] This reaction is stereospecific, and consists of a *trans* addition of a molecule of water, with generation of the R-configuration at both C-9 and C-10.[2-4] The enzyme system responsible for the hydration (as well as the reverse dehydration) can be obtained as a soluble, cell-free preparation.[5] The possibility that the hydration proceeds by way of an intermediate epoxide[6] is ruled out by the fact that the soluble enzyme preparation fails to convert the synthetic epoxystearic acids to 10-hydroxystearic acid. Instead, the enzyme catalyzes the opening of one enantiomer of *d,l-cis*-9,10-epoxystearic acid to optically active *threo*-9,10-dihydroxystearic acid (see Fig. 1), and of one enantiomer

113

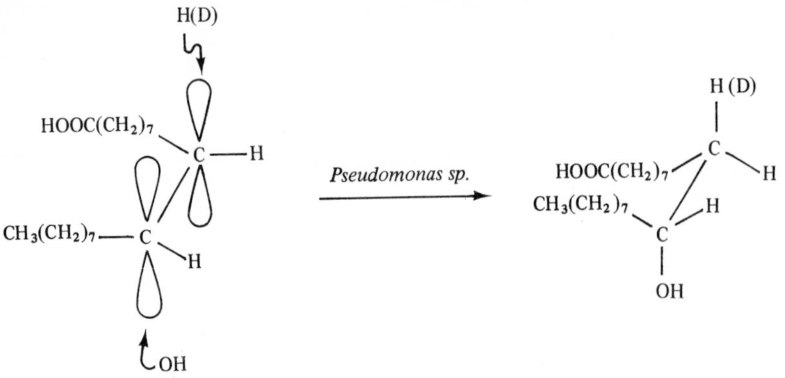

Oleic acid

10-Hydroxystearic acid

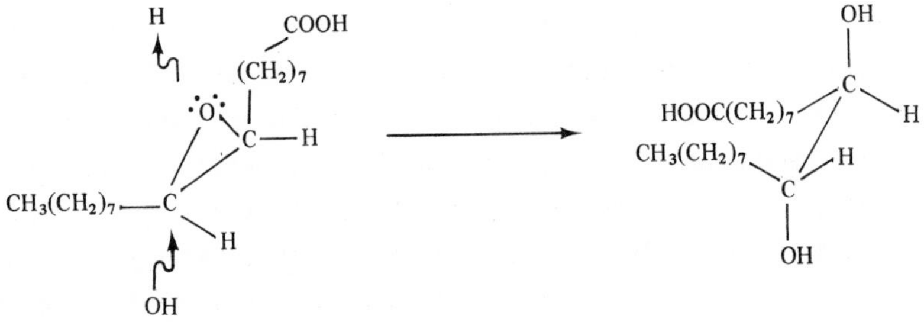

cis-9,10-Epoxystearic acid

threo-9,10-Dihydroxystearic acid

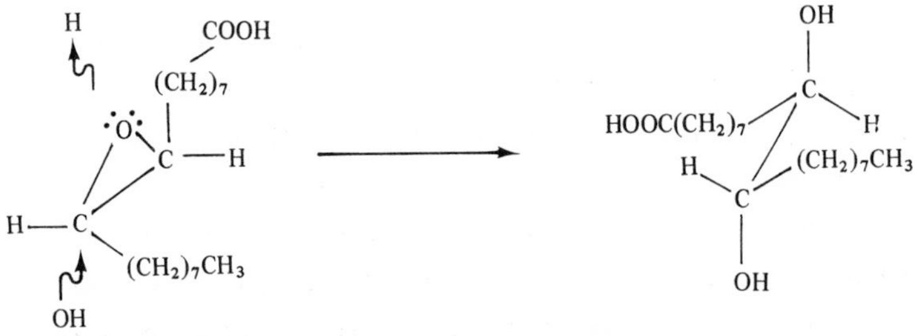

trans-9,10-Epoxystearic acid

erythro-9,10-Dihydroxystearic acid

of d,l-trans-9,10-epoxystearic acid to optically active erythro-9,10-dihydroxystearic acid. The recovered unreacted enantiomeric epoxides are also optically active.[6]

If, as suggested, the same enzyme is catalyzing both the hydration of the olefin and the epoxides, it seems possible that the mechanism of reaction may be similar in both cases and that, therefore,
similar geometries are assumed by the substrate molecules during reaction. The *trans* addition of water to the olefinic bond of oleic
acid, to give the correct absolute configuration of product, may be
depicted as shown in Fig. 1. Assuming a similar mode of addition to
the *cis*-epoxide, it may be expected that hydration of the 9R,10R-en-
antiomer of the racemic substrate will occur, giving an optically active product, also of 9R,10R absolute configuration. If it is further
assumed that the carboxyl end of the chain is the more likely to have
a fixed geometry with respect to the enzyme active site, then the
trans epoxide may be expected to yield the 9R,10S diol from the 9R,
10S enantiomer of racemic substrate.

Several other *cis*-9,10-unsaturated acids, namely palmitoleic and
linoleic, undergo hydration to C-10 mono alcohols by the above enzyme
preparation,[7] but elaidic acid (the *trans*-isomer of oleic acid) fails
to react.

Several reports have described the oxygenation of straight chain
alkenes. As might be expected, oxygenation at the saturated end of
the molecule competes significantly with oxidative attack at the olefinic center. In fact, an early examination of the oxygenation of 1-
octene by *Pseudomonas aeruginosa* led to the tentative conclusion that
oxidation at the saturated terminus of n-alkenes was preferred over
attack at the unsaturated end.[8] The conversion of n-alkenes,
$CH_3(CH_2)_nCH=CH_2$ where n = 11-15, to mixtures of epoxides, diols, and
α-hydroxycarboxylic acids by *Candida lipolytica*, proceeds in unstated
yields.[9,10] Again, oxygenation at the saturated end of the chain is
important. These examples demonstrate the possibility of oxidative
attack at olefinic bonds in n-alkenes, but no results have yet been
described which allow exclusive hydrative attack at olefin in preference to oxygenative attack at the saturated end of the chain.

The hydration of cyclic olefins has been described by several
groups. In general, the investigations into this particular type of
reaction have not been extensive and yields usually have been poor

from a synthetic point of view.   The apparent hydration of cyclopen-
tene, cyclohexene, and cyclo-octene by a *Bacterium* species is followed by dehydrogenation of the alcohols to the ketones.[11] The epoxides may be intermediates in these reactions.[11]

Hydration of the double bond of 1-methylcyclohexene with *Aspergillus niger* occurs in low yield, giving 1-methylcyclohexanol as well
as products of allylic oxidation (see Chapter 2).[12] Similarly, in the
oxygenation of limonene by a soil pseudomonad,[13] products of allylic

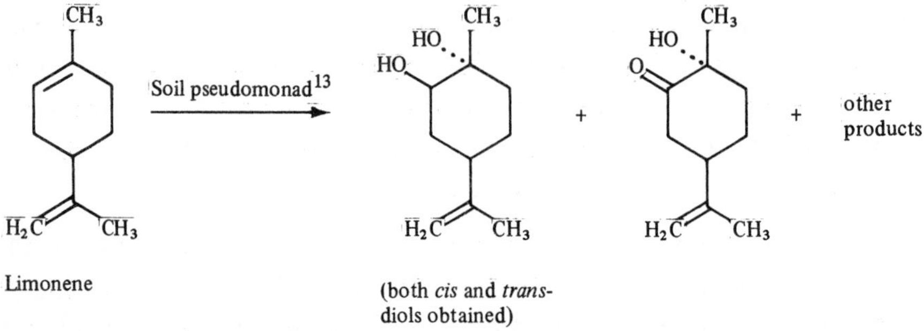

Limonene

(both *cis* and *trans-*
diols obtained)

oxidation and of extensive ring rearrangement dilute the yield of the
cis- and *trans-*diols and a keto-alcohol that represent the olefinic
hydration products.

Several other examples of the hydration of cyclic olefins may be
mentioned.   Both *Nocardia opaca* and *N. restrictus* hydrate the double
bond of the steroid fragment A, giving B as the product.[14] Other com-

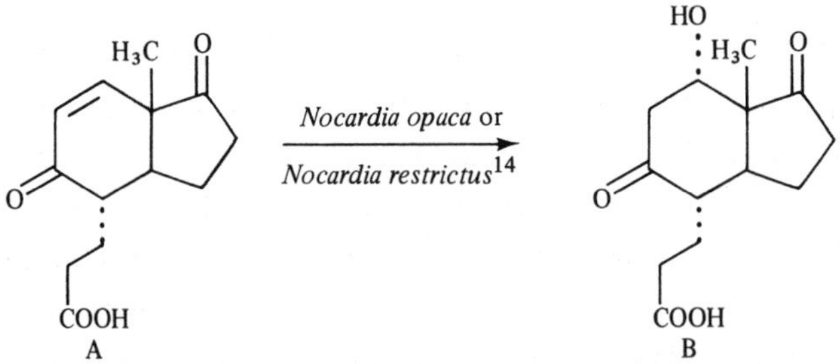

A

B

pounds of very similar structure are not substrates for this hydration reaction, suggesting a specific nature for this particular enzyme system.   Hydration of the 7,8-double bond in abietic acid by an
Alcaligenes species results in formation of the 7-hydroxy derivative
(55 mg from 48 g!).[15]

Curvularia lunata and *Cunninghamella blakesleeana* both 11β-hy-
droxylate C-ring-saturated steroids. When either organism is incubated with 17α,21-dihydroxypregna-4,9(11)-diene-3,20-dione, the corresponding 9β,11β-epoxide is obtained as one of the products.[16] The same

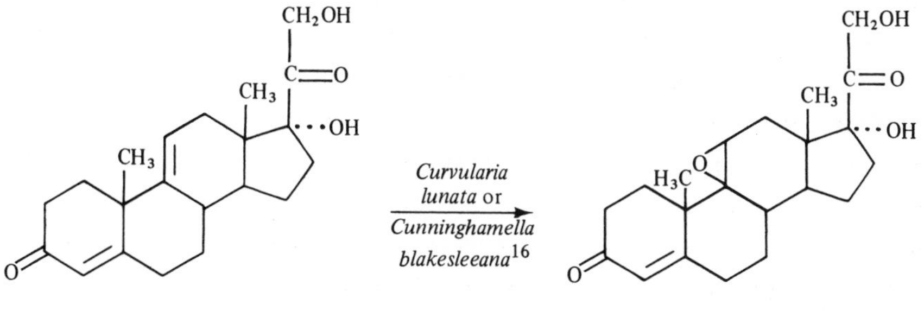

Curvularia
lunata or
Cunninghamella
blakesleeana[16]

17α,21-Dihydroxypregna-
4,9(11)-diene-3,20-dione

C. lunata brings about 11β,12β-epoxidation of Δ[11]-progesterone in
about 1% yield, while another *Curvularia* species, which 9α-hydroxy-
lates C-ring-saturated steroids, gives a 46% yield of 9α,11α-epoxide
from 17α,21-dihydroxypregna-4,9(11)-diene-3,20-dione 21-acetate.[17]

Several microorganisms capable of introducing a 14α-hydroxyl
group into D-ring-saturated steroids are able to convert the Δ[14]-
pregnene derivative A to the 14α,15α-epoxide B.[16]

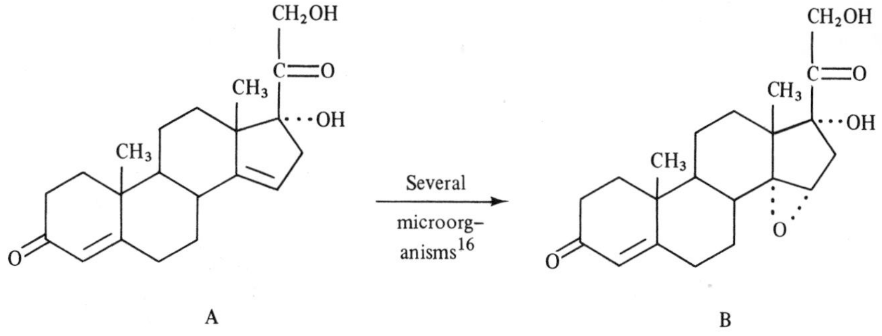

Several

microorg–
anisms[16]

A   B

It will be seen that all of the above steroid epoxidations are
carried out by microorganisms that effect axial hydroxylation at the
same position in the corresponding saturated steroid,[16] while organisms capable of equatorial hydroxylation, or incapable of 14α-hydrox-
ylation, fail to epoxidize Δ[9(11)]- and Δ[14]-olefins, respectively.

Since hydroxylation of a saturated carbon proceeds with displacement of hydrogen without inversion of configuration,[18,19] the two oxidative reactions appear to be geometrically related. A comparison of the geometry of the olefinic π-electron orbital and of the axial C-H bond has been proferred in possible explanation of these observations.[17-20] Implied by this explanation are the electrophilic character of the attacking species during the actual oxygenation reaction and the possible similarity of the enzymic hydroxylation site to the enzymic epoxidation site. The latter implication is supported by results with a cell-free preparation of 9α-hydroxylase from *Nocardia restrictus*, that indicate that portions of this enzyme system are involved in both hydroxylation and epoxidation.[21,22]

A minor reaction pathway in the oxidation of α-olefins by *Pseudomonas aeruginosa* is the epoxidation of the double bond system.[23] Thus, 1-heptene,[23] 1-octene,[8,23] and 1-nonene[23] are epoxidized by cells of the culture grown on the saturated *n*-hydrocarbons. The yields are infinitesimal, and the reader is reminded that the major pathway of oxidation, namely, the oxidation of the terminal methyl group, is also undoubtedly occurring during this time. The β-olefin, 2-octene does not give any epoxide under similar conditions.

Several miscellaneous olefins are epoxidized by the mold *Sporotrichum sulfurescens*.[24] Oxygenation of N-benzoyl-N-methyl-4-methyl-eneadamantan-1-amine gives a hydroxy epoxide (51%) as the major

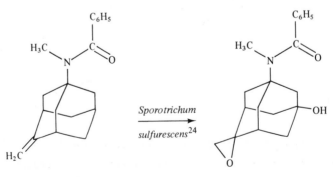

N-Benzoyl-N-methyl-4-
methyleneadamantan-1-amine

product, while N-*p*-toluenesulfonyl-4-methylenepiperidine is oxygenated to give the dihydroxy compound as the isolable product (21%).
A trace of the epoxide, which can be prepared chemically, is found to
be present by gas-liquid chromatographic analyses, suggesting that
the epoxide may be an intermediate in the formation of the diol.[24]

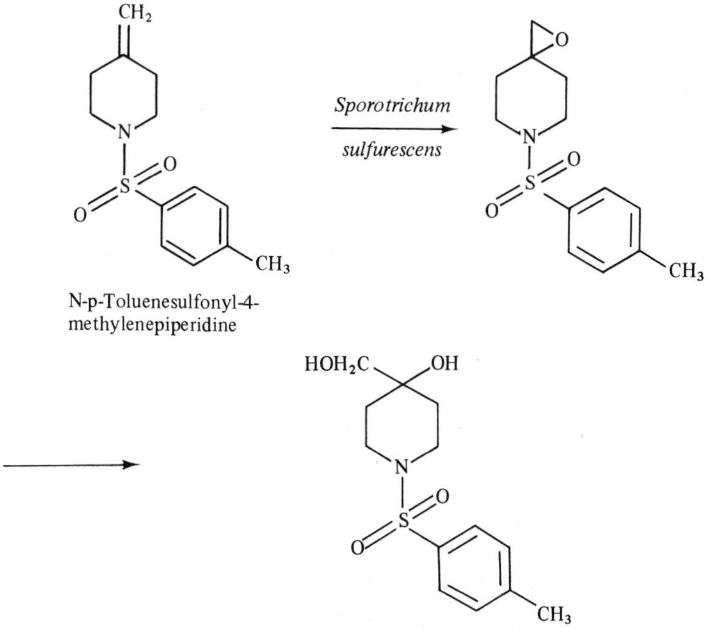

N-p-Toluenesulfonyl-4-
methylenepiperidine

Sporotrichum

sulfurescens

Epoxidation of the insecticide aldrin is catalyzed by several
fungi, leading to the compound dieldrin.[25] The latter may be further

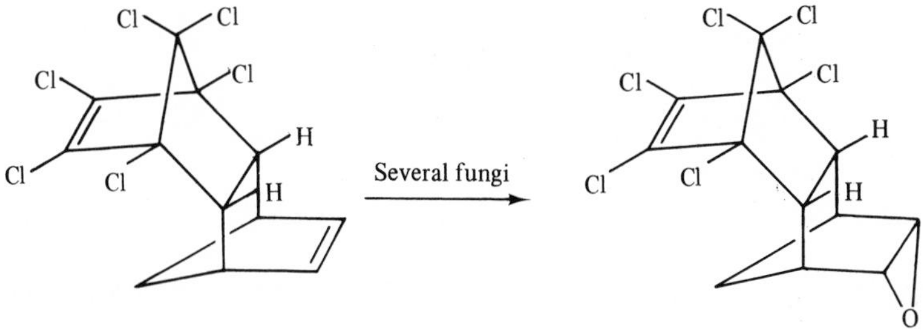

Aldrin

Several fungi

Dieldrin

acted upon by microorganisms, giving several products resulting from
the apparent cleavage of the epoxide.   Other "metabolites" of rather

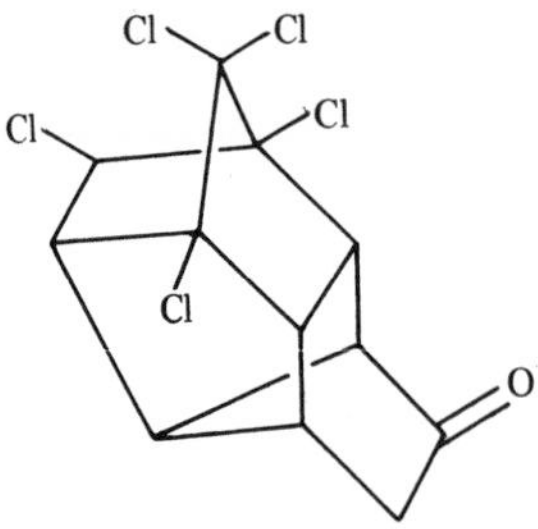

A 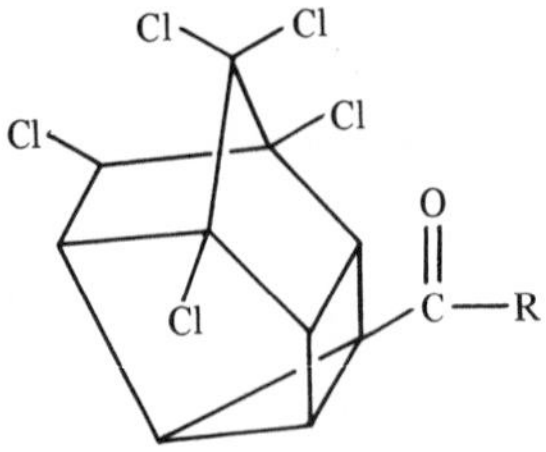   B   R = H or OH

remarkable structure, *i.e.*, A and B, have also been suggested as
arising from microbial reaction.

Diol Formation. Oxidative attack by microorganisms upon aromatic
rings may proceed by various pathways (see Chapters 4 and 5), some of
which appear analogous to the preceding examples of attack upon isolated double bonds. The potential value of this approach to the synthesis of systems in which the aromatic character of a ring has been
interrupted and simultaneously functionalized remains largely un-
tapped.   Indicative of this potential is the ability of a mutant
strain of *Pseudomonas putida* to convert benzene and toluene to *cis*-
1,2-dihydroxycyclohexa-3,5-diene[27] (isolated) and (+)-*cis*-2,3-dihy-
droxy-1-methylcyclohexa-4,6-diene[28] (20% yield), respectively.   Par-

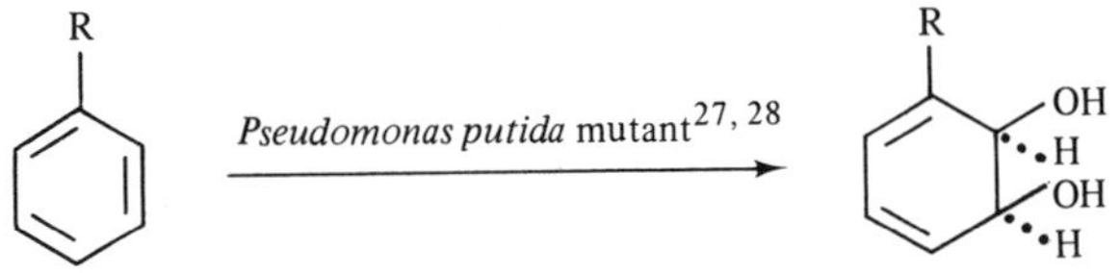

Benzene, R = H
Toluene, R = CH$_3$

ticularly interesting is the optical activity of the product ($[\alpha]_D$
+25°) obtained from toluene.   It has been speculated that an intermediate cyclic peroxide (in preference to an epoxide) may be formed
in these reactions since both oxygens of the diol are derived from
molecular oxygen.[27]   Although satisfactory chemical routes exist for

the synthesis of the *cis*- and *trans*-1,2-dihydroxy-3,5-cyclohexadi-
enes,[29],[30] these methods are not applicable to the preparation of
other substituted derivatives. Thus, for example, (+)-*cis*-4-chloro-
2,3-dihydroxy-1-methylcyclohexa-4,6-diene, isolated from fermentation
of *p*-chlorotoluene with *Ps. putida*,[31] could not be prepared easily by
the chemical methods.

An intermediate cyclic peroxide (A) has also been proposed in
the oxygenation of anthranilic acid by *Pseudomonas fluorescens*,[32]

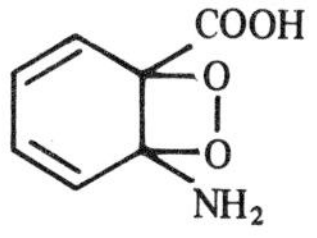

A 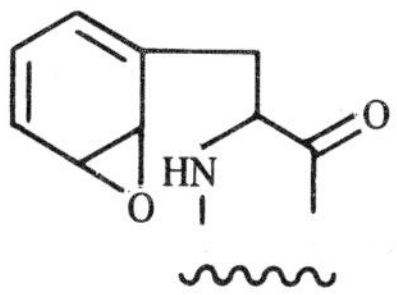   B

while an epoxy intermediate (B) has been proposed in the biosynthesis
of aranotin and related compounds (see Chapter 5) by *Arachniotus
aureus*.[33]

Diols have also been isolated from the oxygenation of naphthalene and several of its derivatives. D-*trans*-1,2-Dihydroxy-1,2-dihy-
dronaphthalene is obtained from the interaction of naphthalene with a
soil bacterium.[34] Analogous products are obtained from 1-chloronaph-
thalene[35] and 2-methylnaphthalene[36] (see Table 1), when these are incubated with *Pseudomonas desmolytica*.[36]

Table 1. Oxygenation of Naphthalenes

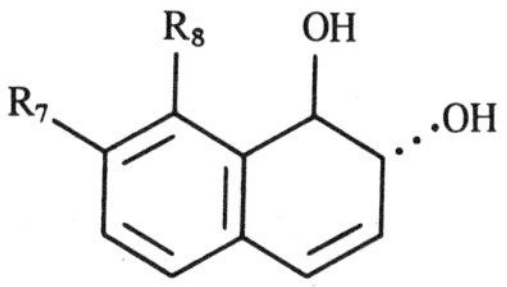

Product	$[\alpha]_D$	Reference
$R_7 = R_8 = H$	+ 161°	34
$R_7 = H$, $R_8 = Cl$	+ 77°	35
$R_7 = Cl$, $R_8 = H$	+ 225°	36
$R_7 = CH_3$, $R_8 = H$	+ 227°	36

A diol (2%; 8% conversion) of undetermined stereochemistry is obtained from the oxygenation of kynurenic acid by *Pseudomonas fluorescens* or by enzyme preparations thereof.[37]   The available evidence

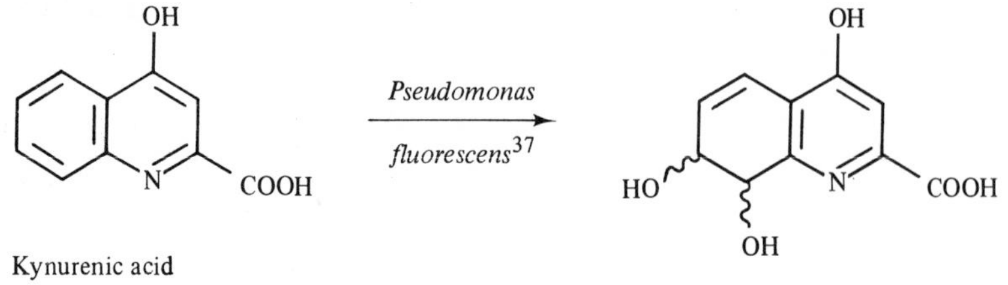

Kynurenic acid

supports the postulation of an epoxide intermediate between kynurenic acid and the diol,[37] and thus the diol probably has a *trans* configuration, since epoxides generally undergo hydrolysis to *trans*-diols.

Various 3-substituted indoles have been found to be transformed to oxindole acetic acid by *Hygrophorus conicus*.[38]   The best yield (43%) is obtained with the substrate tryptamine (R = $CH_2CH_2NH_2$).   The organism *Omphalia flavida* converts indole acetic acid to 3-methyloxindole, however.[39]

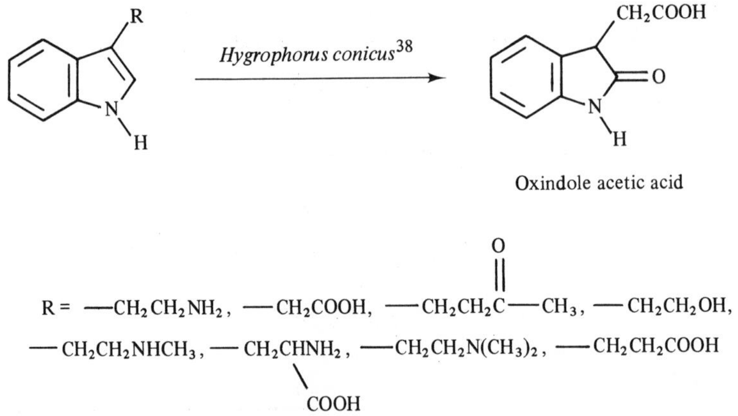

Oxindole acetic acid

$$R = \!-CH_2CH_2NH_2, \; -CH_2COOH, \; -CH_2CH_2\overset{\overset{\displaystyle O}{\|}}{C}\!-CH_3, \; -CH_2CH_2OH,$$

$$-CH_2CH_2NHCH_3, \; -CH_2\underset{\underset{\displaystyle COOH}{\diagdown}}{C}HNH_2, \; -CH_2CH_2N(CH_3)_2, \; -CH_2CH_2COOH$$

Rehydration of 5a,6-anhydrotetracycline derivatives to the tetracyclines has been shown to be a biochemical reaction catalyzed by *Streptomyces aureofaciens* and *S. rimosus* in up to 75% conversion yields.[40]

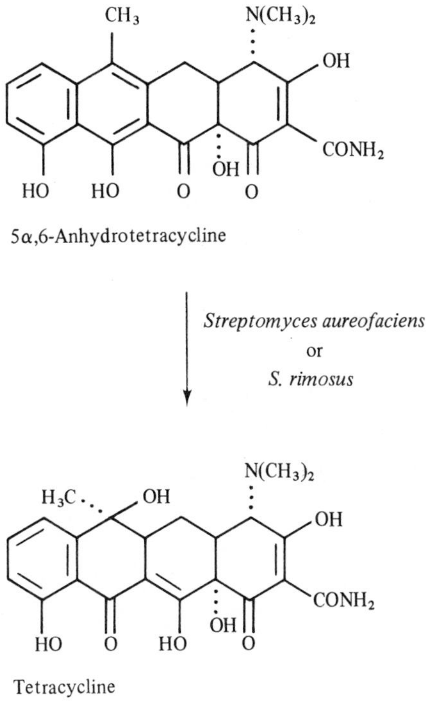

5α,6-Anhydrotetracycline

Streptomyces aureofaciens
or
S. rimosus

Tetracycline

REFERENCES

1. L. L. Wallen, R. G. Benedict, and R. W. Jackson, *Arch. Biochem. Biophys.* *99*, 249 (1962).

2. G. J. Schroepfer, Jr., *J. Amer. Chem. Soc.* *87*, 1411 (1965).

3. G. J. Schroepfer, Jr., and K. Bloch, *J. Biol. Chem.* *240*, 54 (1965).

4. G. J. Schroepfer, Jr., *J. Biol. Chem.* *241*, 5441 (1966).

5. W. G. Niehaus, Jr., and G. J. Schroepfer, Jr., *Biochem. Biophys. Res. Commun.* *21*, 271 (1965).

6. W. G. Niehaus, Jr., and G. J. Schroepfer, Jr., *J. Amer. Chem. Soc.* *89*, 4227 (1967).

7. G. J. Schroepfer, Jr., Abstr. Papers, Amer. Chem. Soc., No. 154, C260 (1967).

8. R. Huybregtse and A. C. Van der Linden, *Antonie van Leeuwenhoek* *30*, 185 (1964).

9. M. J. Klug and A. J. Markovetz, *J. Bacteriol. 93*, 1847 (1967).

10. M. J. Klug and A. J. Markovetz, *J. Bacteriol. 96*, 1115 (1968).

11. J. Ooyama and J. W. Foster, *Antonie van Leeuwenhoek 31*, 45 (1965).

12. K. Ganapathy, K. S. Khanchandani, and P. K. Bhattacharyya, *Indian J. Biochem. 3*, 66 (1966).

13. R. S. Dhavalikar and P. K. Bhattacharyya, *Indian J. Biochem. 3*, 144 (1966).

14. E. Kondo, B. Stein, and C. J. Sih, *Biochim. Biophys. Acta 176*, 135 (1969).

15. B. E. Cross and P. L. Myers, *Biochem. J. 108*, 303 (1968).

16. B. M. Bloom and G. M. Shull, *J. Amer. Chem. Soc. 77*, 5767 (1955).

17. Y. Kurosawa, M. Hayano, and B. M. Bloom, *Agr. Biol. Chem. 25*, 838 (1961).

18. M. Hayano, M. Gut, R. J. Dorfman, O. K. Sebek, and D. H. Peterson, *J. Amer. Chem. Soc. 80*, 2336 (1958).

19. E. J. Corey, G. A. Gregoriou, and D. H. Peterson, *J. Amer. Chem. Soc. 80*, 2338 (1958).

20. M. Hayano, in *Oxygenases* (O. Hayaishi, ed.), Academic, New York, 1962, pp. 122-124.

21. C. J. Sih, *J. Bacteriol. 84*, 382 (1962).

22. F. N. Chang and C. J. Sih, *Biochemistry 3*, 1551 (1964).

23. A. C. Van der Linden, *Biochim. Biophys. Acta 77*, 157 (1963).

24. M. E. Herr, R. A. Johnson, H. C. Murray, and L. M. Reineke, The Upjohn Company, unpublished results.

25. F. Korte, G. Ludwig, and J. Vogel, *Justus Liebigs Ann. Chem. 656*, 135 (1962).

26. F. Matsumura, G. M. Boush, and A. Tai, *Nature 219*, 965 (1968).

27. D. T. Gibson, G. E. Cardini, F. C. Maseles, and R. E. Kallio, *Biochemistry 9*, 1631 (1970).

28. D. T. Gibson, M. Hensley, H. Yoshioka, and T. J. Mabry, *Biochemistry 9*, 1626 (1970).

29. M. Nakajima, J. Tomida, A. Hashizume, and S. Takei, *Chem. Ber. 89*, 2224 (1956).

30. M. Nakajima, J. Tomida, and S. Takei, *Chem. Ber. 92*, 163 (1959).

31. D. T. Gibson, J. R. Koch, C. L. Schuld, and R. E. Kallio, *Biochemistry 7*, 3795 (1968).

32. S. Kobayashi, S. Kuno, N. Itada, O. Hayaishi, S. Kozuka, and S. Oae, *Biochem. Biophys. Res. Commun. 16*, 556 (1964).

33. N. Neuss, R. Nagarajan, B. B. Molloy, and L. L. Huckstep, *Tetrahedron Lett.*, 4467 (1968).

34. N. Walker and G. H. Wiltshire, *J. Gen. Microbiol. 8*, 273 (1953).

35. N. Walker and G. H. Wiltshire, *J. Gen. Microbiol. 12*, 478 (1955).

36. L. Canonica, A. Fiecchi, and V. Treccani, *Rend. Ist. Lomb. Sci. Lett., Part 1, 91*, 119 (1957).

37. H. Taniuchi and O. Hayaishi, *J. Biol. Chem. 238*, 283 (1963).

38. D. J. Siehr, *J. Amer. Chem. Soc. 83*, 2401 (1961).

39. P. M. Ray and K. V. Thimann, *Arch. Biochem. Biophys. 64*, 175 (1956).

40. J. R. D. McCormick, P. A. Miller, S. Johnson, N. Arnold, and N. O. Sjolander, *J. Amer. Chem. Soc. 84*, 3023 (1962).

Chapter 4

AROMATIC RING HYDROXYLATION

Microbiological hydroxylation of the aromatic ring, although covered in a voluminous scientific literature, has to date turned up far less of synthetic value to the organic chemist than has the much smaller amount of research into aliphatics. There are at least two plausible reasons for this situation: (1) the aromatic ring, once hydroxylated, ordinarily becomes extremely susceptible to further oxidation by other enzymes in the cell or in a cell-free preparation, usually with ring cleavage and ultimate degradation to small fragments (this will be discussed in Chapter 5); and (2) the bulk of the reported work has been carried out by microbiologists or biochemists who were far more concerned with exploring the metabolic (degradative) pathways (possibly in order to get rid of unwanted aromatics), than in creating preparative biosynthetic systems. The paucity of even marginally useful synthetic procedures may, however, be viewed optimistically as an indicator of the wealth of synthetic methodology yet to be discovered in this area, and the crystallization of at least one hydroxylase[13] suggests that reactions in a degradative sequence may ultimately be dealt with individually.

Several comprehensive reviews are available (see the General References), and the following discussion is thus intended only to give the reader a brief overview of some of those known reaction types that appear to have some potential for synthetic use. Since no categorization would be completely satisfactory, we have arbitrarily included heterocyclic (heteroaromatic) compounds, and have presented this information roughly in the order of increasing bulk of the substrate, first for the compounds in which hydroxylation occurs on a

127

benzenoid ring and then for those in which it occurs on a heterocyclic ring.

   Benzenoid Substrates.   Oxidation of melilotic acid with an enzyme preparation from an *Arthrobacter* species gives the 3-hydroxy-
lated derivative in 20% yield.[1]   The reaction has a very high sub-

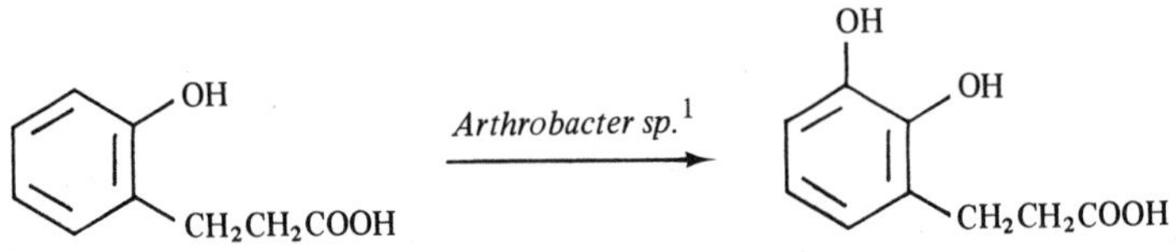

Melilotic acid

strate specificity and requires the addition of a large amount of
NADH -- both factors that could preclude its synthetic application.
Hydroxylations of other phenylalkanecarboxylic acids by other microorganisms may lead to concomitant β-oxidation, and this possibility
should not be overlooked.[2]

   o-Cresol, oxidized by *Pseudomonas aeruginosa*, gives only a trace
of the catechol, owing in part to the susceptibility of the catechol

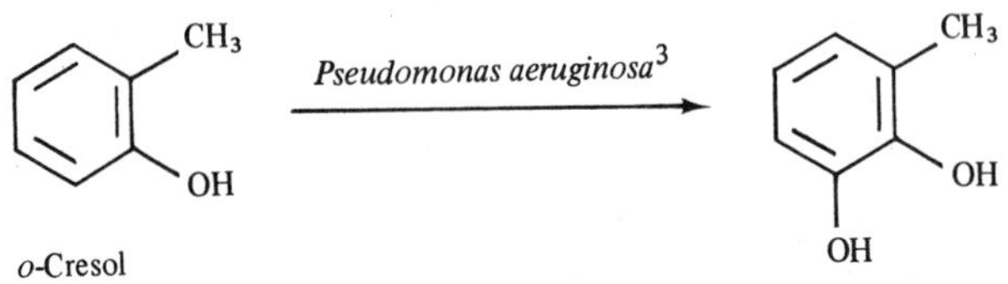

o-Cresol

to further oxidation by the microorganism.[3] A similar situation pre-
vails in the oxidation of 2-fluorobenzoic acid to 3-fluorocatechol
(5% yield) by a *Pseudomonas* species.[4] On the other hand, the oxidation of thymol by *Ps. putida* affords the hydroxyquinone (presumably
via the indicated intermediates) in about 10% yield.   The quinone
apparently serves to withhold trihydroxylated substrate from ring
cleaving enzymes, at least temporarily.[5]

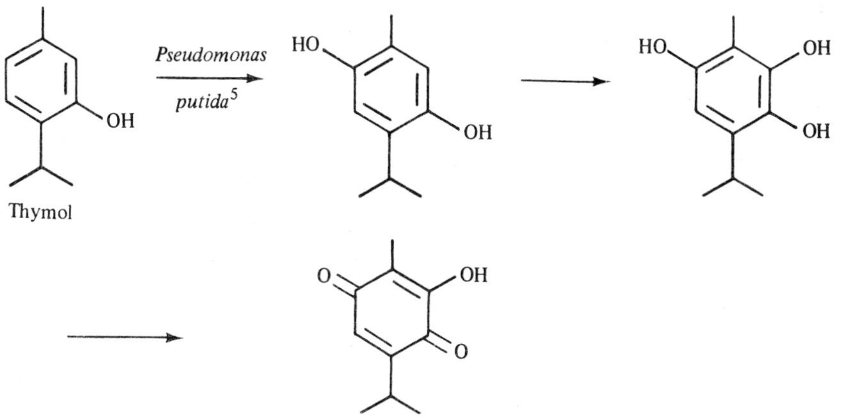

The oxidation of acetanilide by a *Streptomyces* species or by *Amanita muscaria* to give 4-hydroxyacetanilide and 2-hydroxyacetanilide, respectively, apparently proceeds in about 40% assay yield.[6] Similarly, assays of an incubation of phenylalanine with a *Pseudomonas* species (or an enzyme preparation from this organism) indicate about 68% conversion to tyrosine.[7]

The great medical interest in L-DOPA for the treatment of Parkinson's disease stimulated the development of a successful microbiological method for oxidation of suitably derivatized L-tyrosine:

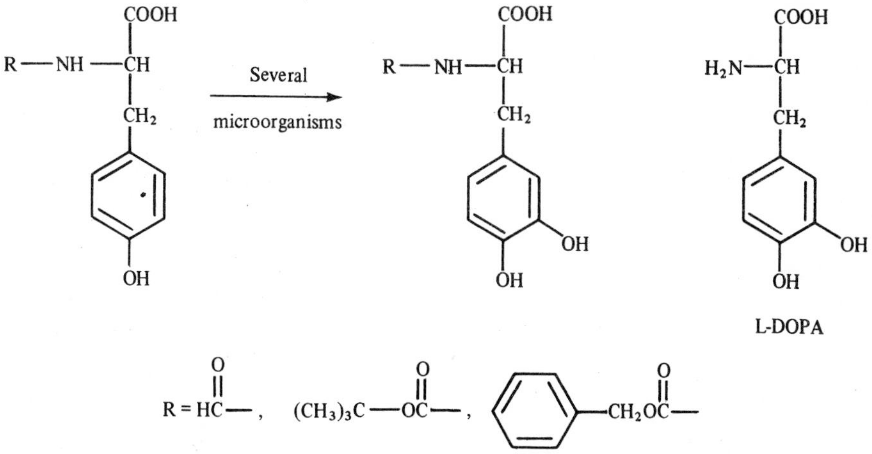

Yields are in the 25-30% range or higher.[8]

When dealing with reactions of this type, the possibility of the occurrence of an NIH shift[9,10] must always be kept in mind.

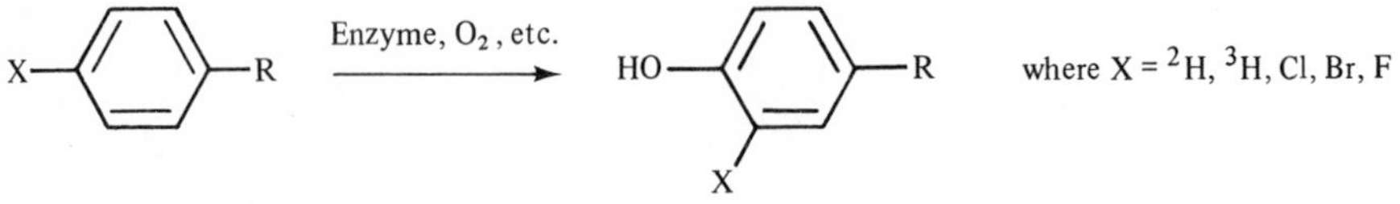

X—⟨ ⟩—R $\xrightarrow{\text{Enzyme, O}_2\text{, etc.}}$ HO—⟨ ⟩—R   where $X = {}^2H, {}^3H, Cl, Br, F$
                                          |
                                          X

As illustrated, the NIH shift occurs during some enzymatic hydroxylations of aromatic rings and results in a shift of the group at the
position of attack to an adjacent carbon rather than in complete replacement from the ring.   The NIH shift is catalyzed by the phenylalanine hydroxylase of a *Pseudomonas* species and by the tryptophan-5-
hydroxylase of *Chromobacterium violaceum*.[9]

It is sometimes possible to remove oxidative reaction products
from the effective fermentation area as they are formed, either to
prevent further oxidation or to remove a toxic product that is inhibitory to the fermentation of the remaining substrate.   In the oxidation sequence: *p*-xylene → *p*-toluic acid → 2,3-dihydroxy-*p*-toluic
acid → α,α-*cis*,*cis*-dimethylmuconic acid, the assay yields of all the
acidic products increase following the addition of anion-exchange
resins to the fermentation to remove the final product.[11]

A few other oxidations of simple monocyclic aromatic systems are
listed below to illustrate the potential scope of these reactions.

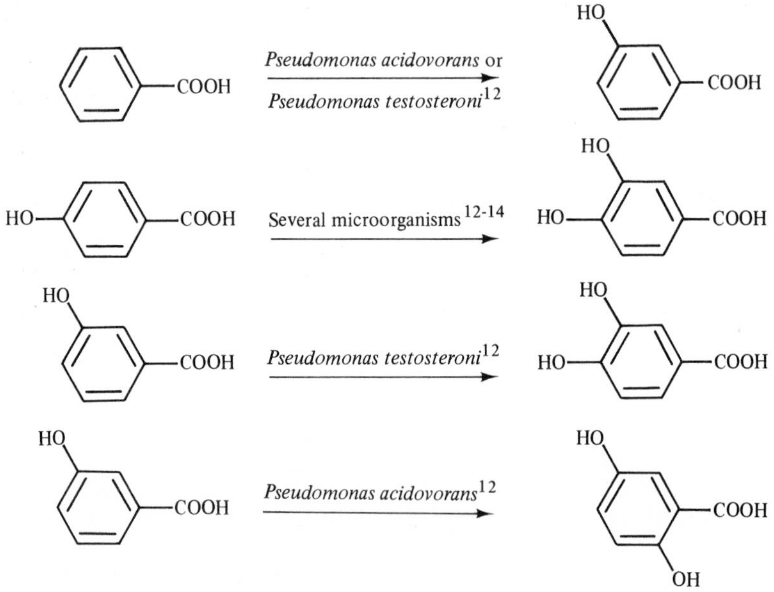

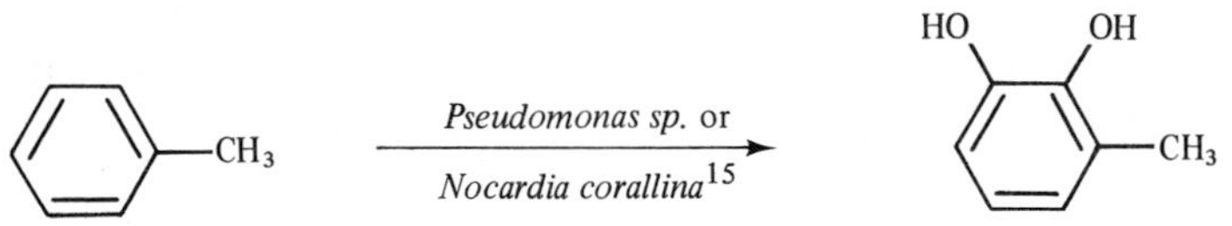

CH_3-benzene $\xrightarrow{\textit{Pseudomonas sp. or Nocardia corallina}[15]}$ 3-methylbenzene-1,2-diol

2-fluoro-4-nitrobenzoic acid $\rightarrow$ 2-fluoro-4-hydroxybenzoic acid $\rightarrow$ 2-fluoro-3,4-dihydroxybenzoic acid

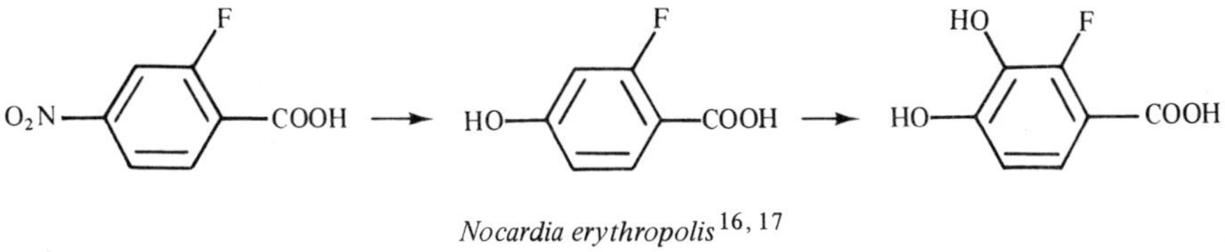

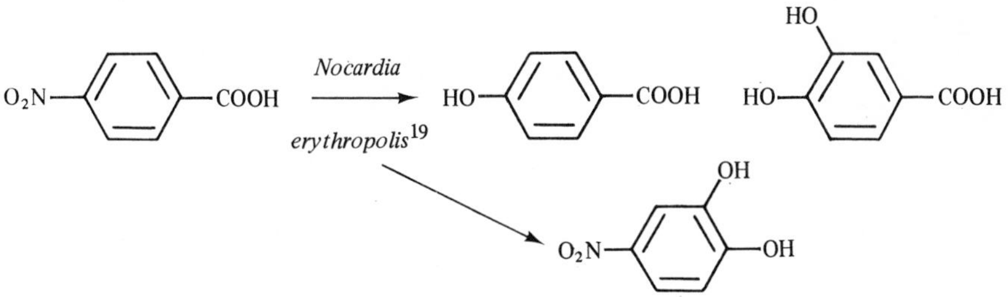

Nocardia erythropolis[16, 17]

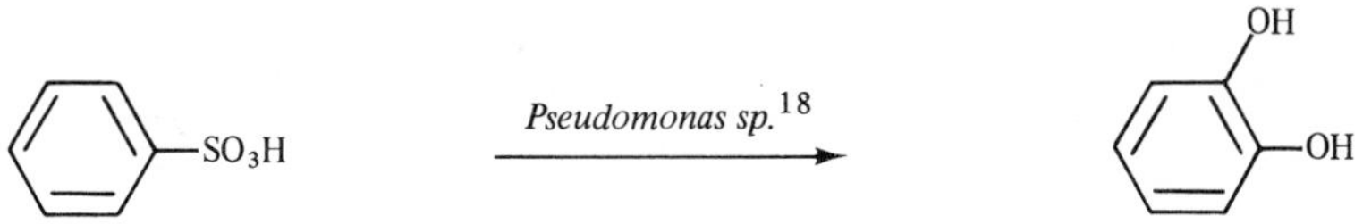

benzenesulfonic acid ($-SO_3H$) $\xrightarrow{\textit{Pseudomonas sp.}[18]}$ benzene-1,2-diol

4-nitrobenzoic acid (O_2N-...$-COOH$) $\xrightarrow[\textit{erythropolis}[19]]{\textit{Nocardia}}$ 4-hydroxybenzoic acid and 3,4-dihydroxybenzoic acid; and 4-nitrobenzene-1,2-diol (O_2N-...$-OH$)

phthalic acid (benzene-1,2-dicarboxylic acid) $\xrightarrow{\textit{Pseudomonas sp.}[20, 21]}$ 4,5-dihydroxyphthalic acid

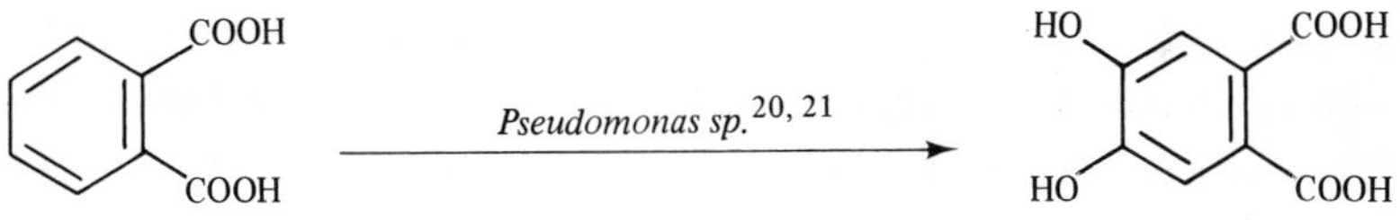

anthranilic acid (2-aminobenzoic acid) $\xrightarrow{\textit{Nocardia opaca}[22]}$ 2-amino-5-hydroxybenzoic acid

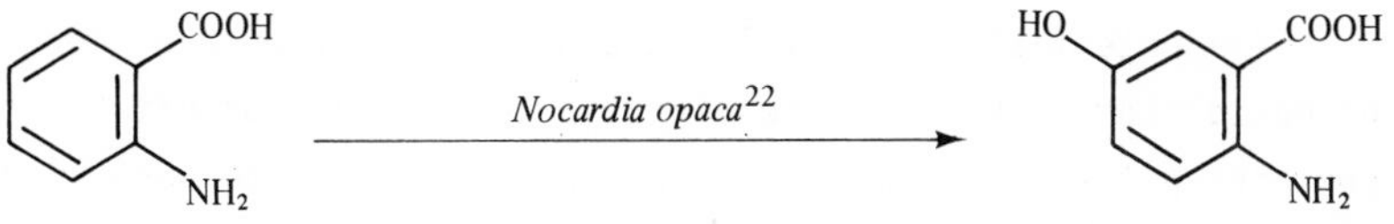

4-hydroxycinnamic acid ($HO-$...$-CH=CHCOOH$) $\xrightarrow[\textit{lepideus}[23, 24]]{\textit{Lentinus}}$ 3,4-dihydroxycinnamic acid ($HO-$...$-CH=CHCOOH$, with HO)

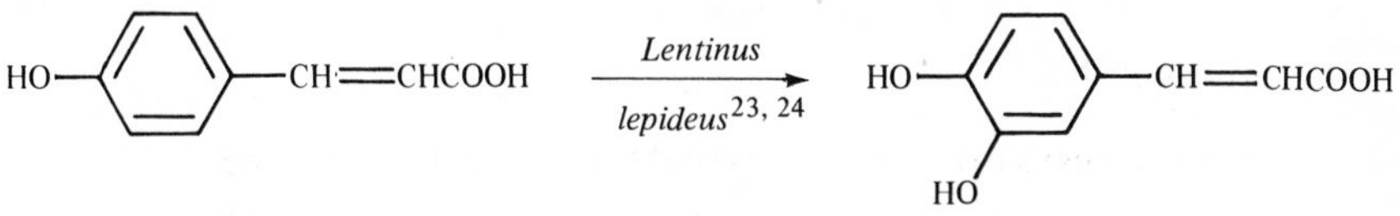

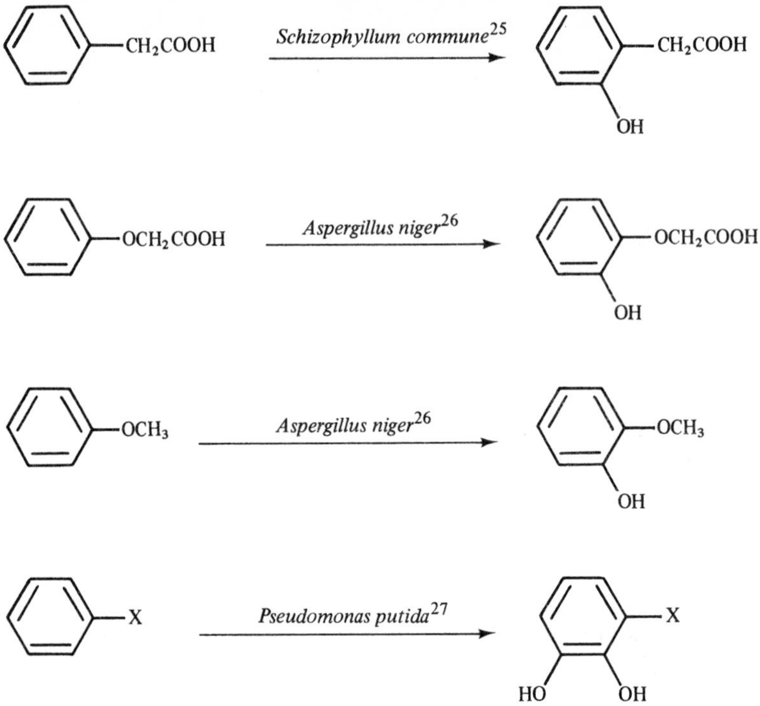

—CH$_2$COOH   *Schizophyllum commune*[25]   —CH$_2$COOH OH

—OCH$_2$COOH   *Aspergillus niger*[26]   —OCH$_2$COOH OH

—OCH$_3$ *Aspergillus niger*[26]   —OCH$_3$ OH

—X   *Pseudomonas putida*[27]   —X HO OH

(X = F, Cl, Br, I)   (All yields ~10% except
                                                               X = F ~1%)

It is worth noting briefly that the degradation of naphthalene by a *Pseudomonas* species involves 1,2-dihydroxynaphthalene as a transitory (but not primary) intermediate.[28] It will be seen from subsequent discussion of aromatic ring cleavages that this is a general reaction of polycyclic aromatic molecules. The preparative potential of these intermediates is noted in Chapter 3.

The potential commercial medicinal utility of the products has prompted some study of the microbial oxidation of natural products containing the indole moiety. In the case of tryptophane, its oxidation to 5-hydroxytryptophane by *Chromobacterium violaceum* is demonstrable, but no product is isolated, owing in part to a very low conversion (with much residual substrate) and in part to the lability of the product.[29]

More successful is the oxidation of yohimbine by *Streptomyces platensis*, to give about a 50% yield of 10-hydroxyyohimbine (based on the 60% of the substrate that is not recovered).[30]   The same micro-

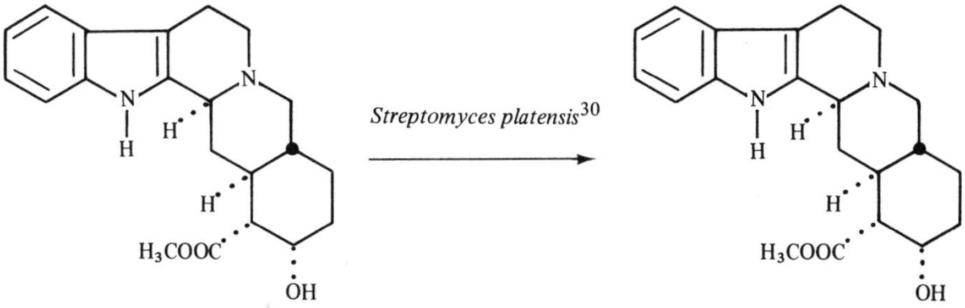

Yohimbine   10-Hydroxyyohimbine

organism hydroxylates the related compounds rauwolscine, 3-epi-α-
yohimbine, and alloyohimbine, but fails to hydroxylate a number of
other substances containing an indole moiety.[30] Other organisms also
oxygenate the aromatic ring of these related alkaloids,[31-34] as indicated in Table 2.

Table 2. Aromatic Hydroxylations of Yohimbe Alkaloids

Substrate	Organism	Product	Yield	Ref.
Yohimbine	*Cunninghamella ber-tholletiae*	11-Hydroxyyohimbine	3%	34
Yohimbine	*C. echinulata*	10-Hydroxyyohimbine	5%	33
α-Yohimbine	*C. bertholletiae*	11-Hydroxy-α-yohimbine	3%	33
α-Yohimbine	*C. echinulata*	10-Hydroxy-α-yohimbine	10%	33
β-Yohimbine	*Cunninghamella* species	No products		33
β-Yohimbine	*Streptomyces rimosus*	10-Hydroxy-β-yohimbine	5%	33
Apoyohimbine	*Cunninghamella blakesleeana*	10-Hydroxy-apoyohimbine	-	31, 33
β-Yohimbine methyl ester	*C. blakesleeana*	10-Hydroxy-apoyohimbine	-	31
3-epi-Apoyohimbine	*C. blakesleeana*	10-Hydroxy-apoyohimbine	-	31

The complex pathway of microbial degradation of riboflavin and related substances by a *Pseudomonas* species has an aromatic hydroxylation as an early step, as illustrated, but this is probably of no synthetic utility.[35-37]

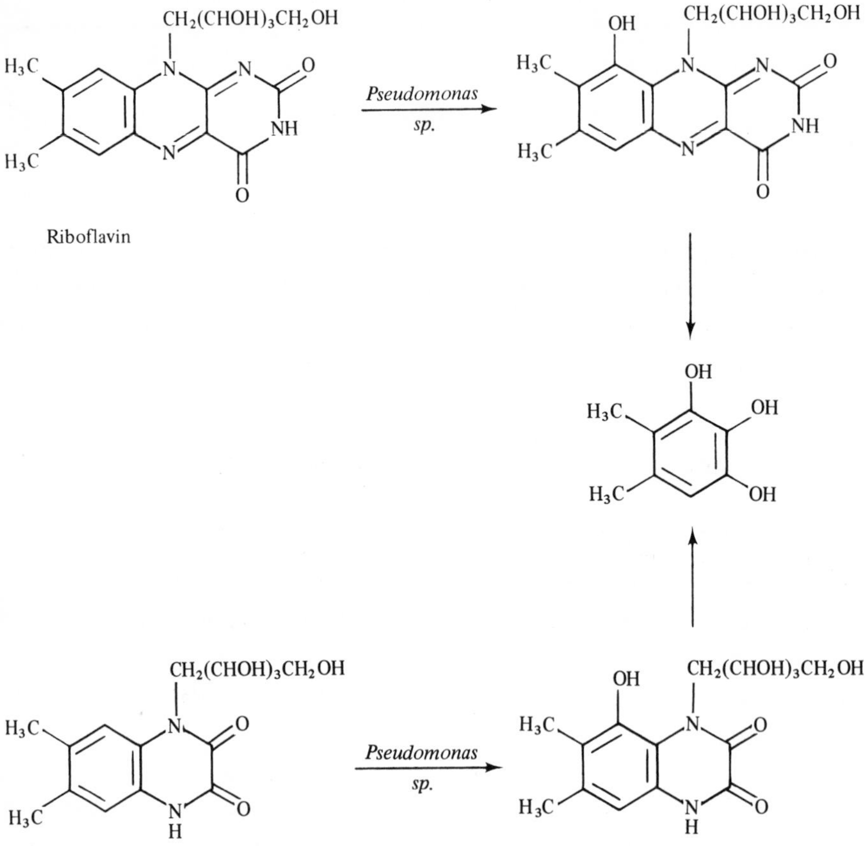

Riboflavin

The pathway of degradation of kynurenic acid by *Pseudomonas fluorescens* includes the 7,8-dihydroxy derivative, obtained in 35% yield by the use of an enzyme preparation from this microorganism.[38]

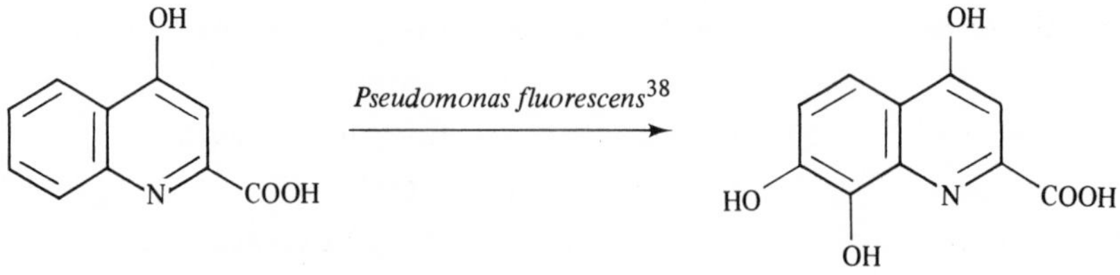

Kynurenic acid

Heterocyclic Substrates.   The hedonistic poet who described tobacco as "a dirty weed," but added, "I like it!" was probably euphor-
ically unaware of such problems as lung cancer and air pollution.
Nor could he have guessed the somewhat different (and possibly less
harmful) enjoyment that countless scientists would derive from the
study of the microbial metabolism of the constituents of tobacco.

Before considering the transformations of the tobacco alkaloids
themselves, it will be helpful to review briefly the microbiological
oxidations of nicotinic acid.   Nicotinic acid is oxidized to 6-hy-
droxynicotinic acid by several pseudomonads, either as the intact organisms (25% yield) or as cell-free extracts.[39]   Although oxygen gas
is required, the introduced hydroxyl receives its oxygen from the
aqueous medium.[40]   The further oxidation of 6-hydroxynicotinic acid
that takes place when intact organisms are used[41] may usually be pre-
vented by the use of cell-free extracts,[42] although in one instance,
this tendency toward continued oxidation has been used to prepare
2,6-dihydroxynicotinic acid (in 3% yield) from nicotinic acid by the
use of a *Bacillus* species cell-free extract.[43]

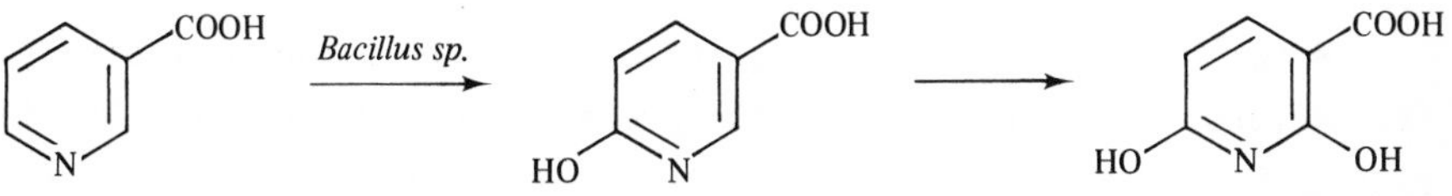

Nicotinic acid

The extensive degradation of the somewhat similar substrate,
pyridine-2,6-dicarboxylic acid, by an *Achromobacter* species includes
3-hydroxypyridine-2,6-dicarboxylic as a trace, unisolated intermediate.[44]

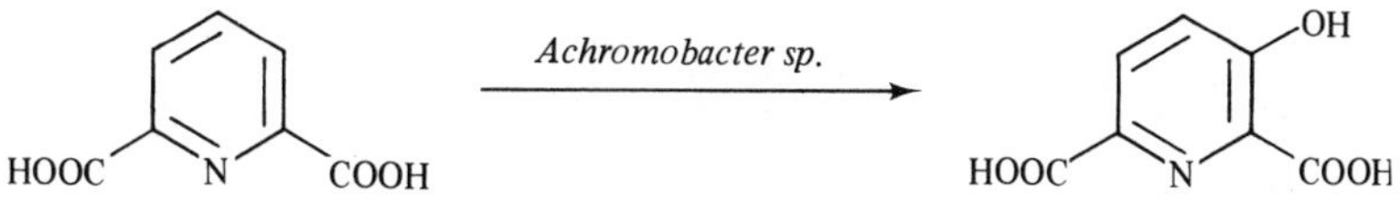

Pyridine-2,6-dicarboxylic acid

The microbial metabolism of the major tobacco alkaloid nicotine has engaged the attention of several research groups.[39,45-52] The introduction of a 6-hydroxyl group, analogous to the above-described oxidation of nicotinic acid, is the first stage in an extended series of degradative reactions. *Arthrobacter oxydans* has been used most commonly, either intact, or as the source of a cell-free enzyme prep-

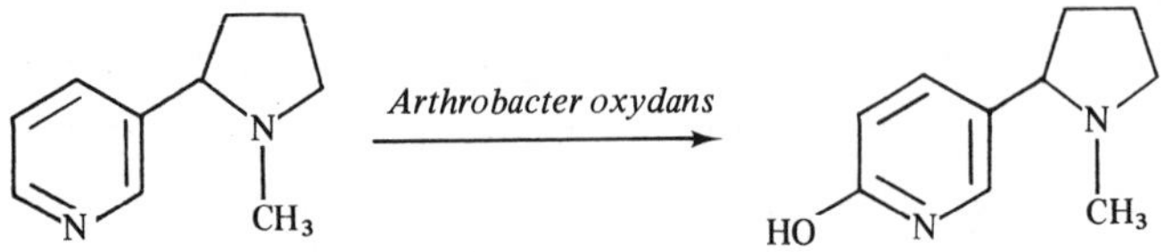

Nicotine

aration,[45] or, in one instance, of purified enzyme.[47] Yield informa-tion is lacking. One enzyme system oxidizes both D- and L-nicotine, as well as nornicotine, nicotine N-oxide, myosmine, and anabasine, but not nicotyrine, 6-hydroxynicotine, 6-hydroxypseudooxynicotine, nicotinic acid, N'-methylnicotinamide, pyridine, quinoline, or α,α'-dipyridyl.[47]

A soil *Pseudomonas* species oxidizes nornicotine and anabasine to the 6-hydroxy dehydro compounds in (crude) yields of less than 10% in each case.[41] (The same microorganisms degrade nicotine in the satu-

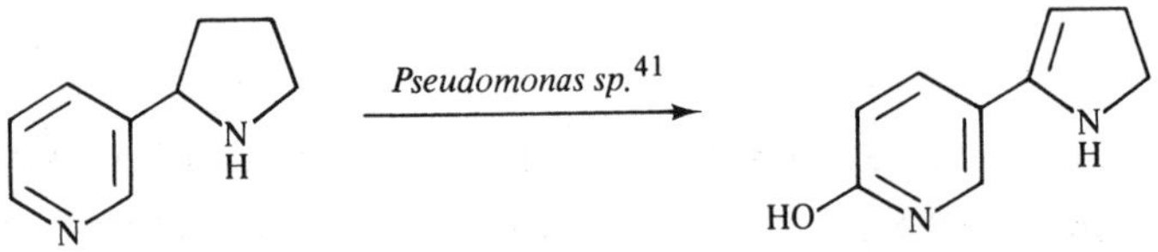

Nornicotine   6-Hydroxymyosmine

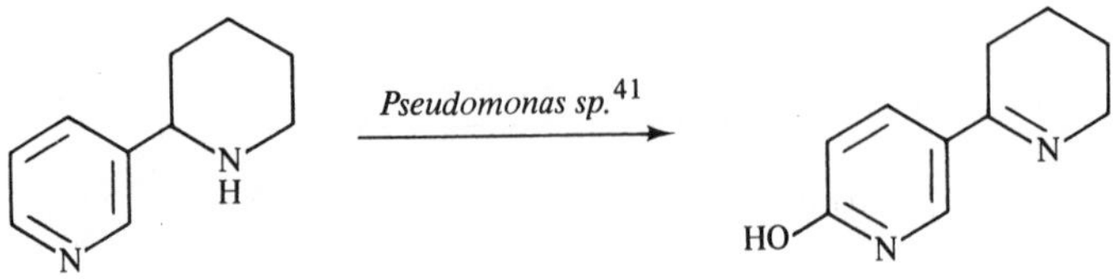

Anabasine

rated ring before introducing a 6-hydroxyl, while another *Pseudomonas* strain does the same to nornicotine, giving the substituted butyric acid in 38% yield.[53])

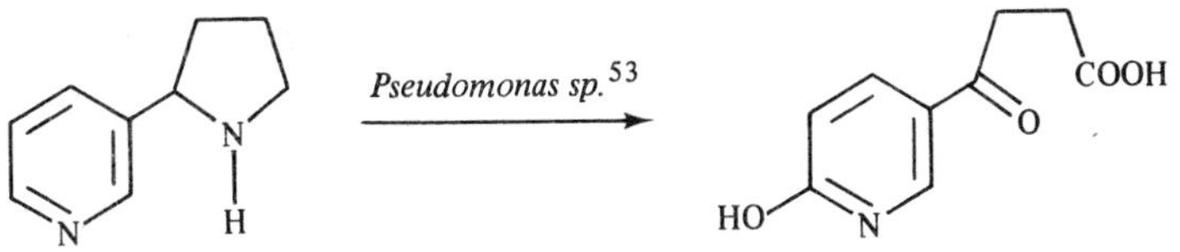

Nornicotine

Miscellaneous. The oxidation of urocanic acid by an enzyme prep- aration from *Pseudomonas aeruginosa* proceeds through a hydroxylated derivative, isolated in 4% yield from its mixture with further degra- dation products.[54,55]

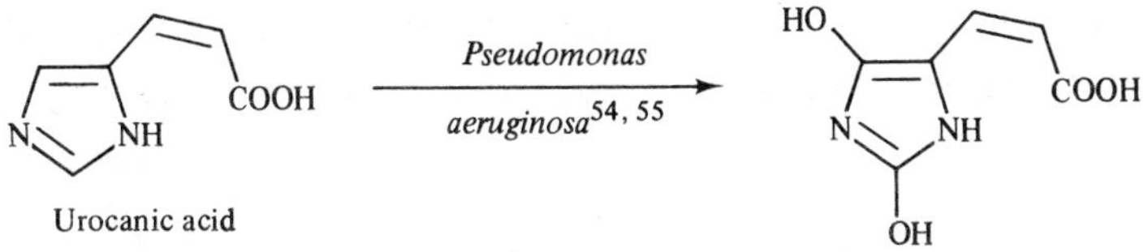

Urocanic acid

The oxidation of xanthopterin by an enzyme preparation from a soil bacterium gives both leucopterin (1% yield) and 6,7-dioxyluma- zine (<1% yield).[56] The enzyme system also oxidizes lumazine and 6-

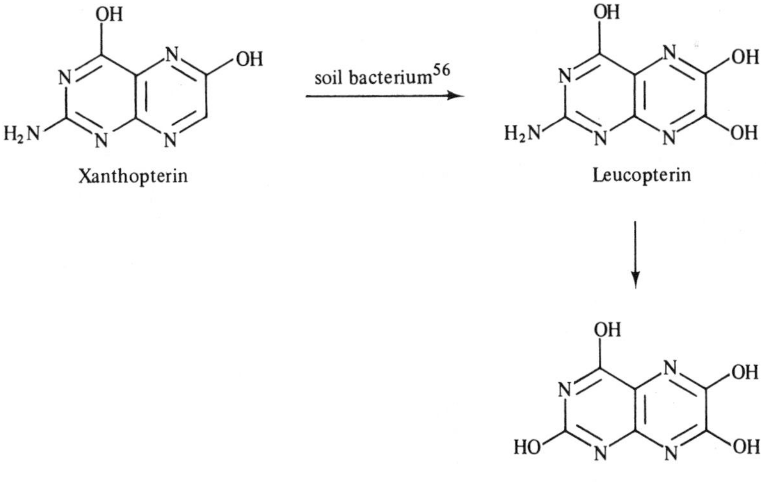

Xanthopterin   Leucopterin

6,7-Dioxylumazine

oxylumazine to the corresponding 7-oxycompounds, but fails to oxidize 7-oxylumazine.[56]

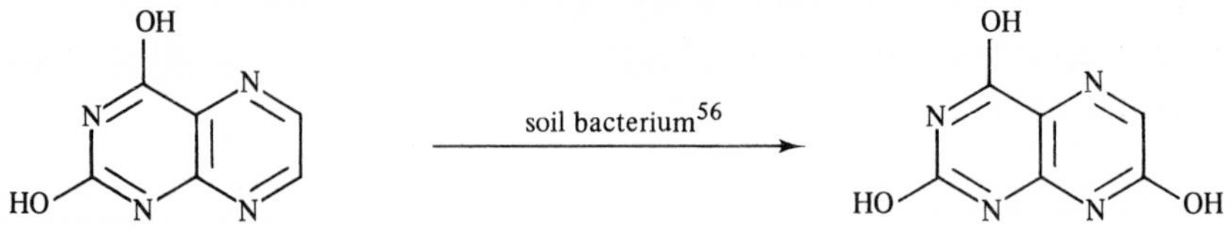

Lumazine

soil bacterium[56]

7-Oxylumazine

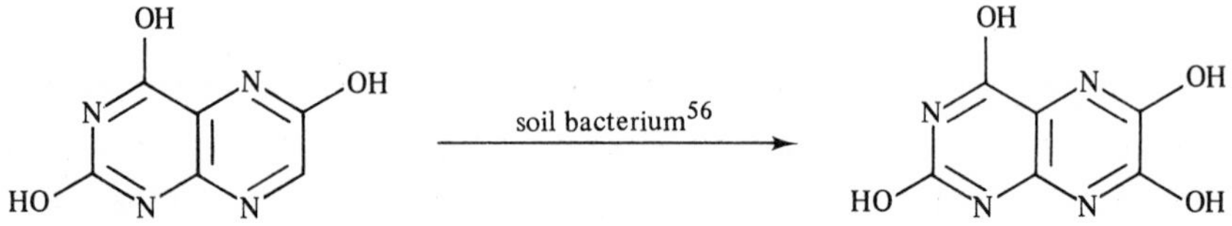

6-Oxylumazine

soil bacterium[56]

6,7-Dioxylumazine

The hydroxylation of several pyrrolopyrimidines at C-8 by *Ps. aeruginosa* can be blocked by a methyl group at that position.[57]

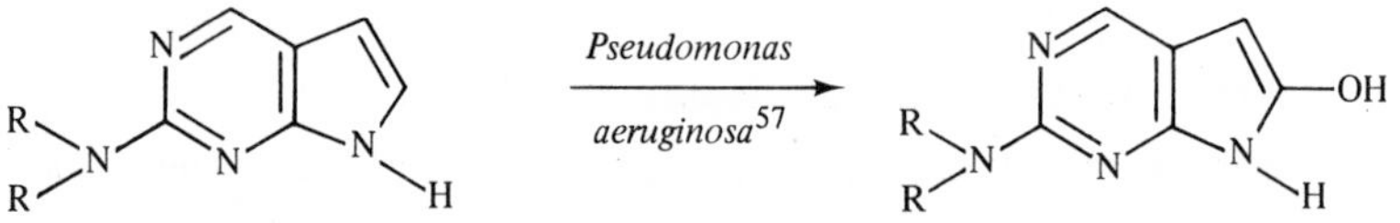

R = H, CH$_3$

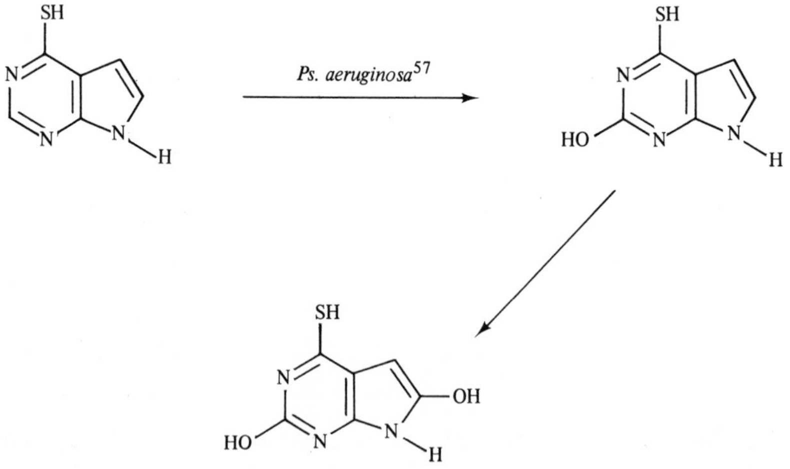

Pseudomonas aeruginosa[57]

Ps. aeruginosa[57]

The oxidation of o-coumaric acid to 4-hydroxycoumarin by several microorganisms[58-60] (including *Aspergillus niger*) would appear at first glance to proceed *via* coumarin.   It has, however, been shown that *A. niger* converts coumarin largely to melilotic acid.[59]   The

o-Coumaric acid   Melilotic acid   Coumarin   4-Hydroxycoumarin

possibility exists that the intermediate product is 2-hydroxybenzoyl-acetic acid, but in the absence of convincing data, the reaction is cited here as a heteroaromatic hydroxylation. When *Fusarium solani* is used, the yield of 4-hydroxycoumarin is 52%.[60]

REFERENCES

1. C. C. Levy and P. Frost, *J. Biol. Chem. 241*, 997 (1966).

2. D. M. Webley, R. B. Duff, and V. C. Farmer, *J. Gen. Microbiol. 13*, 361 (1955).

3. D. W. Ribbons, *J. Gen. Microbiol. 44*, 221 (1966).

4. P. Goldman, G. W. A. Milne, and M. T. Pignataro, *Arch. Biochem. Biophys. 118*, 178 (1967).

5. E. M. Chamberlain and S. Dagley, *Biochem. J. 110*, 755 (1968).

6. R. J. Theriault and T. H. Longfield, *Appl. Microbiol. 15*, 1431 (1967).

7. G. Guroff and T. Ito, *J. Biol. Chem. 240*, 1175 (1965).

8. C. J. Sih, P. Foss, J. Rosazza, and M. Lemberger, *J. Amer. Chem. Soc. 91*, 6204 (1969).

9. G. Guroff, J. W. Daly, D. M. Jerina, J. Renson, B. Witkop, and S. Udenfriend, *Science 157*, 1524 (1967).

10. J. W. Daly, G. Guroff, D. M. Jerina, S. Udenfriend, and B. Witkop, in *Oxidation of Organic Compounds*, Vol. III, American Chemical Society, Washington, D. C., 1968. pp. 279-289.

11. R. L. Raymond, V. W. Jamison, and J. O. Hudson, *Appl. Microbiol. 17*, 512 (1969).

12. M. L. Wheelis, N. J. Palleroni, and R. Y. Stanier, *Arch. Mikrobiol. 59*, 302 (1967).

13. K. Hosokawa and R. Y. Stanier, *J. Biol. Chem. 241*, 2453 (1966).

14. G. D. Hegeman, *Arch. Mikrobiol. 59*, 143 (1967).

15. J. R. Forro, Diss. Abstr. *26*, 6305 (1965-66).

16. R. B. Cain, E. K. Tranter, and J. A. Darrah, *Biochem. J. 106*, 211 (1968).

17. A. Smith, E. K. Tranter, and R. B. Cain, *Biochem. J. 106*, 203 (1968).

18. R. B. Cain and D. R. Farr, in *Biological and Chemical Aspects of Oxygenases*, K. Bloch and O. Hayashi, eds., Maruzen Co., Ltd., Tokyo, 1966. pp. 125-144.

19. N. J. Cartwright and R. B. Cain, *Biochem. J. 71*, 248 (1959).

20. D. W. Ribbons and W. C. Evans, *Biochem. J. 76*, 310 (1960).

21. W. C. Evans, *Biochem. J. 61*, x (1955).

22. R. B. Cain, *Antonie van Leeuwenhoek 34*, 417 (1968).

23. D. M. Power, G. H. N. Towers, and A. C. Neish, *Can. J. Biochem. 43*, 1397 (1965).

24. G. H. N. Towers, in *Perspectives in Phytochemistry*, J. B. Harborne and T. Swain, eds., Academic, New York, 1969, pp. 179-191.

25. K. Moore and G. H. N. Towers, *Can. J. Biochem. 45*, 1659 (1967).

26. S. M. Bocks, J. R. L. Smith, and R. O. C. Norman, *Nature 201*, 398 (1964).

27. D. T. Gibson, J. R. Koch, C. L. Schuld, and R. E. Kallio, *Biochemistry 7*, 3795 (1968).

28. J. F. Murphy and R. W. Stone, *Can. J. Microbiol. 1*, 579 (1955).

29. C. Mitoma, H. Weissbach, and S. Udenfriend, *Arch. Biochem. Biophys. 63*, 122 (1956).

30. Y. H. Loo and M. Reidenberg, *Arch. Biochem. Biophys. 79*, 257 (1959).

31. W. O. Godtfredsen, G. Korsby, H. Lorck, and S. Vandegal, *Experientia 14*, 88 (1958).

32. E. Meyers and S. C. Pan, *J. Bacteriol. 81*, 504 (1961).

33. E. L. Patterson, W. W. Andres, E. F. Krause, R. E. Hartman, and L. A. Mitscher, *Arch. Biochem. Biophys. 103*, 117 (1963).

34. R. E. Hartman, E. F. Krause, W. W. Andres, and E. L. Patterson, *Appl. Microbiol. 12*, 138 (1964).

35. P. Z. Smyrniotis, H. T. Miles, and E. R. Stadtman, *J. Amer. Chem. Soc. 80*, 2541 (1958).

36. H. T. Miles, P. Z. Smyrniotis, and E. R. Stadtman, *J. Amer. Chem. Soc. 81*, 1946 (1959).

37. D. R. Harkness, L. Tsai, and E. R. Stadtman, *Arch. Biochem. Biophys. 108*, 323 (1964).

38. H. Taniuchi and O. Hayaishi, *J. Biol. Chem. 283*, 283 (1963).

39. W. G. Frankenburg and A. A. Vaitekunas, *Arch. Biochem. Biophys. 58*, 509 (1955).

40. A. L. Hunt, D. E. Hughes, and J. M. Lowenstein, *Biochem. J. 69*, 170 (1958).

41. E. Wada, *Arch. Biochem. Biophys. 72*, 145 (1957).

42. D. E. Hughes, *Biochem. J. 60*, 303 (1955).

43. J. C. Ensign and S. C. Rittenberg, *J. Biol. Chem. 239*, 2285 (1964).

44. Y. Kobayashi and K. Arima, *J. Bacteriol. 84*, 765 (1962).

45. K. Decker and H. Bleeg, *Biochim. Biophys. Acta 105*, 313 (1965).

46. K. Decker, H. Eberwein, F. A. Gries, and M. Brühmüller,'*Z. Physiol. Chem. 319*, 279 (1960).

47. L. I. Hochstein and B. P. Dalton, *Biochim. Biophys. Acta 139*, 56 (1967).

48. L. I. Hochstein and S. C. Rittenberg, *J. Biol. Chem. 234*, 151 (1959).

49. L. I. Hochstein and S. C. Rittenberg, *J. Biol. Chem. 234*, 156 (1959).

50. S. H. Richardson and S. C. Rittenberg, *J. Biol. Chem. 236*, 959 (1961).

51. S. H. Richardson and S. C. Rittenberg, *J. Biol. Chem. 236*, 964 (1961).

52. G. D. Griffith, R. U. Byerrum, and W. A. Wood, *Proc. Soc. Exptl. Biol. Med. 108*, 162 (1961).

53. E. Wada, *Arch. Biochem. Biophys. 64*, 244 (1956).

54. K. Ichihara, H. Satani, N. Okada, Y. Takagi, and Y. Sakamoto, *Proc. Japan Acad. 33*, 105 (1957); *Chem. Abstr. 51*, 15673 (1957).

55. K. Uranaka, *Kumamoto Med. J. 11*, 222 (1958); *Chem. Abstr. 53*, 19002 (1959).

56. C. C. Levy and W. S. McNutt, *Biochemistry 1*, 1161 (1962).

57. F. Bergmann, H. Ungar-Warm, H. Kwietny-Govrin, H. Goldberg, and S. Leon, *Biochim. Biophys. Acta 55*, 512 (1962).

58. K. Moore, P. F. Subba Rao, and G. H. N. Towers, *Biochem. J. 106*, 507 (1968).

59. S. M. Bocks, *Phytochemistry 6*, 127 (1967).

60. H. S. Shieh and A. C. Blackwood, *Can. J. Biochem. 45*, 2045 (1967).

General References

L. L. Wallen, F. H. Stodola, and R. W. Jackson, *Type Reactions in Fermentation Chemistry*, United States Department of Agriculture, 1959, pp. 185-189.

D. W. Ribbons, *Ann. Rept. Chem. Soc. London 62*, 445 (1965).

W. C. Evans, *Ann. Rept. Chem. Soc. London 53*, 279 (1956).

C. Arnaudi, L. Canonica, and V. Treccani, *Ric. Sci. 25*, 3244 (1955).

F. C. Happold, *Biochem. Soc. Symposia 5*, 85-96 (1950).

R. L. Raymond, *Process Biochem. 4*, 71 (1969).

D. T. Gibson, *Science 161*, 1093 (1968).

Chapter 5

AROMATIC RING OPENING

As we have noted in the preceding chapter on aromatic ring hydroxylation, that reaction is frequently followed by rapid additional oxidation, which usually results in the opening of the aromatic ring. Despite the great inherent microbiological and biochemical interest in the many studies of aromatic ring cleavages, the usefulness of the reactions as preparative synthetic tools remains (with two notable exceptions) essentially undemonstrated. The fact that several of the responsible enzymes have been prepared in crystalline form may enhance the synthetic potential of this area.

Ring opening reactions are of two types, designated *ortho* cleavage and *meta* cleavage, respectively.

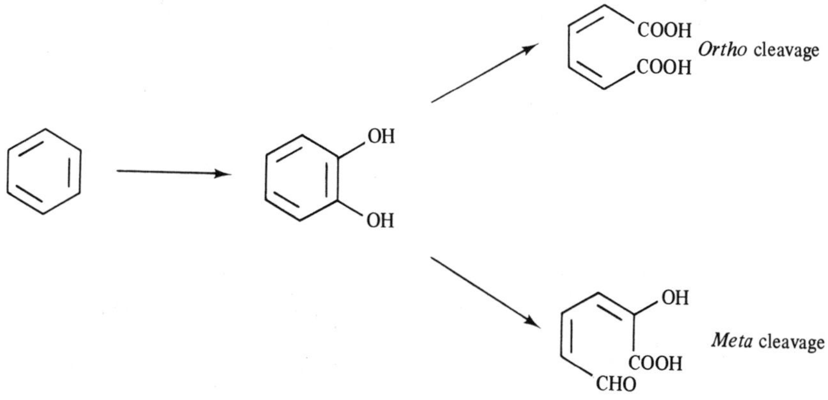

Ortho cleavage of *p*-xylene by *Nocardia corallina*, which has been studied extensively, produces good yields of α,α-dimethyl-*cis*,*cis*-muconic acid.[1-3] (Apparently, the process has been used on a commercial scale.[4]) Since the end product is somewhat toxic to the microorganism, the efficiency of the conversion may be enhanced by the addition of ion exchange resin to the fermentation to absorb some of

the muconic acid.   (It must be noted that some *Nocardia* give other
oxidation products,   notably *p*-toluic acid and 2,3-dihydroxy-*p*-toluic
acid.[3])

The *ortho* cleavage of 2,4-dichlorophenoxyacetic acid (2,4-D) by
a *Pseudomonas* species   gives an   infinitesimal   isolated yield of α-
chloro-*cis*,*cis*-muconic acid[5]:

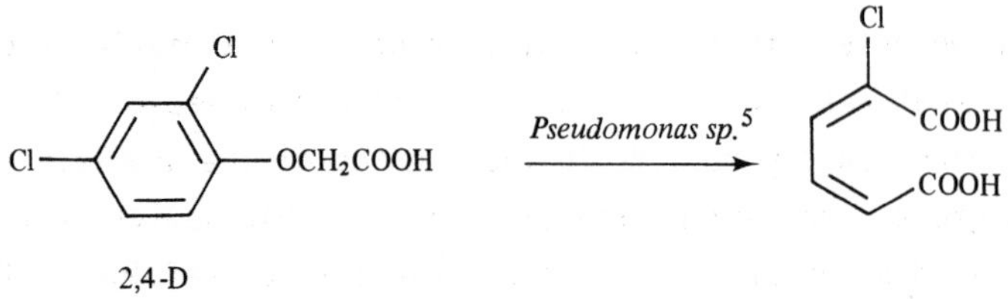

2,4-D

and a similar cleavage of 2-methyl-4-chlorophenoxyacetic acid   leads
to the muconic acid-related lactone[6]:

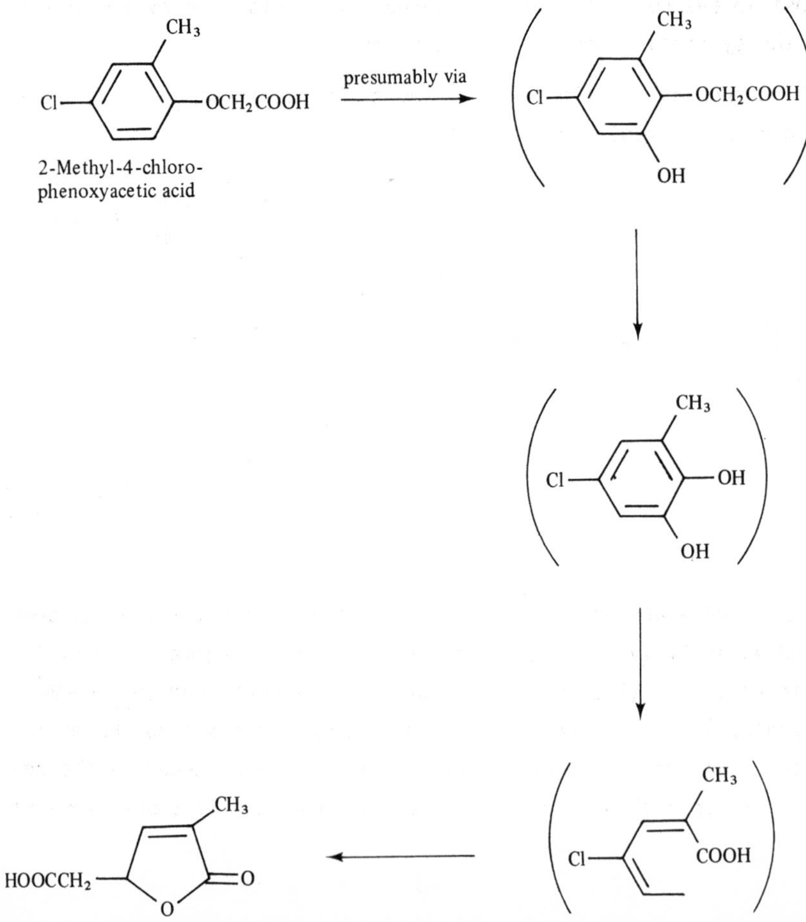

2-Methyl-4-chloro-
phenoxyacetic acid

Cleavage of 2-fluorobenzoic acid proceeds through the intermediate fluorocatechol (5% yield) to α-fluoro-*cis*,*cis*-muconic acid (0.8%

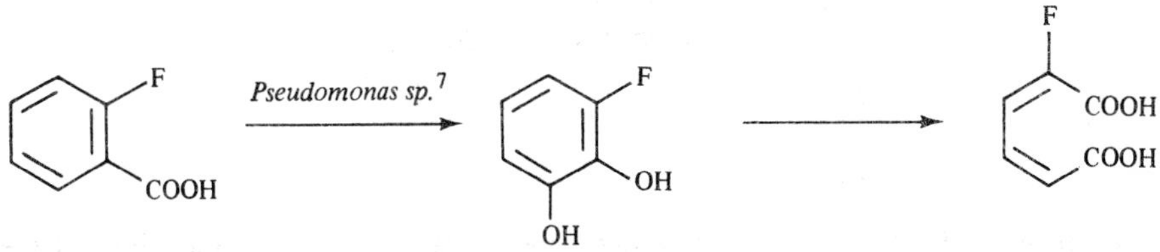

2-Fluorobenzoic acid

yield).[7] Since fluorocatechol is chemically oxidizable to the fluoromuconic acid in 45% yield,[7] the microbial method seems to be of academic interest only.

In contrast to the preceding reactions leading to (substituted) *cis*,*cis*-muconic acids, the action of a *Micrococcus spheroides*-like organism on benzene gives *trans*,*trans*-muconic acid.[8]

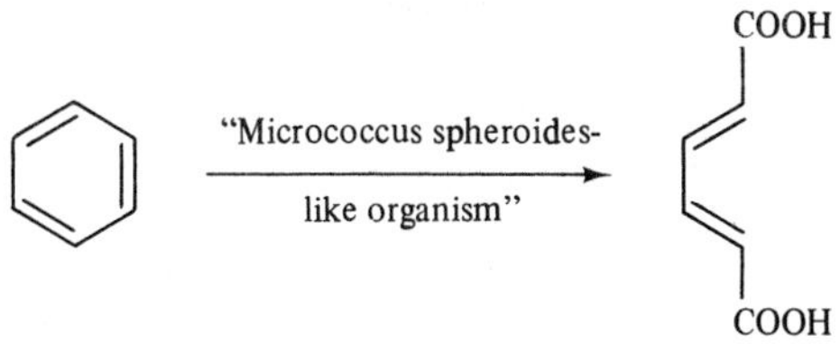

Benzene

In a rather unusual variant of the *ortho* cleavage reaction, 2-fluoro-3,4-dihydroxybenzoic acid is converted by *Nocardia erythropolis* extracts to 2-fluoro-3-oxohexanedioic acid (94% assay yield).[9]

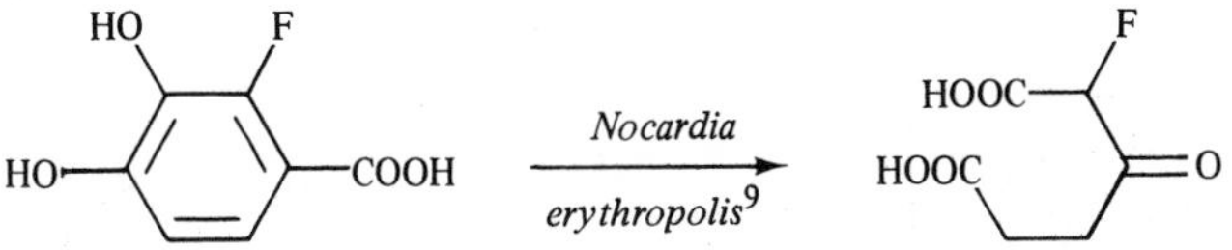

2-Fluoro-3,4-dihydroxybenzoic acid

The *meta* cleavage of 4-methylcresol by *Pseudomonas desmolyticum* gives the substituted muconic semialdehyde (11% yield).[10]

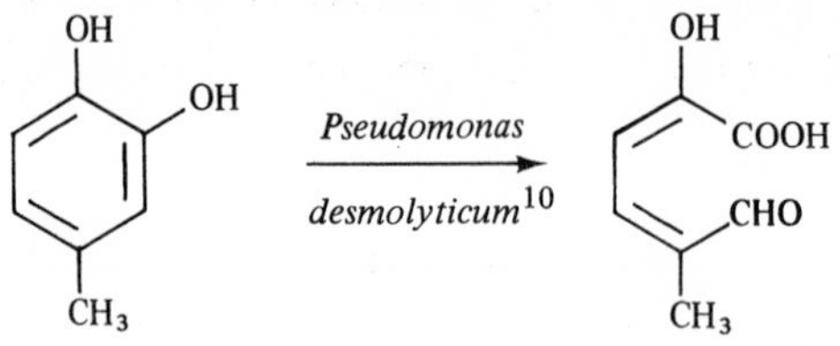

4-Methylcresol

A variety of other *meta* cleavages are listed to illustrate the potential scope of this reaction type. In some cases, the nature of the isolation procedures has led to isomerization to *trans,trans* isomers.

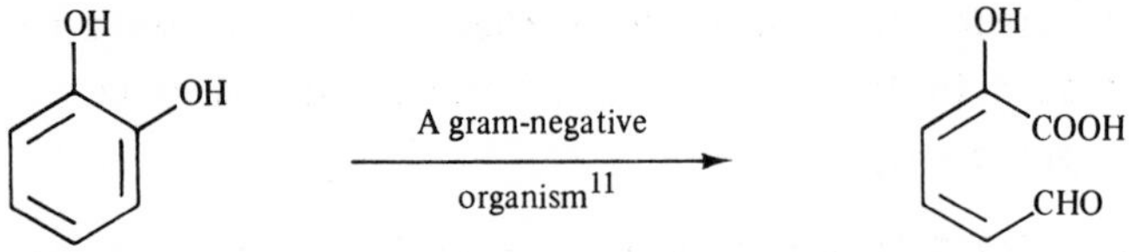

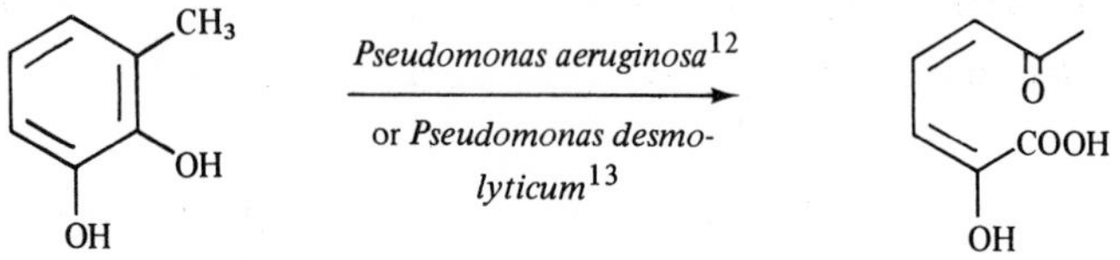

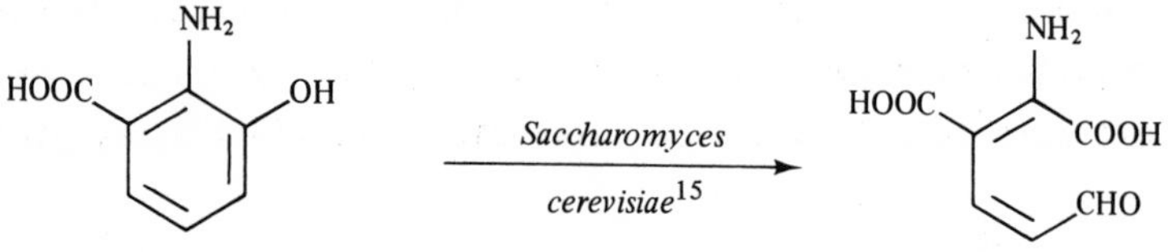

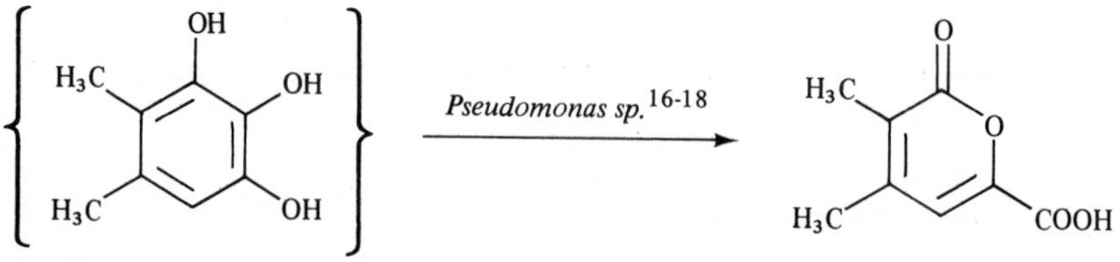

Riboflavin

(21% yield from riboflavin)

No discussion of *meta* cleavage would be complete without mentioning the extensive studies with steroids. For example, the oxidation of 11α-hydroxyprogesterone to a bicyclic acid by *Proactinomyces*

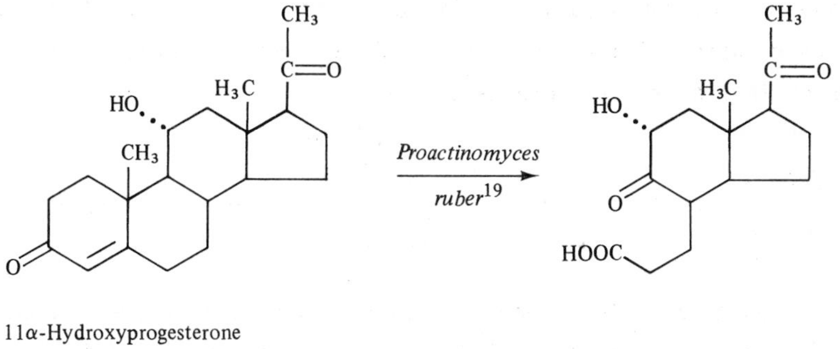

11α-Hydroxyprogesterone

ruber[19] is the net result of a multiplicity of reactions, including dehydrogenation, hydroxylation, retroaldol reaction, and β-oxidation,[20] from which the aromatic ring cleavage portion may be dissected out as shown (in another steroid):

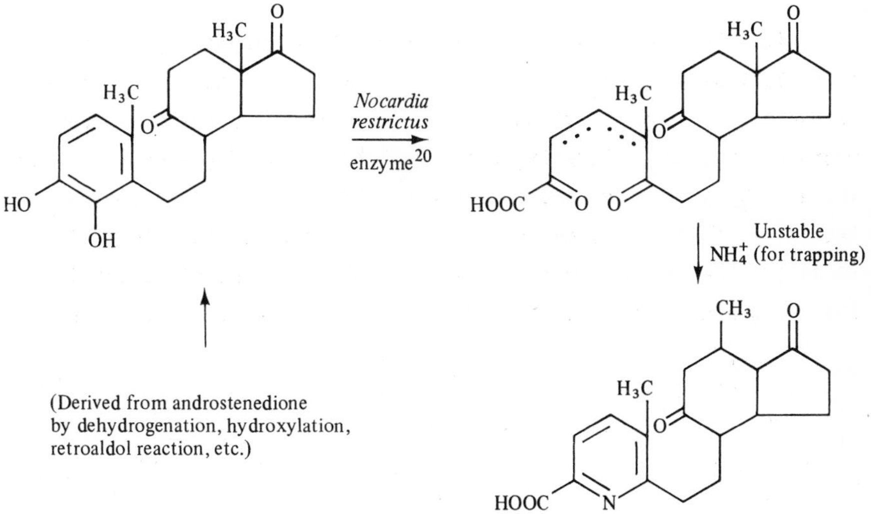

Nocardia restrictus enzyme[20]

Unstable
NH$_4^+$ (for trapping)

(Derived from androstenedione by dehydrogenation, hydroxylation, retroaldol reaction, etc.)

One enzyme responsible for *meta* cleavage has been isolated and crystallized.[21] Although this metapyrocatechase (catechol 2,3-oxygenase) apparently has high substrate specificity (only catechol and homocatechol are cleaved), it might conceivably oxidize other compounds. More important, it suggests that other cleaving enzymes may be amenable to isolation, with the resultant promise of good substrate specificity under clean reaction conditions.

The problem of the disposal of detergent wastes has been studied extensively by microbiologists.   At least one of the reaction types in the degradation of alkylbenzenesulfonates involves ring cleavage, but the synthetic utility remains unproven.[22,23]

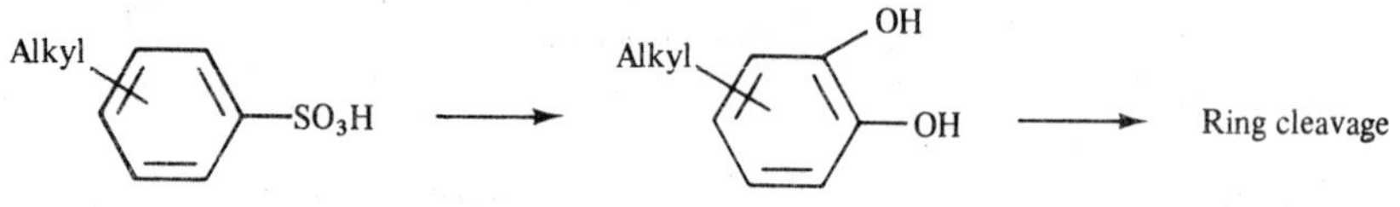

Long-chain alkanes with a phenyl group attached and with a sulfonic acid moiety somewhere in the molecule are known to be degraded by microorganisms.[24,25] Although products have not been characterized (and hence no yields are reported), there are enough differences between isomeric materials with regard to substrate suitability and degradation rate to suggest that this will ultimately be a fruitful area for synthetic work.

With the increased headaches of such ecologically disturbing technological problems as the disposal of detergents and other waste products, it seems inevitable that a microbial approach to aspirin production would be investigated. Naphthalene is degraded to salicylic acid by several *Pseudomonas* species[26-37] in excellent yield, particularly if ion exchange resin is incorporated in the fermentation to absorb the product, which otherwise inhibits complete oxidation of the substrate.[38] The reaction probably proceeds by the pathway shown.

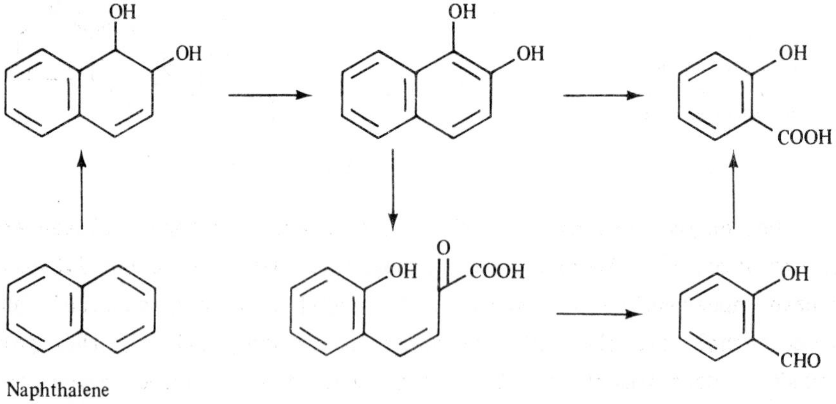

Naphthalene

The first monocyclic intermediate can be isolated by using an enzyme preparation that lacks the aldolase and dehydrogenase needed to degrade it to salicylic acid.

The degradations of some substituted naphthalenes, and of anthracene and phenanthrene are depicted below as an indication of the potential scope of this area.

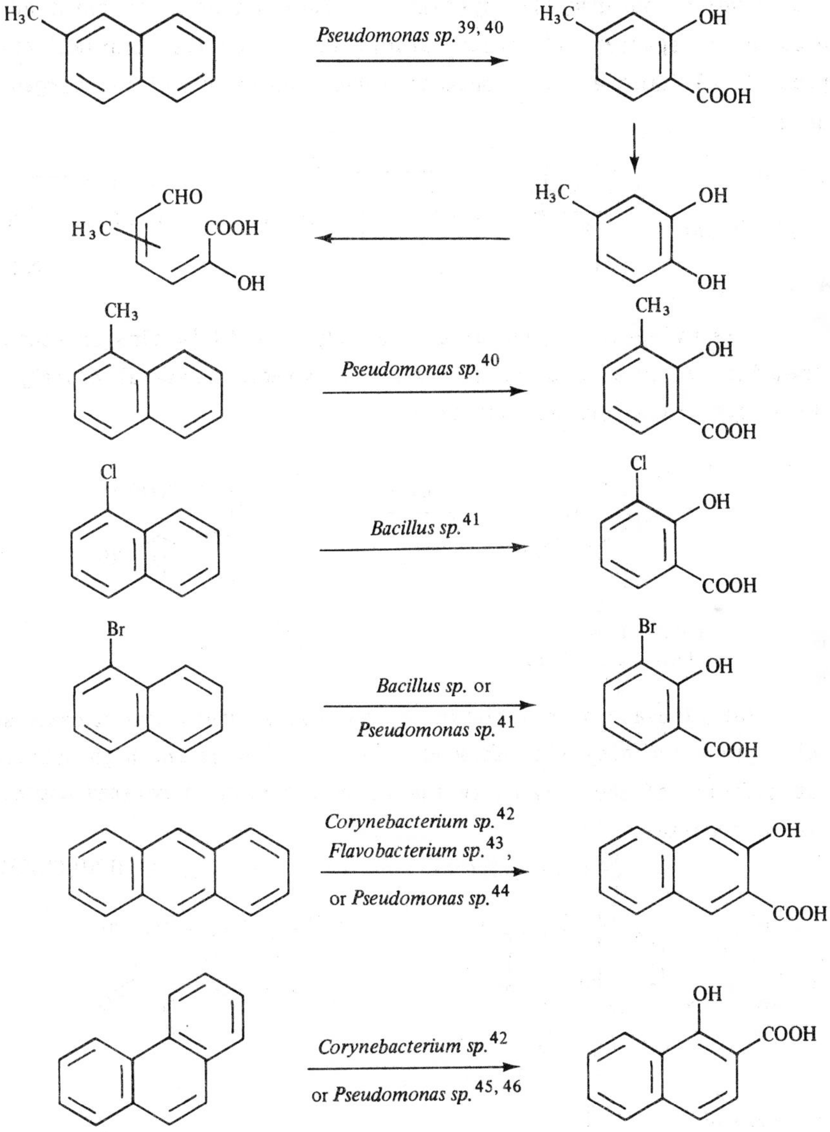

The differences in reactivity toward microbial oxidation that are
seen with alkylated phenanthrenes and naphthalenes may be attributable to substrate binding to the enzyme at its electron-rich K region,
followed by oxidative ring cleavage at another site whose reactivity
is a function of its distance from the binding site.[47]

Several reactions that can be classified as aromatic ring openings involve heterocyclic systems. 2-Furanoic acid is oxidized by
Pseudomonas species to α-ketoglutaric acid[48] or its transamination
product, glutamic acid,[49] depending upon the strain of microorganism
used.

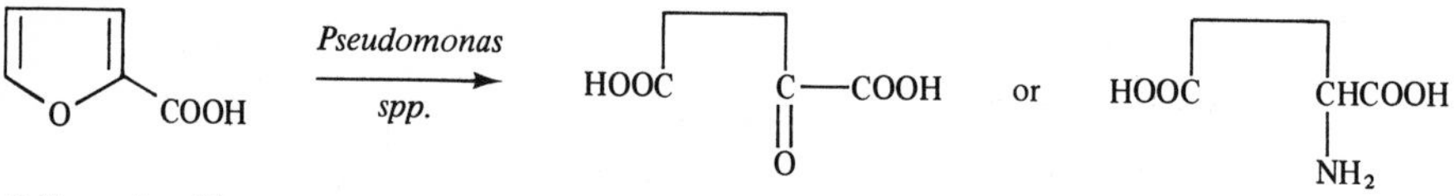

2-Furanoic acid

2-Methyl-3-hydroxypyridine-5-carboxylic acid is cleaved stoichi-
ometrically by a *Pseudomonas* enzyme, as shown. Several closely related substrates are not attacked.[50]

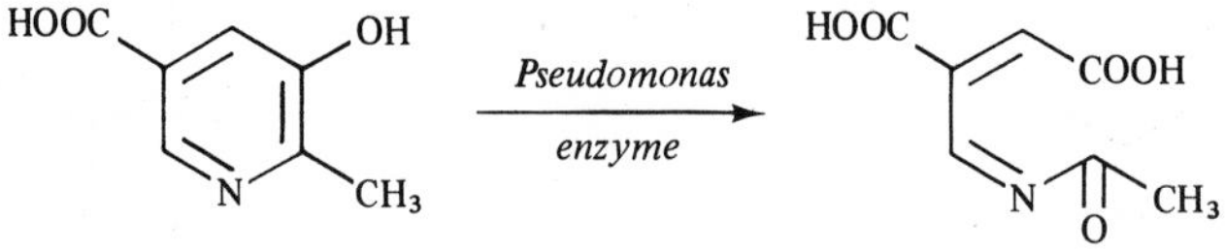

2-Methyl-3-hydroxy-
pyridine-5-carboxylic acid

The cleavage of riboflavin by an enzyme from a *Pseudomonas* species gives the bicyclic compound,[51] but in view of the high substrate
specificity of the enzyme, synthetic application to related compounds
seems doubtful.

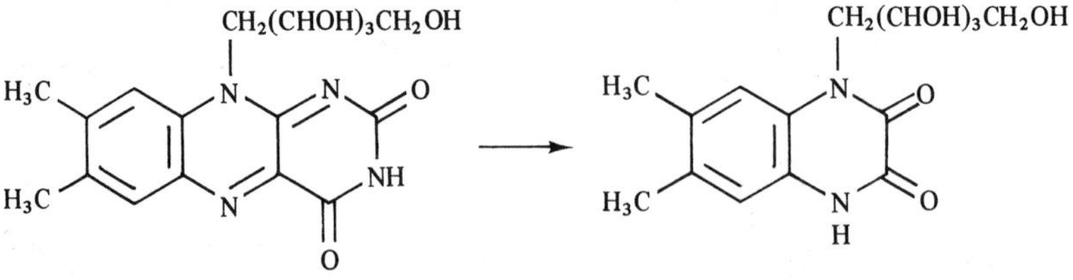

Riboflavin

The well-known enzymatic transformation of L-tryptophane to L-kynurenine, usually followed by further transformation to kynurenic acid, has been used to prepare a ^{14}C-labeled kynurenic acid from ap-

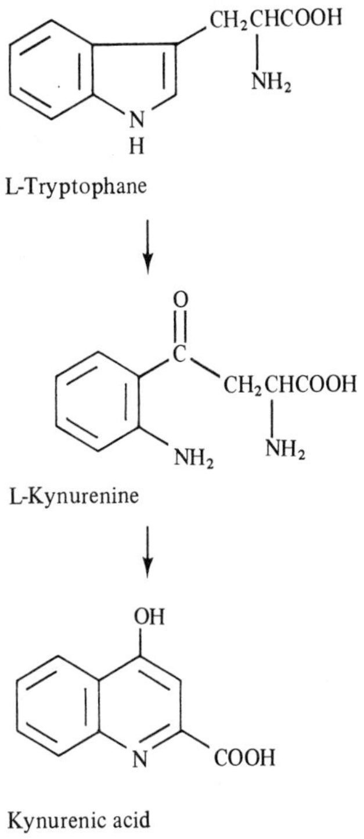

L-Tryptophane

L-Kynurenine

Kynurenic acid

propriately C^{14}-labeled tryptophane.[52] What such a microbiological-biochemical sequence lacks in yield may well be compensated for by the specificity of labeling obtainable.

An unusual example of selective aromatic ring cleavage in the complex polycyclic alkaloid strychnine gives C_{16}-hanssenic acid in 34% yield.[53]   Curiously, the closely related alkaloid brucine is not attacked by the microorganism.

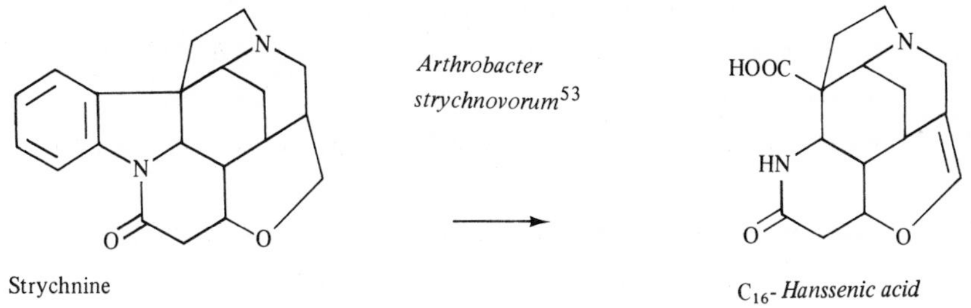

Arthrobacter
strychnovorum[53]

Strychnine

C_{16}- *Hanssenic acid*

Finally, it has been proposed that a biosynthetic pathway leading to the metabolite aranotin in *Arachniotus aureus* involves an aromatic ring enlargement that is technically a ring cleavage without
actual rupture.[54]   It is cited here only for the possibilities that
it suggests.

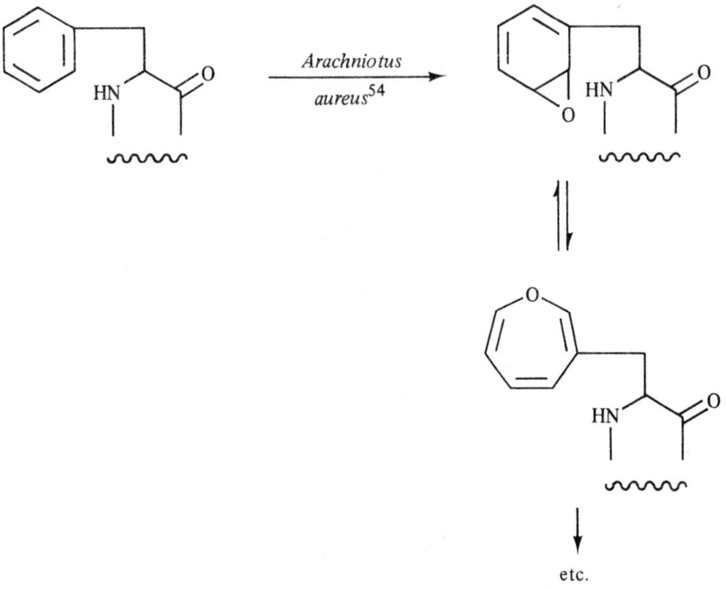

etc.

REFERENCES

1. V. W. Jamison, R. L. Raymond, and J. O. Hudson, *Appl. Microbiol.*
    17, 853 (1969).

2. P. Hosler and R. W. Eltz, in *Fermentation Advances*, (D. Perlman,
    ed.), Academic, New York, 1969, pp. 789-805.

3. R. L. Raymond and V. W. Jamison, *U. S. Patent* 3,383,289 (May 14,
    1968).

4. *Chem. Eng. News*, April 10, 1967, p. 18.

5. H. N. Fernley and W. C. Evans, *Biochem. J. 73*, 22P (1959).

6. J. K. Gaunt and W. C. Evans, *Biochem. J. 79*, 25P (1961).

7. P. Goldman, G. W. A. Milne, and M. T. Pignataro, *Arch. Biochem.*
    Biophys. 118, 178 (1967).

8. A. Kleinzeller and Z. Fencl, *Chem. Listy 46*, 300 (1952); *Chem.*
    Abstr. 47, 4290 (1953).

9. R. B. Cain, E. K. Tranter, and J. A. Darrah, *Biochem. J. 106*,
    211 (1968).

10. L. Canonica, A. Fiecchi, M. Galli Kienle, A. Scala, and V.
    Treccani, *Gazz. Chim. Ital. 96*, 915 (1966).

11. S. Dagley and D. A. Stropher, *Biochem. J. 73*, 16P (1959).

12. D. W. Ribbons, *J. Gen. Microbiol. 44*, 221 (1964).

13. D. Catelani, A. Fiecchi, and E. Galli, *Experientia 24*, 113
    (1968).

14. D. W. Ribbons, *Biochem. J. 99*, 30P (1966).

15. F. Lingens and H. D. Heilmann, *Naturwiss. 54*, 369 (1967).

16. P. Z. Smyrniotis, H. T. Miles, and E. R. Stadtman, *J. Amer.
    Chem. Soc. 80*, 2541 (1958).

17. H. T. Miles, P. Z. Smyrniotis, and E. R. Stadtman, *J. Amer.
    Chem. Soc. 81*, 1946 (1959).

18. D. R. Harkness, L. Tsai, and E. R. Stadtman, *Arch. Biochem.
    Biophys. 108*, 323 (1964).

19. A. Capek, O. Hanc, and M. Tadra, *Naturwiss. 54*, 70 (1967).

20. C. J. Sih, K. C. Wang, D. T. Gibson, and H. W. Whitlock, Jr., *J.
    Amer. Chem. Soc. 87*, 1386 (1965).

21. M. Nozaki, H. Kagamiyama, and O. Hayaishi, *Biochem. Z. 338*, 582
    (1963).

22. R. B. Cain and D. R. Farr, in *Biological and Chemical Aspects
    of Oxygenases*, (K. Bloch and O. Hayaishi, eds.), Kyoto, 1966,
    pp. 125-144.

23. D. R. Farr and R. B. Cain, *J. Gen. Microbiol. 41*, xv (1965).

24. R. D. Swisher, in *Developments in Industrial Microbiology*, vol.
    9, American Institute of Biological Sciences, Washington, D. C.,
    1968, pp. 270-279.

25. W. J. Payne and V. E. Feisal, *Appl. Microbiol. 11*, 339 (1963).

26. R. J. Strawinski and R. W. Stone, *J. Bacteriol. 45*, 16 (1943).

27. N. Walker and G. H. Wiltshire, *J. Gen. Microbiol. 8*, 273 (1953).

28. V. Treccani, N. Walker, and G. H. Wiltshire, *J. Gen. Microbiol.
    11*, 341 (1954).

29. R. J. Strawinski and R. W. Stone, *Can. J. Microbiol. 1*, 206
    (1955).

30. J. F. Murphy and R. W. Stone, *Can. J. Microbiol.* 1, 579 (1955).

31. R. E. Klausmeier and R. J. Strawinski, *J. Bacteriol.* 73, 461 (1957).

32. H. N. Fernley and W. C. Evans, *Nature 182*, 373 (1958).

33. J. I. Davies and W. C. Evans, *Biochem. J.* 91, 251 (1964).

34. I. Hill and A. Gordon, Abstr. Papers, Am. Chem. Soc., No. 150, 21U (1965).

35. V. V. Modi and R. N. Patel, *Appl. Microbiol. 16*, 172 (1968).

36. D. F. Wessley, Abstr. Papers, Am. Chem. Soc., No. 154, Q13 (1967).

37. M. H. Rogoff, Abstr. Papers, Am. Chem. Soc., No. 154, Q14 (1967).

38. H. Tone, A. Kitai, and A. Ozaki, *Biotech. Bioeng. 10*, 689 (1968).

39. V. Treccani, A. Fiecchi, G. Baggi, and E. Galli, *Ann. Microbiol. Enzimol. 15*, 11 (1965).

40. M. H. Rogoff and I. Wender, *J. Bacteriol. 77*, 783 (1959).

41. N. Walker and G. H. Wiltshire, *J. Gen. Microbiol. 12*, 478 (1955).

42. I. D. Hill, *U. S. Patent* 3,318,781 (May 9, 1967).

43. C. Colla, C. Biaggi, and V. Treccani, *Atti Accad. Naz. Lincei Cl. Sci. Fis. Mat. Nat. Rend. 23*, 66 (1957); *Chem. Abstr. 53*, 2350 (1959).

44. M. H. Rogoff and I. Wender, *J. Bacteriol. 74*, 108 (1957).

45. M. H. Rogoff and I. Wender, *J. Bacteriol. 73*, 264 (1957).

46. W. C. Evans, H. N. Fernley, and E. Griffiths, *Biochem. J. 95*, 819 (1965).

47. M. H. Rogoff, *J. Bacteriol. 83*, 998 (1962).

48. A. H. Jones and P. W. Trudgill, *Biochem. J. 105*, 31P (1967).

49. A. Kakinuma and S. Yamatodani, *Nature 201*, 420 (1964).

50. E. J. Nyns, T. K. Sundaram, D. Zach, and E. E. Snell, *Arch. Intern. Physiol-Biochim. 71*, 639 (1963).

51. D. R. Harkness and E. R. Stadtman, *J. Biol. Chem. 240*, 4089 (1965).

52. T. Tanaka and E. J. Behrman, *Anal. Biochem. 1*, 181 (1960).

53. H. Niemer and H. Bucherer, *Z. Physiol. Chem. 328*, 108 (1962).

54. N. Neuss, R. Nagarajan, B. B. Molloy, and L. L. Huckstep, *Tetrahedron Lett.*, 4467 (1968).

General References

O. Hayashi and M. Noyaki, *Science 164*, 389 (1969).

R. L. Raymond, *Process Biochemistry 4*, 71 (1969).

D. W. Ribbons, *Ann. Rept. Chem. Soc. London 62*, 445 (1965).

W. C. Evans, *Ann. Rept. Chem. Soc. London 53*, 279 (1956).

W. C. Evans, in *Fermentation Advances*, (D. Perlman, ed.), Academic, New York, 1969, p. 649.

D. T. Gibson, *Science 161*, 1093 (1968).

Chapter 6

MICROBIOLOGICAL BAEYER-VILLIGER OXIDATION

Oxygenation at the ß-carbon of the chain is a major pathway of alkane metabolism by microorganisms. The alcohols or ketones initially formed[1] sometimes undergo further degradation to primary alcohols with a chain length two carbons shorter than that of the substrate,[2] suggesting that a microbiological Baeyer-Villiger oxidation has taken place. The trace of 1-undecanol acetate isolated[3] from the conversion of 2-tridecanone to 1-undecanol by *Pseudomonas multivorans* lends credence to this mechanism, but definitive proof is lacking. (A similar chain-shortening of an N-alkylbenzamide may be found in Chapter 1, Table 12).

Analogous microbiological Baeyer-Villiger oxidations of cyclic ketones are known. The oxidation of 2-heptylcyclopentanone or of 2-pentylcyclopentanone by *Pseudomonas oleovorans* gives the corresponding lactones in low yields (cyclopentanone itself goes to glutaric acid), with some suggestion that stereochemical selectivity may be involved.[4] A number of additional examples of microbiological Baeyer-

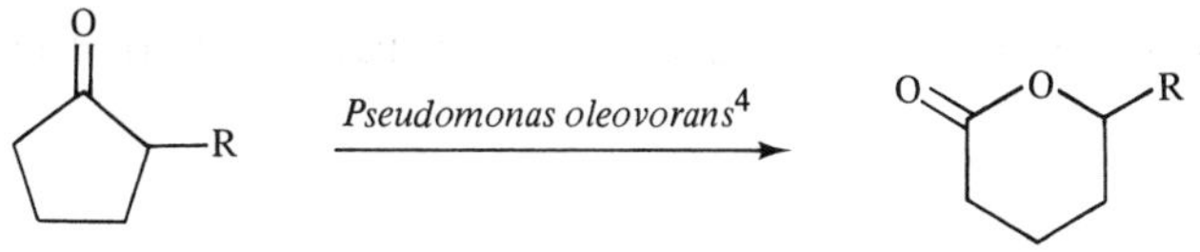

2-Alkylcyclopentanones

Villiger oxidations of simple cyclic ketones are found in the extensive research on the degradation of camphor and other terpenes. Fenchone, when oxidized by a *Corynebacterium* species, gives 1,2-fencholide and 2,3-fencholide in the ratio 9:1. (By contrast, the chemical oxidation[5] gives a 4:6 ratio.) The yield of mixed lactones by the microbiological method is 42%.[6] The addition of 2,2'-bipyridine to

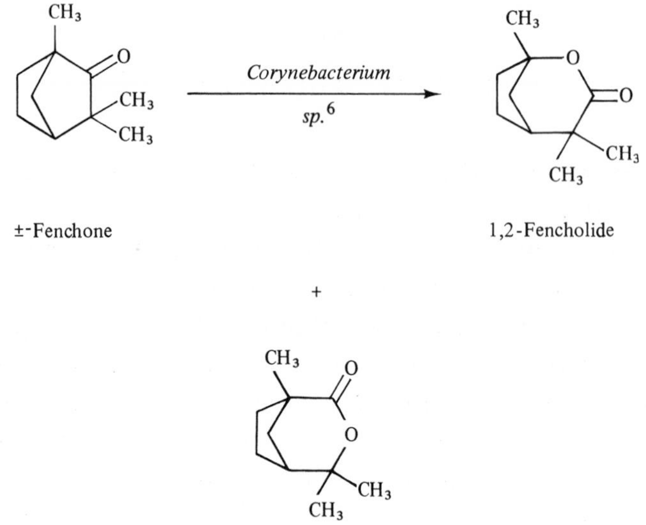

CH₃

±-Fenchone

Corynebacterium
sp.[6]

1,2-Fencholide

+

2,3-Fencholide

resting cells of *Pseudomonas* species permits the accumulation (*i.e.*, prevents further degradation) of 4,5,5-trimethyl-6-carboxymethyl-dihydropyran-2-one from the oxidation of 2,3,3-trimethyl-4-carboxymethylcyclopent-2-en-1-one.[7] Several other microbiological Baeyer-Villi-

Pseudomonas
sp.[7]

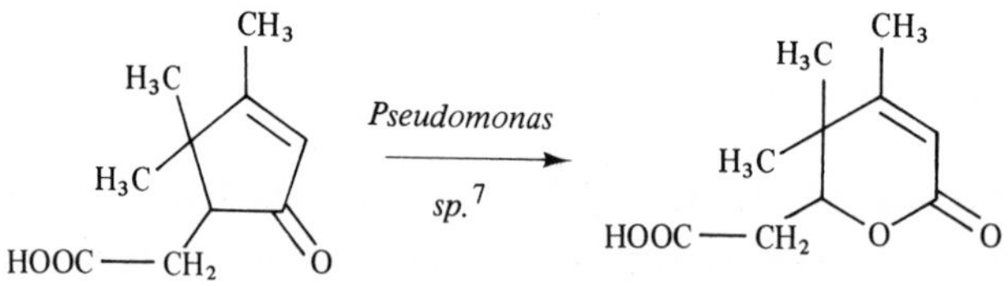

ger reactions in the camphor series,[8-11] for which yield data are generally lacking, are illustrated in Fig. 1.

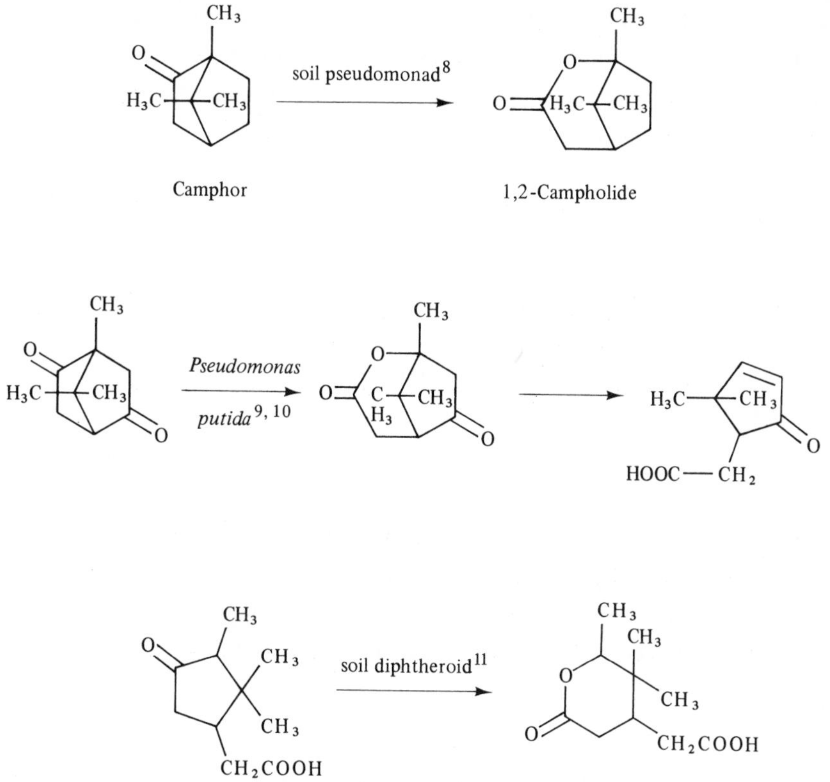

Figure 1. Camphor Series Baeyer-Villiger Oxidations

The bulk of the research on microbiological Baeyer-Villiger reactions has been carried out with steroids. The reaction was discovered at almost the same time as the hydroxylation of steroids but has not had the same commercial importance as the latter reaction. Nonetheless, it is a useful preparative reaction, whose scope still remains largely unexplored.

There are many examples of the oxidative removal of a side-chain (usually acetyl, but also hydroxyacetyl) from position 17 of a steroid[12] to generate a 17-hydroxyl or 17-keto group (with, in some instances, other reactions occurring at other parts of the molecule.) It has been shown for several microorganisms that these reactions involve an intermediate ester that ordinarily undergoes rapid hydrolysis by the esterases present in the fermentation environment. The

tendency of the initially formed ester to undergo hydrolysis and further degradation can be blocked by using the potent esterase inhibitor diisopropylfluorophosphate.   The bulk of the reported work[13-17]
has been concerned with the oxidation of progesterone to testosterone

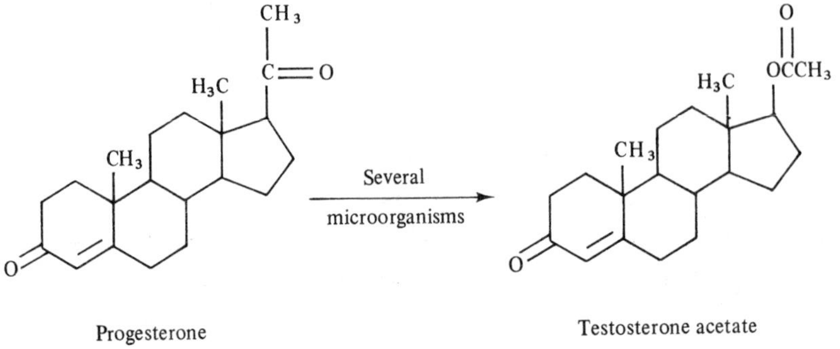

Progesterone   Testosterone acetate

acetate. By the use of a purified enzyme from *Cylindrocarpon radicicola*, yields as high as 17% can be isolated.[18] The specificity of the
C. radicicola enzyme appears to be broad, since several other steroids (illustrated only by their D-rings) are oxidized[19]:

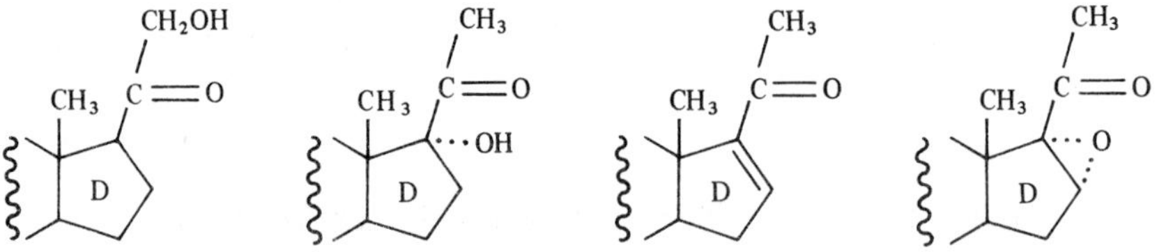

Since the 20-keto group apparently is required in the substrate,
the microbiological Baeyer-Villiger reaction can be blocked by conversion of the ketone to a ketal.   Thus:

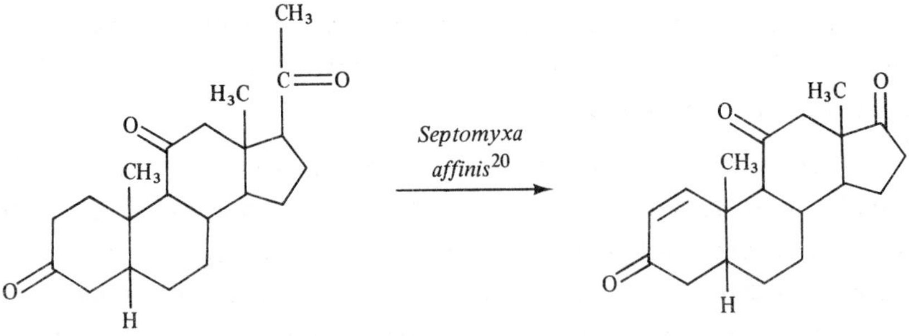

but:

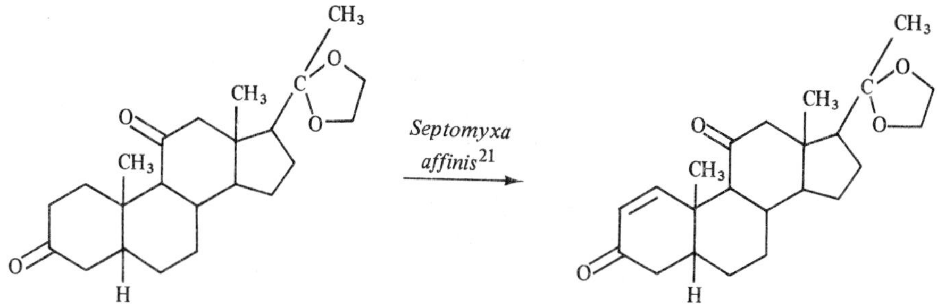

The microorganisms capable of removing side-chains from steroids generally are able to carry out an additional Baeyer-Villiger oxidation on the 17-ketones to give lactones.[22-28] The microbiological double Baeyer-Villiger oxidation of progesterone by *Penicillium chrysogenum*, for example, gives testololactone in 70% yield.[22]

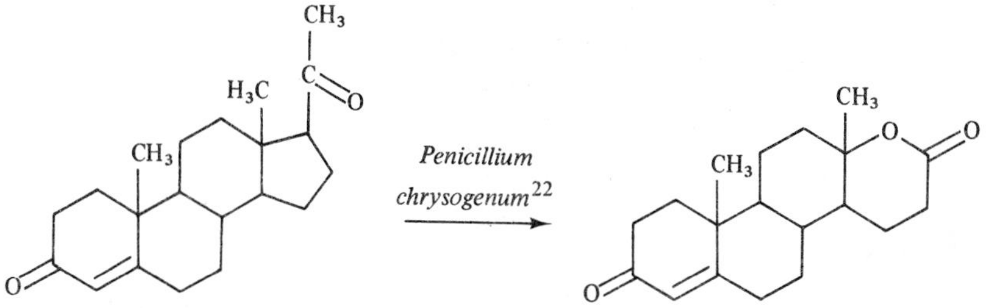

Progesterone

By the use of the enzyme system from *Penicillium lilacinum*,[29] it has been shown that atmospheric oxygen is incorporated during the oxidation of androst-4-ene-3,17-dione to testololactone (80% yield).

The microbiological Baeyer-Villiger reactions of steroids are not limited to C-17 side-chains and D-rings. Eburicoic acid undergoes cleavage of the A-ring by *Glomerella fusarioides*, to give 4-hydroxy-3,4-*seco*-eburica-8,24(28)-diene-3,21-dioic acid in 10% yield.[30,31]

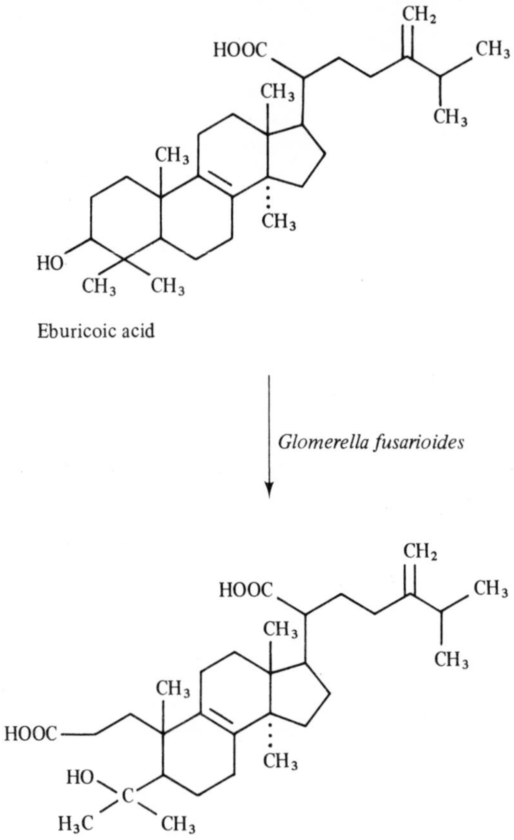

Eburicoic acid

Glomerella fusarioides

REFERENCES

1. H. B. Lukins and J. W. Foster, *J. Bacteriol.* *85*, 1074 (1963).
2. F. W. Forney, A. J. Markovetz, and R. E. Kallio, *J. Bacteriol.* *93*, 649 (1967).
3. F. W. Forney and A. J. Markovitz, *J. Bacteriol.* *96*, 1055 (1968).
4. R. Shaw, *Nature 209*, 1369 (1966).
5. R. R. Savers and G. P. Ahearn, *J. Amer. Chem. Soc. 83*, 2759 (1961).
6. P. J. Chapman, G. Meerman, and I. C. Gunsalus, *Biochem. Biophys. Res. Commun. 20*, 104 (1965).
7. W. H. Bradshaw, H. E. Conrad, E. J. Corey, I. C. Gunsalus, and D. Lednicer, *J. Amer. Chem. Soc. 81*, 5507 (1959).

8. H. E. Conrad, J. Hedegaard, I. C. Gunsalus, E. J. Corey, and H. Uda, *Tetrahedron Lett.*, 561 (1965).

9. H. E. Conrad, R. DuBus, and I. C. Gunsalus, *Biochem. Biophys. Res. Commun. 6*, 293 (1961).

10. P. W. Trudgill, R. DuBus, and I. C. Gunsalus, *J. Biol. Chem. 241*, 4288 (1966).

11. P. J. Chapman, G. Meerman, I. C. Gunsalus, R. Srinivasan, and K. L. Rinehart, Jr., *J. Amer. Chem. Soc. 88*, 618 (1966).

12. W. Charney and H. L. Herzog, *Microbiological Transformations of Steroids*, Academic, New York, 1967.

13. G. S. Fonken, H. C. Murray, and L. M. Reineke, *J. Amer. Chem. Soc. 82*, 5507 (1960).

14. K. Singh and S. Rakhit, *Biochim. Biophys. Acta 144*, 139 (1967).

15. K. Carlstrom, *Acta Chem. Scand. 21*, 1297 (1967).

16. N. Nakano, H. Sato, and B. I. Tamaoki, *Biochim. Biophys. Acta 164*, 585 (1968).

17. K. Carlstrom, *Acta Chem. Scand. 20*, 2620 (1966).

18. M. A. Rahim and C. J. Sih, *J. Biol. Chem. 241*, 3615 (1966).

19. C. J. Sih and M. A. Rahim, *Proc. 2nd Intern. Congress on Hormonal Steroids* (L. Martini, F. Fraschini, and M. Motta, eds.), Excerpta Medica Foundation, Milan, 1967, pp. 303-310.

20. R. C. Meeks, P. D. Meister, S. H. Eppstein, J. P. Rosselet, A. Weintraub, H. C. Murray, O. K. Sebek, L. M. Reineke, and D. H. Peterson, *Chem. Ind.*, 391 (1958).

21. G. S. Fonken and H. C. Murray, *J. Org. Chem. 27*, 1102 (1962).

22. J. Fried, R. W. Thoma, and A. Klingsberg, *J. Amer. Chem. Soc. 75*, 5764 (1953).

23. D. H. Peterson, S. H. Eppstein, P. D. Meister, H. C. Murray, H. M. Leigh, A. Weintraub, and L. M. Reineke, *J. Amer. Chem. Soc. 75*, 5768 (1953).

24. A. Bodanszky, J. Kallonitsch, and G. Wix, *Experientia 11*, 384 (1955).

25. A. Capek, O. Hanc, K. Macek, M. Tadra, and E. Riedl-Tumova, *Naturwiss. 43*, 471 (1956).

26. S. A. Szpilfogel, M. S. DeWinter, and W. J. Alsche, *Rec. Trav. Chim. 75*, 402 (1956).

27. G. E. Peterson, R. W. Thoma, D. Perlman, and J. Fried, *J. Bacteriol. 74*, 684 (1957).

28. G. M. Shull, *Trans. N. Y. Acad. Sci. 19*, 147 (1956).

29. R. L. Prairie and P. Talalay, *Biochemistry 2*, 203 (1963).

30. A. I. Laskin, P. Grabowich, C. deL. Meyers, and J. Fried, *J. Med. Chem. 7*, 406 (1964).

31. A. I. Laskin and J. Fried, *U. S. Patent* 3,377,255 (April 9, 1968).

Chapter 7

β-OXIDATION

The sequence of reactions making up the pathway by which fatty acids are generally metabolized in biological systems has been widely studied. The over-all sequence is commonly known as β-oxidation. The

$$R \!-\! CH_2CH_2 \!-\! \overset{\displaystyle O}{\overset{\displaystyle \|}{C}} \!-\! OH \longrightarrow R \!-\! CH_2CH_2 \!-\! \overset{\displaystyle O}{\overset{\displaystyle \|}{C}} \!-\! S \!-\! \text{Coenzyme A}$$
$$\text{(CoA)}$$

$$\downarrow -H_2$$

$$R \!-\! \overset{\displaystyle OH}{\overset{\displaystyle |}{C}}HCH_2 \!-\! \overset{\displaystyle O}{\overset{\displaystyle \|}{C}} \!-\! S \!-\! CoA \quad \overset{H_2O}{\longleftarrow} \quad R \!-\! CH \!=\! CH \!-\! \overset{\displaystyle O}{\overset{\displaystyle \|}{C}} \!-\! S \!-\! CoA$$

$$\downarrow -H_2$$

$$R \!-\! \overset{\displaystyle O}{\overset{\displaystyle \|}{C}}CH_2 \!-\! \overset{\displaystyle O}{\overset{\displaystyle \|}{C}} \!-\! S \!-\! CoA \longrightarrow R \!-\! \overset{\displaystyle O}{\overset{\displaystyle \|}{C}} \!-\! S \!-\! CoA \;+\; CH_3\overset{\displaystyle O}{\overset{\displaystyle \|}{C}} \!-\! S \!-\! CoA$$

$$\downarrow$$

etc.

five distinct reactions of this sequence, each of which is catalyzed by a specific enzyme, include three (*i.e.*, dehydrogenation [Chapter 8], hydration [Chapter 3], and alcohol dehydrogenation [Chapter 9]) that have been considered in other contexts elsewhere in this book. The entire sequence serves as an important source of energy to the biological machinery. It is estimated, for example, that metabolism of one mole of palmitoyl-CoA produces 1310 kcal of energy in the form of high energy phosphate ester bonds.[1] β-Oxidation is a perfectly normal metabolic pathway found in most microorganisms. In the face of the voracious appetites of microorganisms for energy, it is not surprising that β-oxidation is often more detrimental than useful to the purpose of preparing chemical intermediates by microbial reactions.

However, if the β-oxidation process is interrupted, or blocked in some manner, then it potentially becomes very useful, since these enzyme systems are so commonly found.

A fascinating example of the blocking of β-oxidation by the microorganism itself is found in the cases of *Torulopsis apicola* and *T. gropengiesseri*, which convert long-chain fatty acids into glycolipids (see Chapter 1).[2,3] Both aliphatic hydrocarbons and fatty acids are incorporated into the glycolipid, which is excreted from the cells. Fatty acids and hydrocarbons exceeding a certain maximum chain length are shortened, presumably *via* β-oxidation, to the optimum length before being incorporated into the glycolipid. From the point of view of the microorganism, this excretion of hydrocarbons as an extracellular material seems wasteful of the potential energy stored in the hydrocarbons. The apparent short-sightedness of the microorganism proves to be a boon to the organic chemist, since chemical hydrolysis of the glycolipid provides, in many cases, a good source of hydrocarbon intermediates that are functionalized at both termini.

The technique of co-oxidation[4] has been successfully applied to the inhibition of rampant β-oxidation in the conversion of alkylbenzenes to useful chemical intermediates. In this technique, a second substrate, which may be structurally related to the substrate of interest, is used to initiate and maintain the growth of the culture while also inducing the oxidative enzymes of the microorganism. In this way, the substrate of interest undergoes partial degradation, but complete utilization is inhibited by the second substrate, which can be present in excess. That this partial degradation involves β-oxidation in the cases of alkylbenzenes is strongly suggested by the structures of the products. Since β-oxidation shortens alkyl chains by two-carbon increments, the fact that, for example, arylacetic acids are obtained from alkylaromatics of even-numbered alkyl chain length and 3-arylpropionic acids or benzoic acids[5] are obtained from those of odd numbered alkyl chain lengths supports this contention. Several examples are illustrative of both the cooxidation method and the β-oxidation pathway. Amylbenzene has been converted to *trans*-cinnamic acid by *Cellulomonas galba, Pseudomonas ligustri, Ps. pseudomaleii, Ps. orvilla, Brevibacterium healii,* and an *Alcaligenes* species.[6] Under the best conditions reported, *trans*-cinnamic acid is obtained in 88 to 100% conversion at levels of 5 g./1. with *Cellulomonas galba.* 5-Phenylvaleric acid is detectable in all bioconversions, and is the major product from the *Alcaligenes* fermentation.

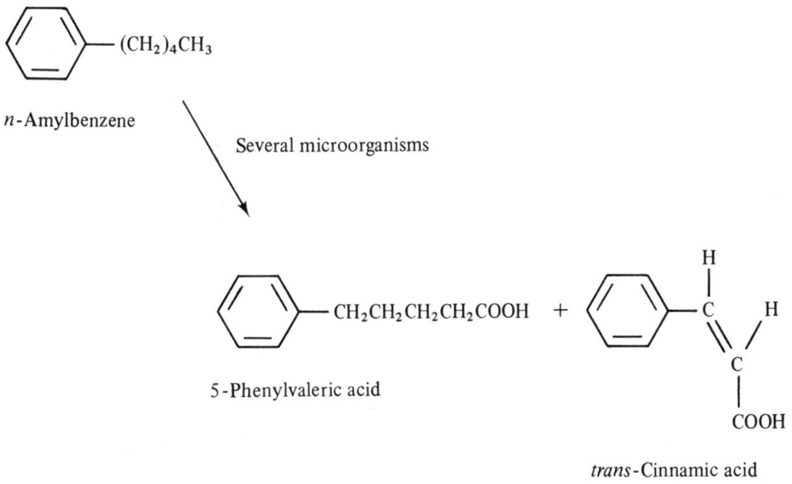

n-Amylbenzene

Several microorganisms

5-Phenylvaleric acid

trans-Cinnamic acid

Isolation of 5-phenylvaleric acid supports the idea that *trans*-cinnamic acid is reached by a β-oxidation route. In these fermentations, *n*-hexadecane is used as a co-oxidant substrate. The use of a mixed substrate, consisting of alkylbenzenes of odd-numbered alkyl chain lengths (C_3 - C_9), does not result in as good a conversion to *trans*-cinnamic acid. Co-oxidation of *n*-hexadecane also serves to support the conversion of 3-alkyltoluenes to 3-*m*-tolylpropionic acid and 3-*m*-tolylacrylic acid by *Micrococcus cerificans* in 80% yield.

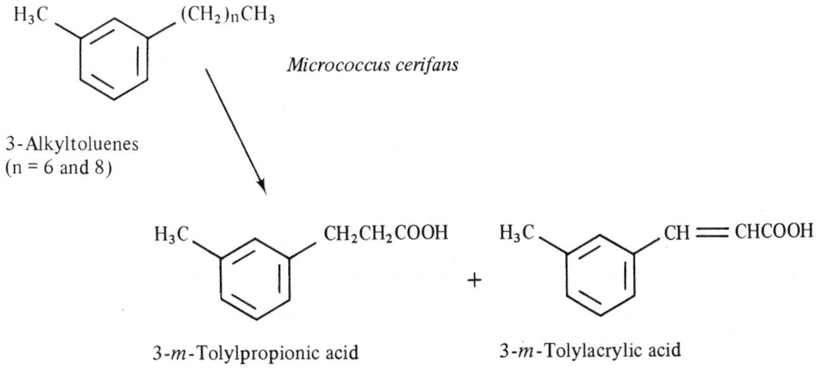

3-Alkyltoluenes
(n = 6 and 8)

Micrococcus cerifans

3-*m*-Tolylpropionic acid

3-*m*-Tolylacrylic acid

Certain alkylbenzenes of long chain length support the growth of *Nocardia* strains.[7,8]   As the alkyl chain undergoes shortening by β-oxidation to a length of two or three carbons, further metabolism is apparently slowed by the phenyl group. Thus, an 80% yield of phenyl-acetic acid is obtained from *n*-dodecylbenzene, while a mixture of acids of which the principal component is phenylacrylic acid is obtained in small amounts from *n*-nonylbenzene.   Short chain alkylben-zenes, however, are not utilized by the culture until a second hydro-carbon (*n*-hexadecane or *n*-octadecane) is added to the fermentation.[7] Then phenylacetic acid is obtained from *n*-butylbenzene (and from eth-ylbenzene) and phenylacrylic acid from *n*-propylbenzene.   Similarly, cyclohexaneacetic acid (26-41%, depending on the *Nocardia* strain used) is obtained from *n*-butylcyclohexane. [*p*-Isopropyltoluene is oxygenated on the isolated methyl group, giving *p*-isopropylbenzoic acid (4%).[7] This reaction is discussed in the chapter on allylic oxi-dations.] Related to these β-oxidations of alkylaromatics is the con-version, by *Sporobolomyces roseus*, of cinnamic acid to benzoic acid as one step in the metabolism of phenylalanine.[9]

Growth of *Pseudomonas aeruginosa* on a hydrocarbon closely re-lated to the desired substrate leads to isolation of intermediates from the subsequent oxidation of the compound of interest by the or-ganism.[10]   Propionic acid (32%) is obtained from heptane with *Ps. aeruginosa* cells grown on hexane, and isovaleric acid (17%) from 2-methylhexane with cells grown on heptane.   Chloramphenicol markedly inhibits the formation of enzyme systems capable of total degradation of the second substrates, and yields are increased to 60% for propi-onic acid and to 35% for isovaleric acid when this antibiotic is added to the fermentation with the second substrate.   Acrylic acid, which presumably acts as an inhibitor of β-oxidation, also increases yields of oxygenated acyclic alkanes obtained from fermentations with *Ps. aeruginosa*.[11]

Several examples illustrate the potential use of β-oxidation to shorten the chain length of various carboxylic acids. Strains of the genera *Rhodotorula, Penicillium,* and *Endomycopsis* shorten the carbox-ylic acid side chain of biotin by two carbons.[12]   In particular, an

Endomycopsis species produces the vitamers A and B in 29% and 13%

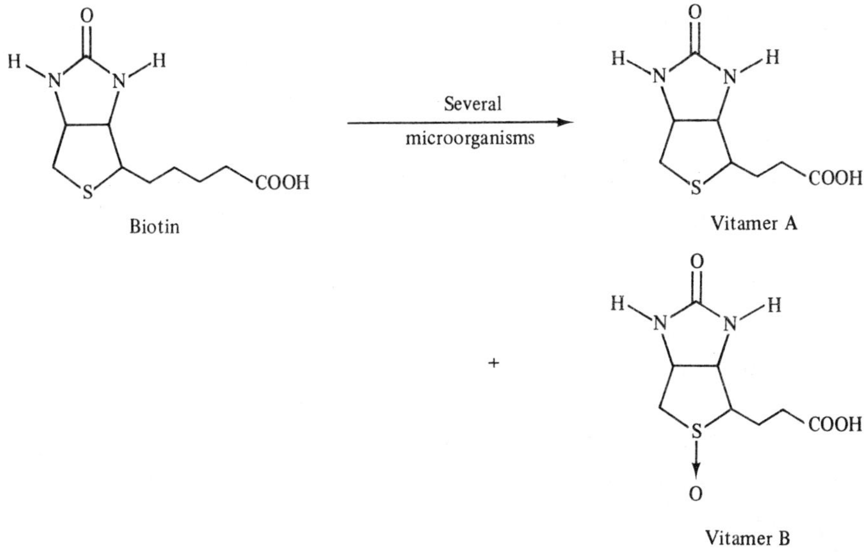

Biotin

Vitamer A

+

Vitamer B

yield, respectively. The action of *Pseudomonas citronellolis* on geraniol is somewhat more complex, being a combination of β-oxidation and a carboxylation reaction.[13]

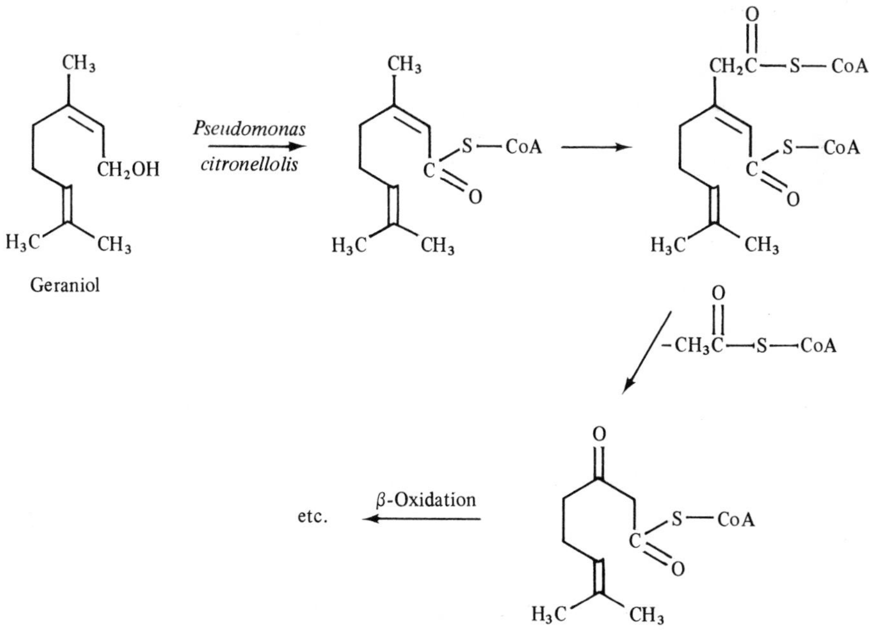

REFERENCES

1. A. White, P. Handler, and E. L. Smith, *Principles of Biochemistry*, 3rd ed., McGraw-Hill, New York, 1964, p. 442.
2. A. P. Tulloch, J. F. T. Spencer, and P. A. J. Gorin, *Canad. J. Chem. 40*, 1326 (1962).
3. D. F. Jones and R. Howe, *J. Chem. Soc. (C)*, 2801 (1968).
4. R. L. Raymond, V. W. Jamison, and J. O. Hudson, *Appl. Microbiol. 15*, 857 (1967).
5. D. M. Webley, R. B. Duff, and V. C. Farmer, *J. Gen. Microbiol. 13*, 361 (1955).
6. J. D. Douros, Jr. and J. W. Frankenfeld, *Appl. Microbiol. 16*, 320 (1968).
7. J. B. Davis and R. L. Raymond, *Appl. Microbiol. 9*, 383 (1961).
8. D. M. Webley, R. B. Duff, and V. C. Farmer, *Nature 178*, 1467 (1956).
9. K. Moore, P. V. Subba Rao, and G. H. N. Towers, *Biochem. J. 106*, 507 (1968).
10. G. J. E. Thijsse and A. C. van der Linden, *Antonie van Leeuwenhoek 29*, 89 (1963).
11. R. Huybregtse and A. C. van der Linden, *Antonie van Leeuwenhoek 30*, 185 (1964).
12. H.-C. Yang, M. Kusumoto, S. Iwahara, T. Tochikura, and K. Ogata, *Agr. Biol. Chem. 32*, 399 (1968).
13. W. Seubert and E. Fass, *Biochem. Z. 341*, 35 (1964).

Chapter 8

ALKYL DEHYDROGENATION

Most of the examples of alkyl dehydrogenation that may be useful to the synthetic organic chemist occur in two areas: (1) steroids, and (2) fatty acids.

Steroids. The medicinally important anti-inflammatory steroids prednisone, prednisolone, and their analogs and derivatives are formed by microbiological removal of the 1α- and 2β-hydrogen atoms from appropriate substrates. It is, therefore, not surprising that the

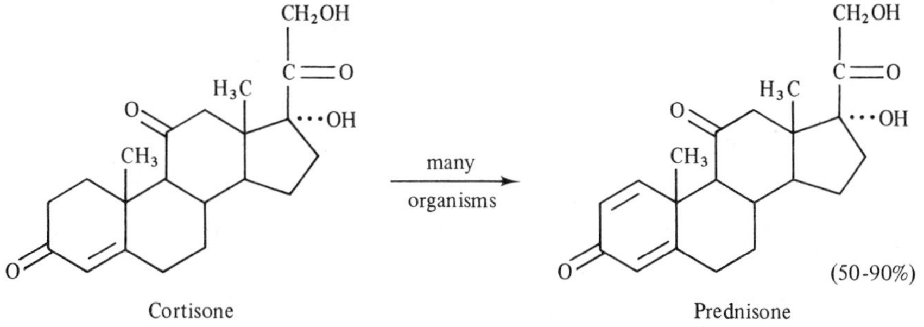

Cortisone	many organisms	Prednisone (50-90%)

bulk of the reported steroid dehydrogenations are 1,2-dehydrogenations, and that many of these proceed in good (60-95%) yields. Almost invariably, the substrate contains a 3-keto-4-ene system, which leads to a 3-keto-1,4-diene product, but more highly saturated substrates have been used in order to obtain 3-keto-1-ene's. Multiple reactions are common with dehydrogenating microorganisms, and may re-

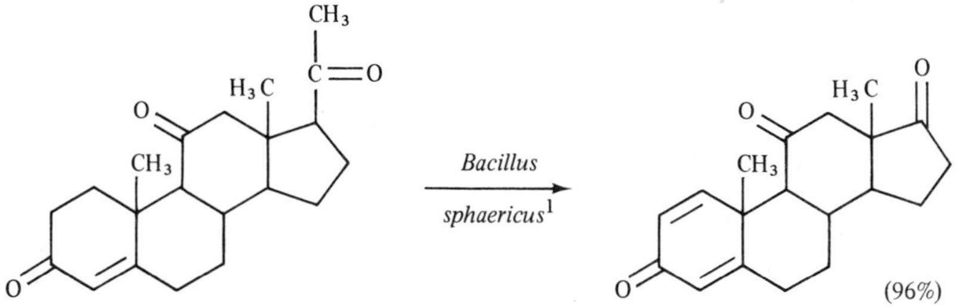

	Bacillus sphaericus[1]	(96%)

171

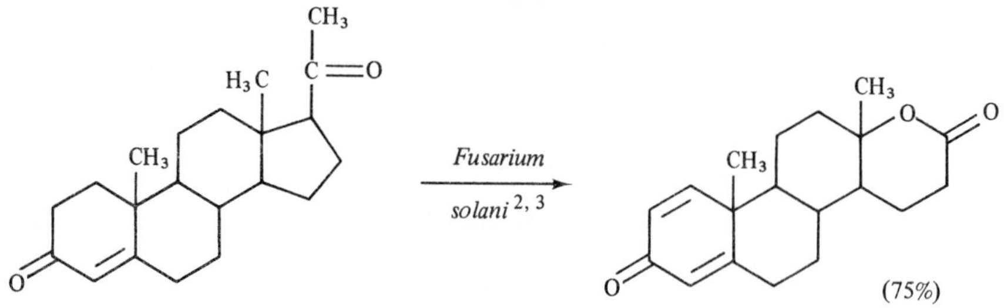

(75%)

sult in products only remotely resembling the substrate, unless some blocking method is available. In the example using *Septomyxa affinis*, the 20-keto group of the substrate has been converted to a cyclic ke-

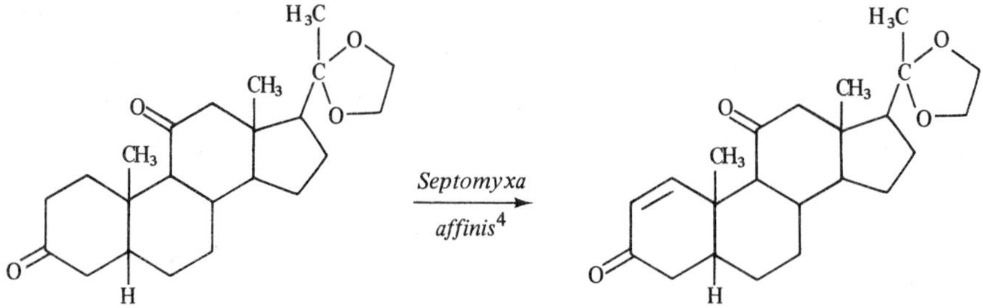

tal to block side-chain cleavage by the microorganism.[4]   In a some-what different approach, the use of the chelating agent 8-hydroxy-quinoline, which blocks hydroxylation enzymes, permits *Mycobacterium phlei* to convert androst-4-ene-3,17-dione to androsta-1,4-diene-3,17-dione in 80% yield, while avoiding the 9α-hydroxylation of this prod-

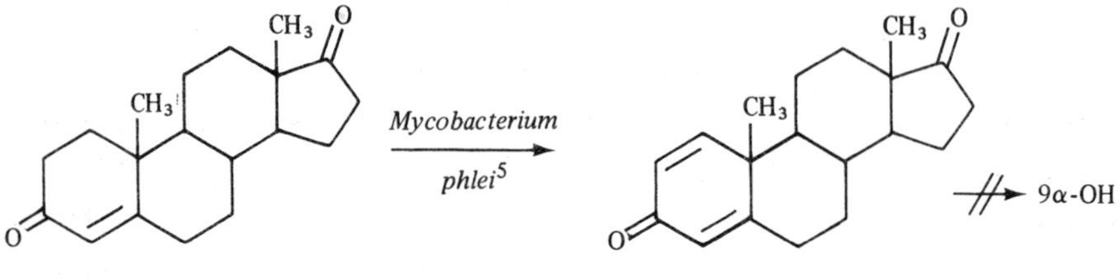

Androst-4-ene-3,17-dione

uct that would result in further skeletal degradation.[5] 2,2'-Dipyri-dyl has a similar effect on the complex transformation of cholesterol to androst-4-ene-3,17-dione and androsta-1,4-diene-3,17-dione.[6]

When the substrate molecule contains appropriate structural fea-tures, 1,2-dehydrogenation may lead to an aromatic steroid, either by enolization of a dienone or by retroaldol elimination of a substituent.

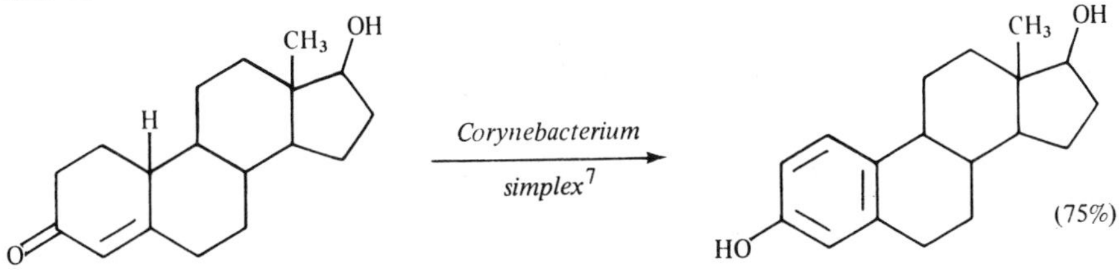
CH3
OH
H
O
Corynebacterium
simplex7
HO
(75%)

19-Nortestosterone

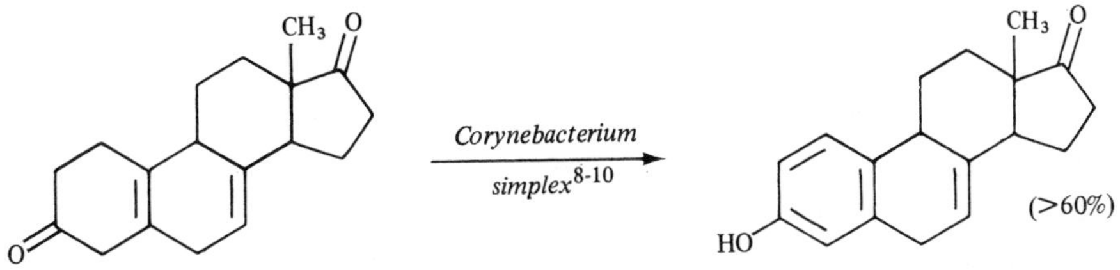
CH3
O
Corynebacterium
simplex8-10
HO
(>60%)

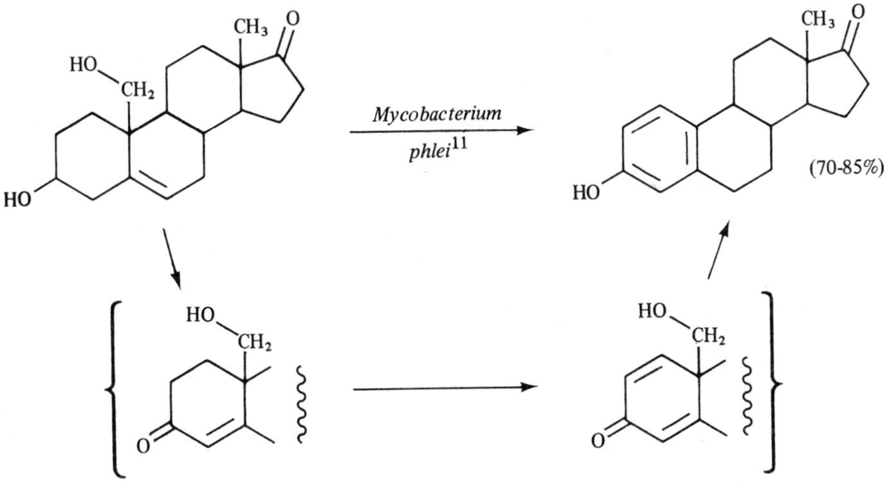
CH3
O
HO
CH2
Mycobacterium
phlei11
(70-85%)

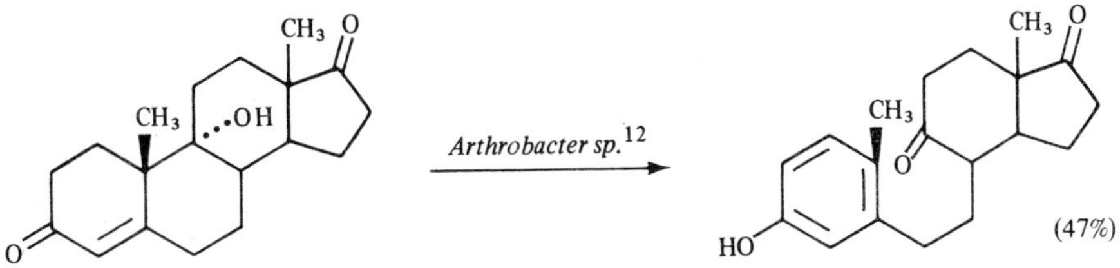
CH3
O
••OH
Arthrobacter sp.12
HO
(47%)

The ability of microorganisms to carry out the efficient 1,2-dehydrogenation of many steroids has been used to effect the resolution of synthetic products having a cyclohexenone system typical of steroids. Acetone powder preparations of *Arthrobacter simplex* cells dehydrogenate the "natural" enantiomer (usually the 10-*R* configuration in steroids) of racemates such as A, and leave unchanged the en-

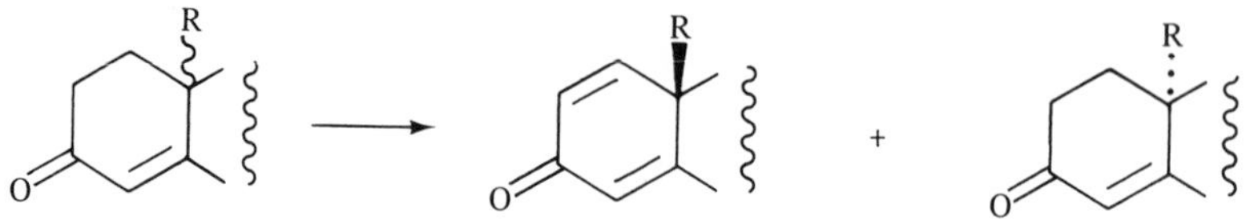

A

antiomer of "unnatural" configuration at the carbon bearing the R-group.[13] When R=H, further dehydrogenation occurs and the aromatic system is obtained. Racemic, synthetic molecules ranging in size from the tetracyclic steroids to the bicyclic octalins have been separated by this method.[13] This fine stereochemical discrimination is not present in all microorganisms, however, and racemic steroids seem to respond to the 1,2-dehydrogenating microorganism, *Corynebacterium hoagii*, for example, in much the same way that natural D-steroids do.[14]

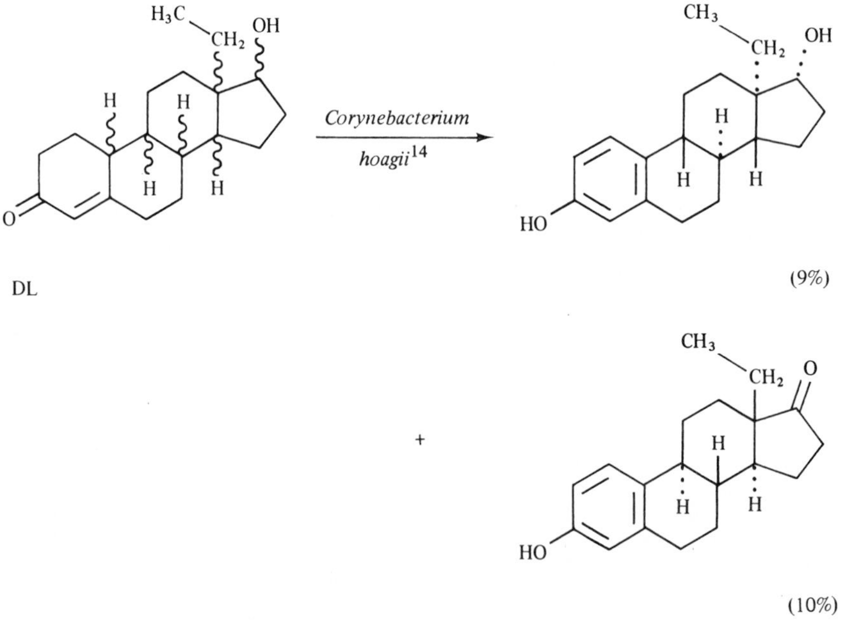

DL

(9%)

(10%)

Introduction of a double bond at the 4,5-positions in steroids has been observed with androstanes, pregnanes, and bile acids.   This reaction is usually accompanied by 1,2-dehydrogenation and appears to be useful only as a route to 1,4-dien-3-ones.

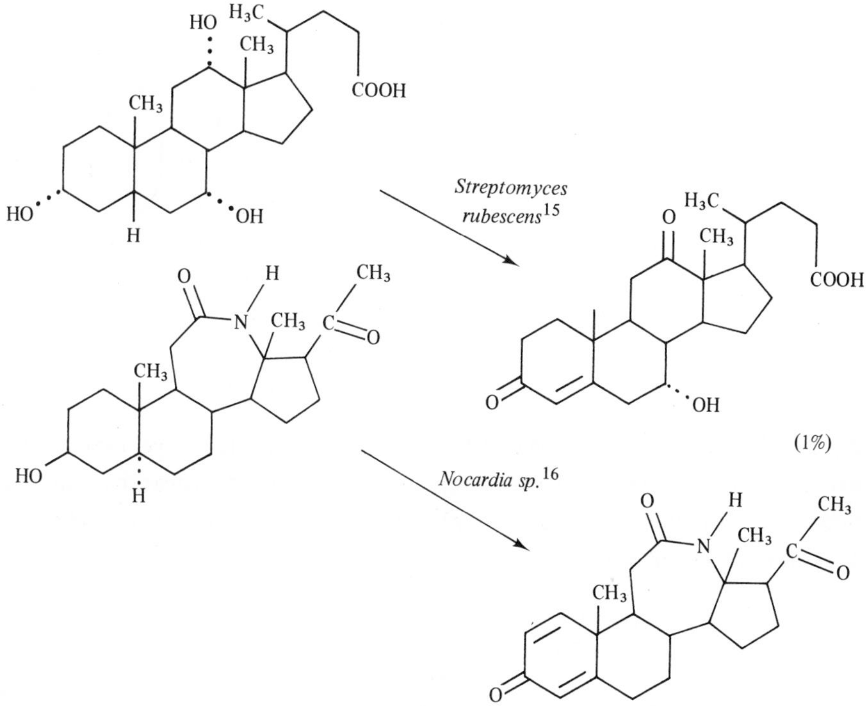

(1%)

Ergosterol biosynthesis in *Saccharomyces cerevisiae* apparently involves *cis*-elimination of the 5α,6α-hydrogens of ergosta-7,22-dien-3β-ol,[17] while the geometry of the oxidation of cholesterol to 7-de-

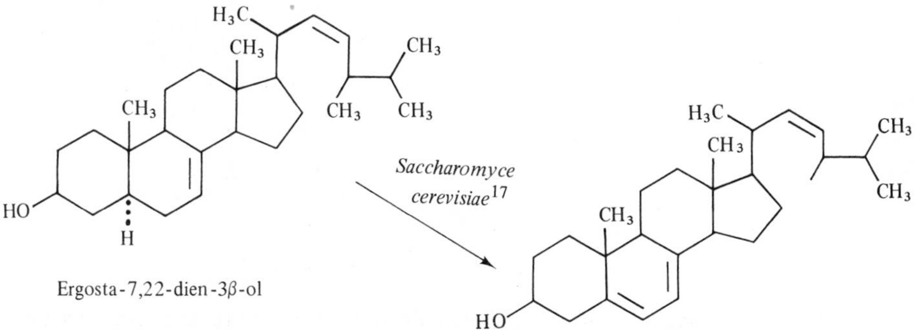

Ergosta-7,22-dien-3β-ol

Ergosterol

hydrocholesterol by an *Azotobacter* species is not known.[18,19]   Dehy-

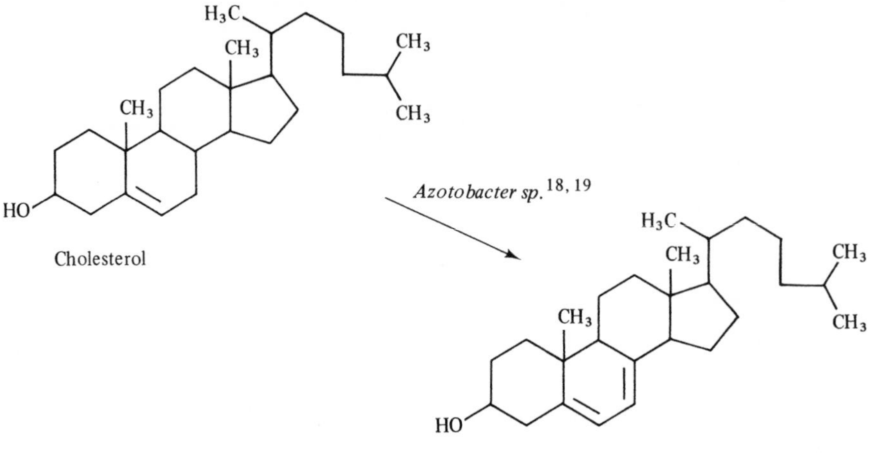

Cholesterol

Azotobacter sp.[18, 19]

7-Dehydrocholesterol

drogenations at other positions (9,11; 14,15; 16,17) are rare, and
probably result from hydroxylation followed by dehydration during
product isolation.   This may also be the case in the illustrated
transformation of cholic acid.[20]

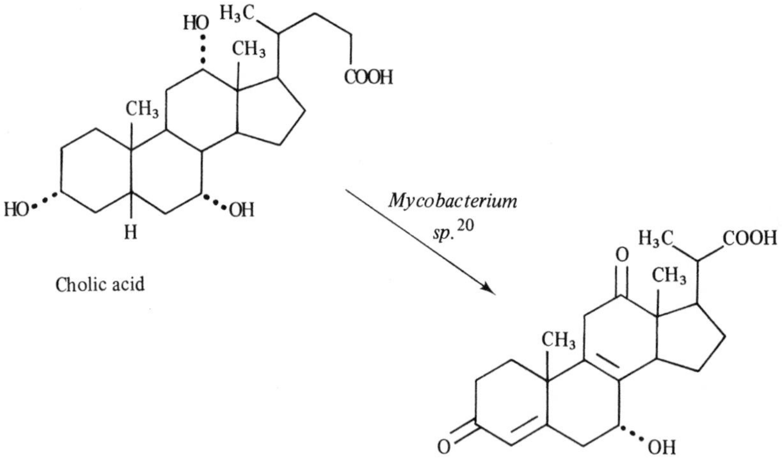

Cholic acid

Mycobacterium sp.[20]

(+ other products)

In a few cases, the 22,23-dehydrogenation that appears to be
part of the over-all biosynthesis of sterols[21] has been observed as

an isolated reaction.[22]   The oxidation of cholesterol by *Tetrahymena*
pyriformis gives 7,22-bisdehydrocholesterol in 29% yield.[23]

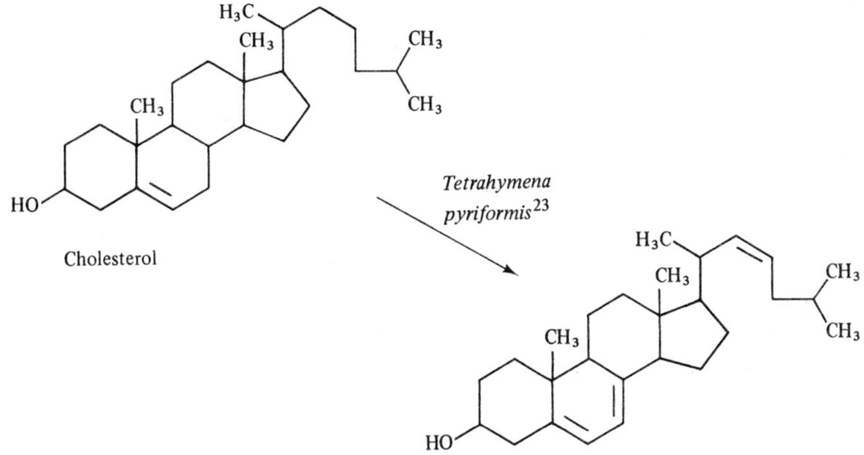

Cholesterol

Tetrahymena pyriformis[23]

Fatty Acids. Studies of the dehydrogenation of fatty acids have
consisted largely of tracer experiments to determine metabolic pathways. The examples are listed chiefly to indicate a synthetic potential.

$CH_3(CH_2)_{16}COOH$ $\xrightarrow{\textit{Corynebacterium diphtheriae}[24]}$ $CH_3(CH_2)_7\overset{H}{\underset{}{C}}=\overset{H}{\underset{}{C}}(CH_2)_7COOH$
Stearic acid   or *Micrococcus lysodeikticus*[25]   Oleic acid

(This reaction is stereospecific. The 9H is removed first.)

$CH_3(CH_2)_{14}COOH$ $\xrightarrow{\textit{Corynebacterium diphtheriae}[25]}$ $CH_3(CH_2)_5CH=CH(CH_2)_7COOH$
Palmitic acid   or *Micrococcus lysodeikticus*[25]

$CH_3(CH_2)_{16}COOH$ $\xrightarrow{\textit{Bacillus megaterium}[25]}$ $CH_3(CH_2)_9CH=CH(CH_2)_5COOH$
Stearic acid

$CH_3(CH_2)_{14}COOH$ $\xrightarrow{\textit{Bacillus megaterium}[25,26]}$ $CH_3(CH_2)_9CH=CH(CH_2)_3COOH$
Palmitic acid

(Shorter fatty acids, *e.g.*, myristic and lauric, are first built up
 to palmitic acid, which is then dehydrogenated as shown.)

$CH_3(CH_2)_{14}COOH$ $\xrightarrow{\text{several bacilli}[26]}$ $CH_3(CH_2)_6CH=CH(CH_2)_6COOH$
Palmitic acid $CH_3(CH_2)_5CH=CH(CH_2)_7COOH$
                                        $CH_3(CH_2)_4CH=CH(CH_2)_8COOH$

Many microorganisms dehydrogenate stearic acid and palmitic acids to oleic and palmitoleic acids,[27] respectively, and in some cases, bring about further dehydrogenation of unsaturated fatty acids, some examples of which are listed.

$$CH_3(CH_2)_7 \overset{H}{C} = \overset{H}{C} (CH_2)_7 COOH$$

Oleic acid

Tricholoma grammapodium[28]

$$CH_3(CH_2)_4 CH=CHCH_2 CH=CH(CH_2)_7 COOH$$

Linoleic acid

$$CH_3(CH_2)_7 \overset{H}{C} = \overset{H}{C} (CH_2)_7 COOH$$

Oleic acid

Tricholoma grammapodium[27]

$$CH_3(CH_2)_7 CH=CHCH_2 CH=CH(CH_2)_4 COOH$$

The conversion of oleic acid to linoleic acid by *Tricholoma grammapodium* also gives crepenynic acid, which presumably arises through further dehydrogenation of the Δ^{12} double bond of linoleic acid.[28]

$$CH_3(CH_2)_4 CH=CHCH_2 CH=CH(CH_2)_7 COOH$$

Linoleic acid

Tricholoma grammapodium[28]

$$CH_3(CH_2)_4 C{\equiv}CCH_2 CH=CH(CH_2)_7 COOH$$

Crepenynic acid

In an *Acanthamoeba* species, the fatty acid biosynthetic pathway includes a chain lengthening of linoleic acid to 11,14-eicosadienoic acid, which is then dehydrogenated further to 8,11,14-eicosatrienoic acid and 5,8,11,14-eicosatetraenoic acid.[29]

The biosynthesis of brefeldin A from palmitic acid by *Penicillium cyaneum* is postulated to proceed through the polyunsaturated fatty acid[30]:

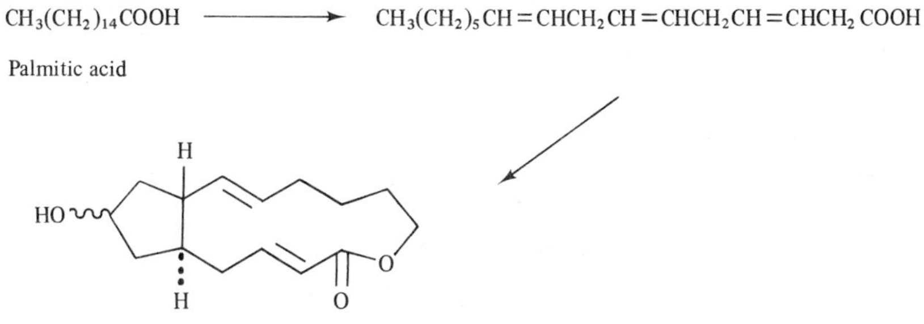

$$CH_3(CH_2)_{14}COOH \longrightarrow CH_3(CH_2)_5 CH=CHCH_2 CH=CHCH_2CH=CHCH_2 COOH$$

Palmitic acid

Brefeldin A

Also related to fatty acid dehydrogenations are several examples of dehydrogenation of alkanes, a less common pathway of alkane metabolism.   Formation of a terminal olefin was first claimed in 1962 in the oxidation of heptane by *Pseudomonas aeruginosa*,[31,32] but subsequent efforts to extend this work have been unsuccessful.   More recently, a minuscule quantity of 1-hexadecene was isolated and identified from the growth of a *Nocardia* species on hexadecane.[33] Finally, a significant quantity (2-8%) of internal olefins has been isolated from the oxidation of $C_{14}-C_{20}$ alkanes by *Nocardia salmonicolor*.[34]

In an unusual situation involving an obligate cooxidant (see Chapter 7, β-Oxidation), propylbenzene is oxidized by a *Nocardia* species to cinnamic acid, a reaction that involves, in part, a formal dehydrogenation.[35]

Miscellaneous. A few other examples of dehydrogenations follow.

In the discussion of aromatic ring opening reactions (Chapter 5) the pathway of degradation of benzene was shown to proceed by way of catechol.   *cis*-Benzene glycol is formed first and is then stoichio-

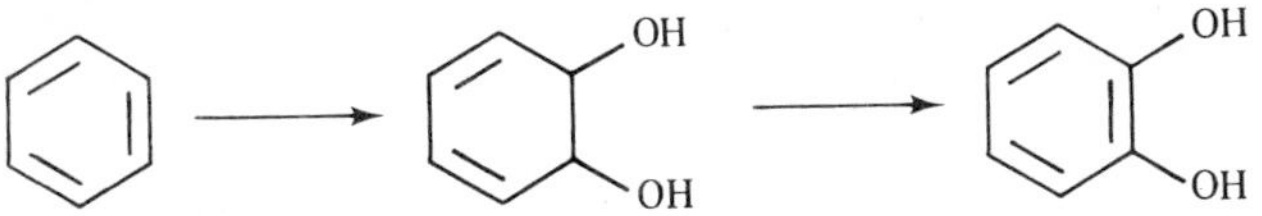

cis-Benzene glycol

metrically dehydrogenated to catechol (*trans*-benzene glycol is not a substrate) by enzymes of *Pseudomonas putida*.[36]

A microorganism isolated from tobacco seed brings about dehydrogenation of nicotine to nicotyrine through the intermediate N-methylmyosmine.[37] It is interesting that this oxidation proceeds even in a nitrogen atmosphere.

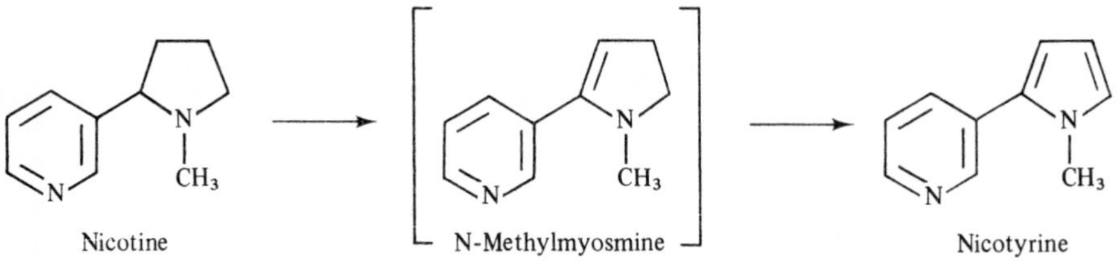

| Nicotine | N-Methylmyosmine | Nicotyrine |

Sparteine and lupanine undergo bacterial dehydrogenation. In the latter case, the dehydrolupanine is merely postulated as the intermediate leading to hydroxylupanine[38] (see Chapter 1), but the postulated structure, which violates Bredt's Rule, is probably incorrect.

A broad study of steroid degradation, particularly by *Nocardia restrictus*, has shown that a steroid-derived ketoacid undergoes dehydrogenation (and other reactions) as illustrated.[39]

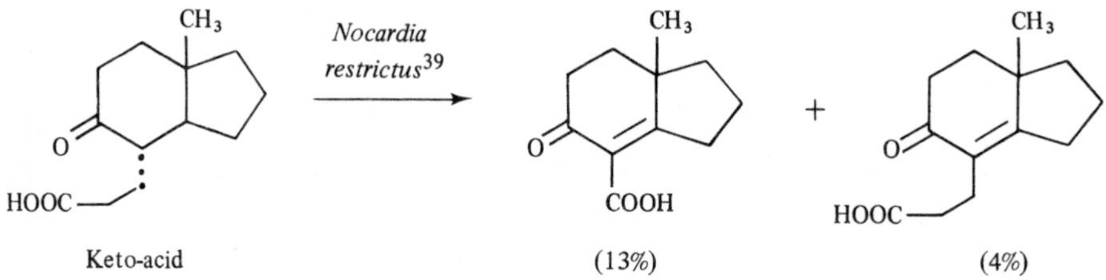

| Keto-acid | (13%) | (4%) |

Dehydrogenations have not been extensively observed (or perhaps not studied) among terpenoids, but are not unknown. Oleanolic acid is oxidized by *Cunninghamella blakesleeana* to a mixture of products,

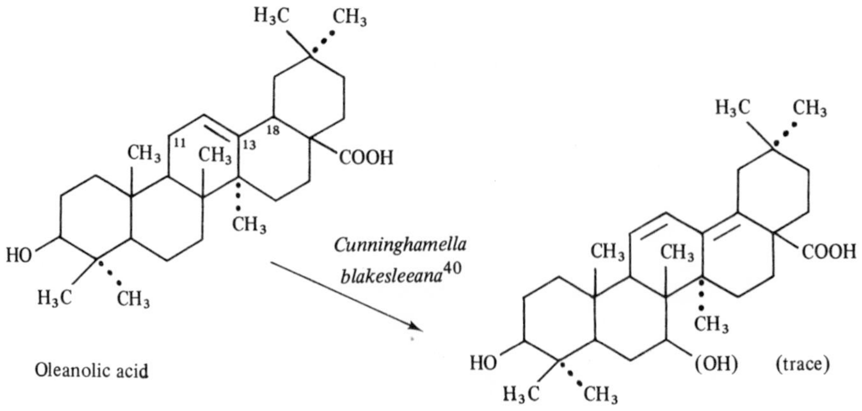

Oleanolic acid   *Cunninghamella blakesleeana*[40]   (trace)

among which are 3ß-hydroxyoleana-11,13(18)-dien-28-oic acid and its 7ß-hydroxy derivative.[40]

The complex transformation of gibberellin A_{12} (or the corresponding diol obtained by reduction of the carboxyl groups) to gibberellic acid involves a dehydrogenation.[41]

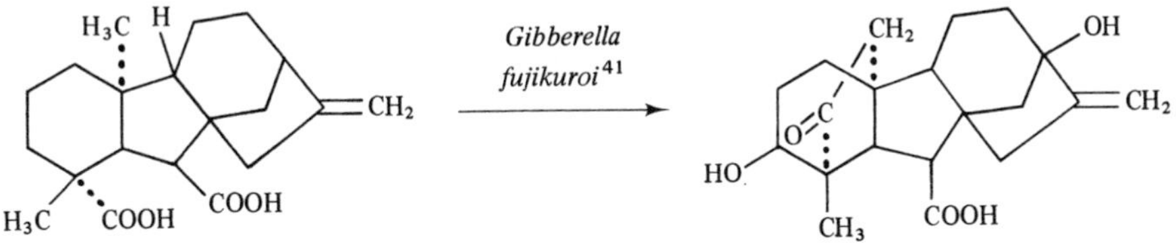

Gibberella fujikuroi[41]

Gibberellin A_{12} Gibberellic acid

In a curious example of the unexpected, the dehydrogenation of imidazolepropionic acid to urocanic acid by an unidentified Gram-negative organism results in product accumulation only if the cells are not first adapted to the substrate.[42]

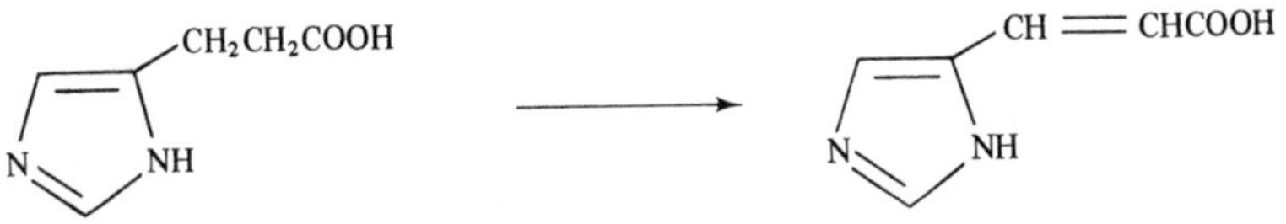

Imidazolepropionic acid   Urocanic acid

All of the preceding examples have dealt with dehydrogenations of carbon-carbon systems. The analogous reaction is known for the carbon-nitrogen system, as illustrated by the degradation of spermine

$$H_2N-(CH_2)_3-NH-(CH_2)_4NH(CH_2)_3NH_2$$

Spermine

$\downarrow$

$$H_2N-(CH_2)_3-NH-(CH_2)_4N=CH-(CH_2)_2NH_2$$

$\downarrow$

$$H_2N-(CH_2)_3-NH-(CH_2)_4NH \quad [+ \ OCH(CH_2)_2NH_2]$$

Spermidine

$\downarrow$

$$H_2N-(CH_2)_3-N=CH(CH_2)_3NH_2$$

$\downarrow$

etc.

and spermidine by *Pseudomonas aeruginosa* or *Serratia marcescens*,[43,44] and of piperidine-2-carboxylic acid by a *Pseudomonas* species.[45]

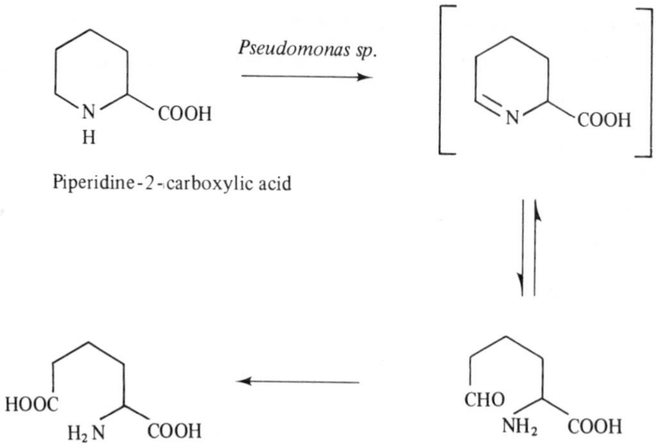

Piperidine-2-carboxylic acid

REFERENCES

1. E. Caspi, W. Schmid, and B. T. Khan, *Tetrahedron 18*, 767 (1962).
2. E. Kondo, *Ann. Rept. Shionogi Res. Lab. 10*, 95 (1960).
3. M. Nishikawa, S. Noguchi, and T. Hasegawa, *Chem. Pharm. Bull. 3*, 322 (1955).
4. G. S. Fonken and H. C. Murray, *J. Org. Chem. 27*, 1102 (1962).
5. G. Wix, K. G. Büki, E. Tömörkeny, and G. Ambrus, *Steroids 11*, 401 (1968).
6. K. Arima, G. Tamura, M. Nagasawa, and M. Bal, *U. S. Patent* 3,388,042 (June 11, 1968).
7. W. Charney, A. Nobile, C. Federbush, D. Sutter, P. L. Pearlman, H. L. Herzog, C. C. Payne, M. E. Tully, M. J. Gentles, and E. B. Hershberg, *Tetrahedron 18*, 591 (1962).
8. C. Vezina, D. J. Marshall, and R. Deghenghi, *U. S. Patent* 3,386,890 (June 4, 1968).
9. J. A. Zderic, A. Bowers, H. Carpio, and C. Djerassi, *J. Amer. Chem. Soc. 80*, 2596 (1958).
10. J. A. Zderic, H. Carpio, A. Bowers, and C. Djerassi, *Steroids 1*, 233 (1963).
11. C. Casas-Campillo, *U. S. Patent* 3,379,621 (April 23, 1968).
12. R. M. Dodson and R. D. Muir, *J. Amer. Chem. Soc. 83*, 4631 (1961).

13. J. Fried, M. J. Green, and G. V. Nair, *J. Amer. Chem. Soc. 92*, 4136 (1970).

14. G. Greenspan, L. L. Smith, T. J. Foell, and R. Rees, *U. S. Patent* 3,344,038 (September 26, 1967).

15. S. Hayakawa, Y. Saburi, and K. Tamaki, *J. Biochem. (Tokyo) 45*, 419 (1958).

16. R. H. Mazur and R. D. Muir, *J. Org. Chem. 28*, 2442 (1963).

17. M. Akhtar and M. A. Parvez, *Biochem. J. 108*, 527 (1968).

18. J. Horvath and A. Kramli, *Nature 160*, 639 (1947).

19. J. Horvath and A. Kramli, *Arch. Biol. Hung. 18*, 19 (1948).

20. L. O. Severina, I. V. Torgov, and G. K. Skryabin, *Dokl. Akad. Nauk SSSR 173*, 1200 (1967).

21. M. Akhtar, M. A. Parvez, and P. F. Hung, *Biochem. J. 106*, 623 (1968).

22. R. Ellouz and M. Lenfant, *Tetrahedron Lett.*, 609 (1969).

23. R. B. Mallory, R. L. Conner, J. R. Landrey, and C. W. L. Iyengar, *Tetrahedron Lett.*, 6103 (1968).

24. G. J. Schroepfer, Jr. and K. Bloch, *J. Biol. Chem. 240*, 54 (1965).

25. A. J. Fulco, R. Levy, and K. Bloch, *J. Biol. Chem. 239*, 988 (1964).

26. A. J. Fulco, *Biochim. Biophys. Acta 144*, 701 (1967).

27. J. Erwin and K. Bloch, *Science 143*, 1006 (1964).

28. J. D. Bu'Lock and G. N. Smith, *Biochem. J. 98*, 6P (1966).

29. E. D. Korn, *J. Biol. Chem. 239*, 396 (1964).

30. J. D. Bu'Lock and P. T. Clay, *Chem. Commun.*, 237 (1969).

31. J. Couteau, E. Azoulay, and J. C. Senez, *Nature 194*, 576 (1962).

32. J. Couteau, E. Azoulay, and J. C. Senez, *Bull. Soc. Chim. Biol. 44*, 671 (1962).

33. F. Wagner, W. Zahn, and U. Buhring, *Angew. Chem. Intern. Ed. Engl. 6*, 359 (1967).

34. B. J. Abbott and L. E. Casida, Jr., *J. Bacteriol. 96*, 925 (1968).

35. J. B. Davis and R. L. Raymond, *Appl. Microbiol. 9*, 383 (1961).

36. D. T. Gibson, J. R. Koch, and R. E. Kallio, *Biochemistry 7*, 2653 (1968).

37. F. Kuffner, H. Klaushofer, and T. Kirchenmayer, *Abh. Deut. Akad. Wiss. Berlin, Kl. Chem., Geol. Biol.*, 103 (1963).

38. H. Rybicka, *Acta Agrobot. 16*, 23 (1964).

39. S. S. Lee and C. J. Sih, *Biochemistry 6*, 1395 (1967).

40. H. Hikino, S. Nabetani, and T. Takemoto, *Yakugaku Zasshi 89*, 809 (1969).

41. B. E. Cross and K. Norton, *Chem. Commun.*, 535 (1965).

42. H. Hassall and F. Rabie, *Biochim. Biophys. Acta 115*, 521 (1966).

43. S. Razin, I. Gery, and U. Bachrach, *Biochem. J. 71*, 551 (1959).

44. U. Bachrach, S. Persky, and S. Razin, *Biochem. J. 76*, 306 (1960).

45. L. V. Basso, D. R. Rao, and V. W. Rodwell, *J. Biol. Chem. 237*, 2239 (1962).

General Reference

W. Charney and H. L. Herzog, *Microbiological Transformations of Steroids*, Academic, New York, 1967.

Chapter 9

ALCOHOL DEHYDROGENATION

The oxidations of alcohols to ketones or aldehydes, and of aldehydes to acids, are common chemical (and microbiological) reactions
that rarely present insurmountable problems to the synthetic organic
chemist. Microbiological methods will therefore be discussed only
sparingly and should in general be considered only when greater reaction specificity is required than is obtainable by purely chemical
means. Situations where this might be the case would be:

(1) Selective oxidation of a portion of a molecule in the presence of other functional groups that might be undesirably
affected by chemical reagents.

(2) Selective oxidation of only one of a mixture of several isomeric materials.

(3) Selective alteration of the composition of a complex mixture, to facilitate the isolation of a desired component.

(4) Oxidation with retention of a stereochemistry that might be
altered by chemical reagents.

Because of the diversity of substrates that have been subjected
to the types of reactions under discussion, any classification scheme
is somewhat arbitrary. Our examples are grouped thus:

(1) Primary alcohol → aldehyde → acid
   (a) No aromatic ring system present
   (b) Aromatic or heteroaromatic ring present

(2) Secondary alcohol → ketone
   (a) Hydroxyl not on a ring (including sugars, which are
       treated as if they were in the open-chain form)
   (b) Hydroxyl on a ring

185

(1) Primary Alcohol → Aldehyde → Acid

(a) No Aromatic Ring Present. The salad dressing responsible
for transforming a mere bundle of greens into a mouth-watering
prelude to the main course owes much of its charm to the product
of the microbiological oxidation of ethanol. Vinegar production
is one of the largest scale manifestations of the reaction type
under discussion, and to the extent that a well-fed organic
chemist is better able to perform synthetic work, the inclusion
of this reaction is justified.[1] Although a number of other hy-
drocarbon-derived alcohols or aldehydes ranging from three to
sixteen carbons have been oxidized by a variety of microorganisms, the results have not been of sufficient preparative value
to warrant detailed reporting.[2-6] Conceivably, the different
substrate specificities of the several microorganisms or their
enzymes could be useful.[7-9]

The bulk of the research effort[10] in this area has been fo-
cused on oxidative transformations of sugars, and the commercial
production of gluconic acid by microbiological oxidation of glucose attests to the success of the method.[11] Similar oxidations
of other sugars are known. Arabinose, xylose, ribose, mannose,
galactose,[12] and 2-deoxyglucose all give the corresponding
acids, usually with a variety of microorganisms (*Acetobacter*
species and *Pseudomonas* species predominating). This reaction
would perhaps be considered useful by the organic chemist only
if great substrate selectivity were desired as, for example, in
the case of modifying the composition of a mixture of sugars.
In some instances, enzymes or enzyme systems with high substrate
specificity have been isolated.[13] *Pseudomonas aeruginosa* has
yielded two enzyme systems, one of which oxidizes both D-glucose
and 2-deoxy-D-glucose to the gluconic acids, while the other oxidizes only the 2-deoxy sugar.[14]

The oxidation of sugars to ketoacids is a fairly common
microbiological reaction. Glucose is converted to 2-ketogluconic
acid in yields exceeding 75% by a variety of microorganisms,[15]
or to 5-ketogluconic acid (65% yield) by several *Acetobacter*

species,[16] and galactose is converted to 2-ketogalactonic acid by *Pseudomonas fluorescens*[17] or *Acetobacter aerogenes*.[18] Another example of two oxidations in the same molecule is the conversion of glucose to saccharic acid by *Aspergillus niger*.[19]

$$
\begin{array}{ccc}
\text{CHO} & & \text{COOH} \\
| & & | \\
\text{HCOH} & & \text{HCOH} \\
| & & | \\
\text{HOCH} & \xrightarrow{\textit{Aspergillus niger}^{19}} & \text{HOCH} \\
| & & | \\
\text{HCOH} & & \text{HCOH} \\
| & & | \\
\text{HCOH} & & \text{HCOH} \\
| & & | \\
\text{CH}_2\text{OH} & & \text{COOH}
\end{array}
$$

D-Glucose   Saccharic acid

Although they are not generally considered to be sugars, pentaerythritol and pantoic acid have enough features reminiscent of sugar alcohols and sugar acids, respectively, to lead to the suspicion that they might undergo oxidations similar to the ones discussed above. Pantoic acid is indeed converted to a carboxyaldehyde by a *Pseudomonas* species or an enzyme derived from

$$
\text{HO—CH}_2\text{—}\underset{\underset{\text{H}_3\text{C}}{|}}{\overset{\overset{\text{CH}_3}{|}}{\text{C}}}\text{—}\underset{\underset{\text{OH}}{|}}{\text{CH}}\text{—COOH} \xrightarrow{\textit{Pseudomonas sp.}^{20\text{-}22}} \text{O}{=}\text{CH—}\underset{\underset{\text{H}_3\text{C}}{|}}{\overset{\overset{\text{CH}_3}{|}}{\text{C}}}\text{—}\underset{\underset{\text{OH}}{|}}{\text{CH}}\text{—COOH}
$$

Pantoic acid

it,[20-22] but the ease of total chemical synthesis of the product makes this microbiological reaction of only academic interest. Not so the biological oxidation of pentaerythritol to the acid by a *Flavobacterium* species,[23] a reaction that is said to pro-

$$
\text{HOCH}_2\text{—}\underset{\underset{\text{CH}_2\text{OH}}{|}}{\overset{\overset{\text{CH}_2\text{OH}}{|}}{\text{C}}}\text{—CH}_2\text{OH} \xrightarrow{\textit{Flavobacterium sp.}^{23}} \text{HOCH}_2\text{—}\underset{\underset{\text{CH}_2\text{OH}}{|}}{\overset{\overset{\text{CH}_2\text{OH}}{|}}{\text{C}}}\text{—COOH}
$$

Pentaerythritol   "Tris acid"

ceed in "almost quantitative yield" and which is apparently un-
dergoing commercial development.

Acyclic terpenes undergo microbiological oxidation. Citronellal may be converted to citronellic acid,[24] while the acid

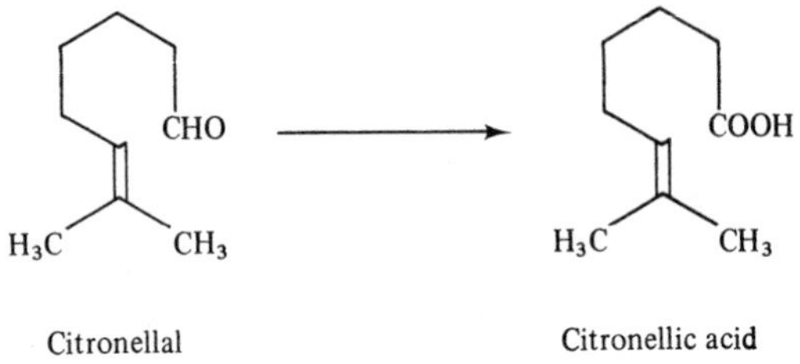

Citronellal   Citronellic acid

similarly derived from geraniol undergoes further oxidation by
the citronellal-induced enzymes of *Pseudomonas citronellolis*,
following a degradative pathway similar to that of β-oxidation
of fatty acids.[25]   This work has not been directed   toward syn-

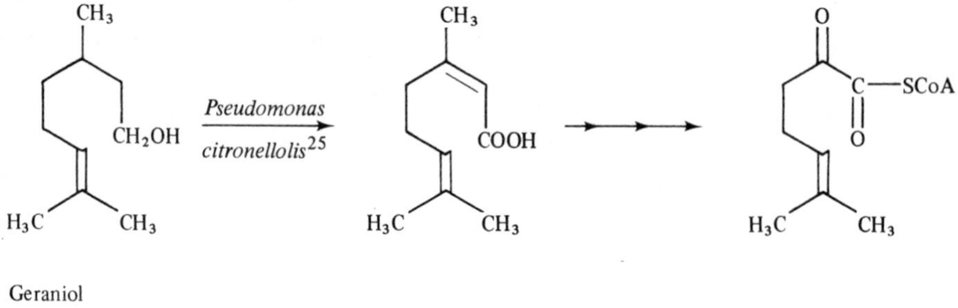

Geraniol

thetic accumulation of the intermediate products.   Furthermore,
these compounds are readily accessible to the chemist without
recourse to microorganisms.

In the area of cyclic terpenes, extensive research into the
metabolic pathway of limonene degradation by a soil pseudomonad
has resulted in isolations of enzyme systems for oxidizing perillyl alcohol and perillyl aldehyde.   Both enzyme systems have

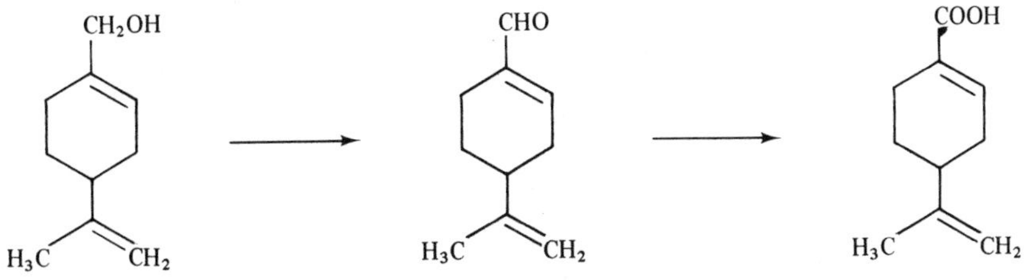

Perillyl alcohol   Perillyl aldehyde

the capability to oxidize some other substrates, but their structural specificity patterns do not coincide. No synthetic uses are detailed.[26-28]

It is worth noting, despite the absence of yield information, that some of the complex biotransformations in the gibberellic acid area involve oxidations of alcohols to acids.[29,30]

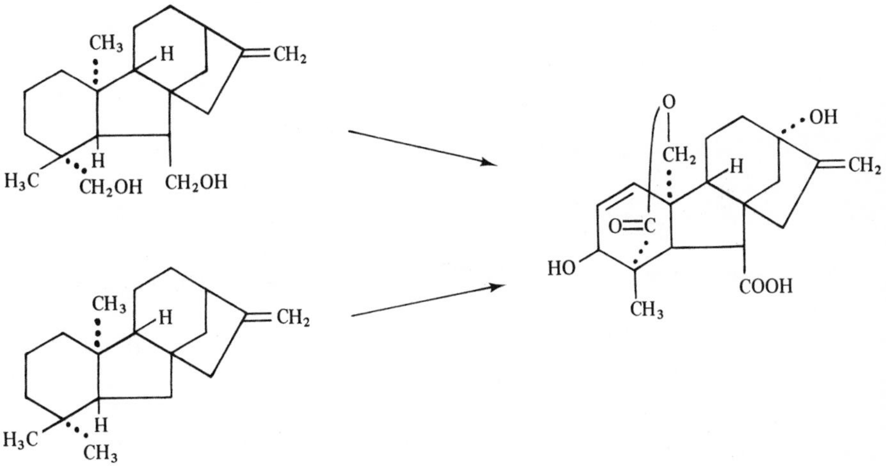

Kaurene

A few miscellaneous examples, all lacking yield data and of no demonstrated synthetic utility, are listed as illustrative of the potential scope of this reaction type:

$H_3C(C\equiv C)_3$ — $CH\equiv CH$ — $COOCH_3$

↓ *Merulius lacrymans*[31]

HO_2C — $(C\equiv C)_3$ — CH_2CH_2 — $COOH$

$H_3C(C\equiv C)_3\ CH_2CH\equiv CH$ — $(CH_2)_3COOH$

↓ *Poria sinusa*[32]

HO_2C — $(CH\equiv CH)_3$ — $CH_2CH\equiv CH(CH_2)_3COOH$

$BrCH_2CH_2CH_2OH$ $\xrightarrow{\textit{Pseudomonas sp.}^{33}}$ $BrCH_2CH_2COOH$

$HOCH_2(CH_2)_{2\ or\ 3}$ — CH — $COOH$
                                  |
                                 NH_2

↓ *Neurospora crassa*[34, 35]

$O\equiv CH$ — $(CH_2)_{2\ or\ 3}$ — $CHCOOH$
                                  |
                                 NH_2

$O\equiv CH$ — $(CH_2)_3$ — $CH\equiv O$ $\xrightarrow[\textit{putida} \text{ enzyme}^{36}]{\textit{Pseudomonas}}$ HO_2C — $(CH_2)_3CHCOOH$
                          | |
                         NH_2 NH_2

$\left[\text{(structure)} \right]$ $\longrightarrow$ $HOCH_2(CH_2)_3COOH$ $\xrightarrow[\textit{spp.}^{37}]{\textit{Pseudomonas}}$ $HO_2C(CH_2)_4COOH$

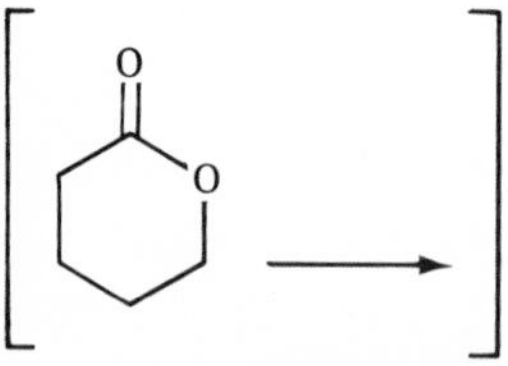

(b) <u>Aromatic Ring System Present</u>. The oxidation of a benzyl alcohol to the aldehyde and thence to the acid is part of a common
microbiological sequence that frequently begins with the alkyl
group. The potential synthetic utility would appear to reside in

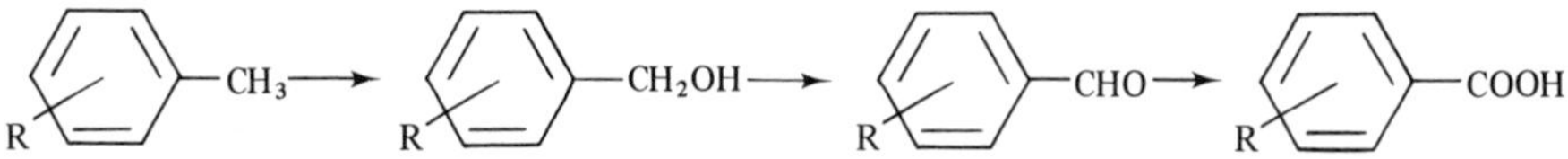

the varying substrate specificities of the microorganisms or their enzymes,[38-41] since purely chemical methods can usually cope with these transformations.

The exquisite selectivity of a microbial oxidation is illustrated by the oxidation of 3-(3,4-methylenedioxybenzyl)-1,5-pentanediol to the hydroxy aldehyde. Of the several microorgan-

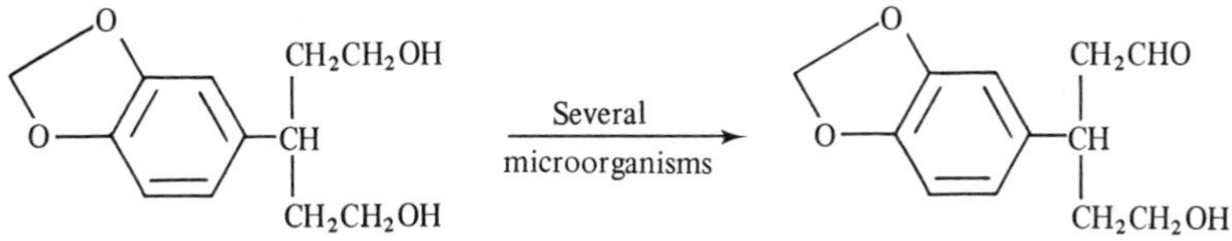

3-(3,4-Methylenedioxybenzyl)-
1,5-pentandiol

isms capable of oxidizing this substrate, *Achromobacter parvulus* gives a 25% yield of the (+)-aldehyde, while *Acetobacter aerogenes* gives an 11% yield of the (-)-aldehyde.[42] 2-(3,4-Methylenedioxybenzyl)-1,3-propanediol is not oxidized under these conditions.

Aside from the conversions of thienylglyoxal to the keto-acid (quantitative)[43] and of phenyl glyoxal to *l*-mandelic acid

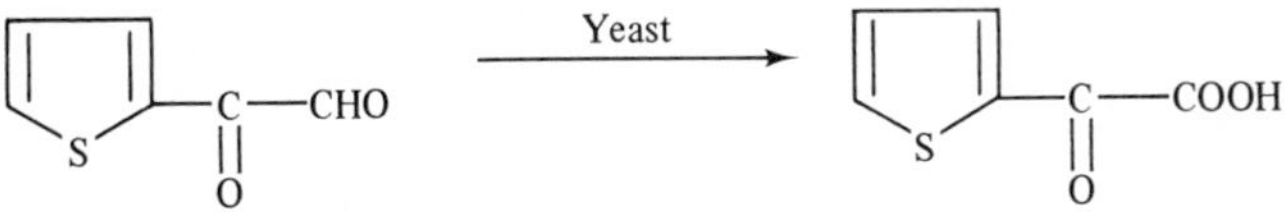

Thienylglyoxal

(∼26% yield), both by yeast, the following illustrations of this reaction type are again offered for speculative purposes only.

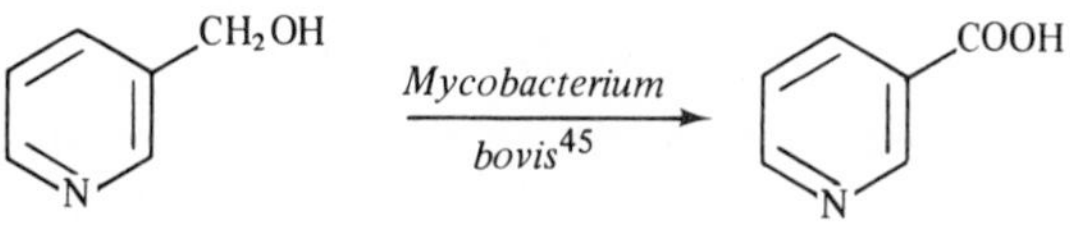

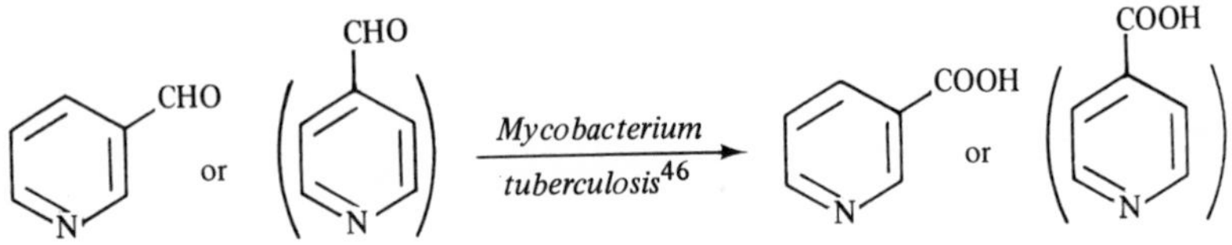

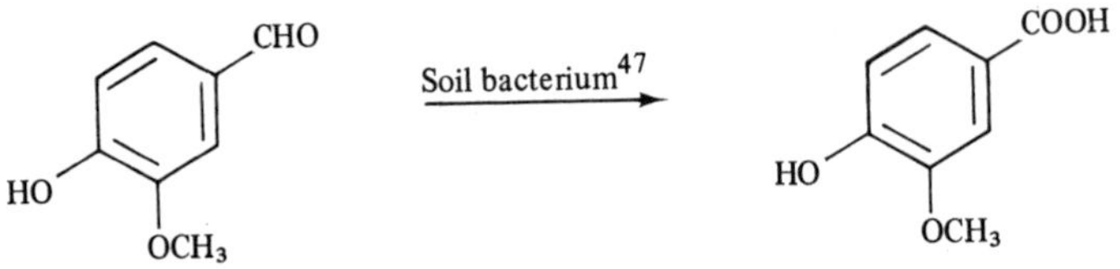

Vanillin → Vanillic acid

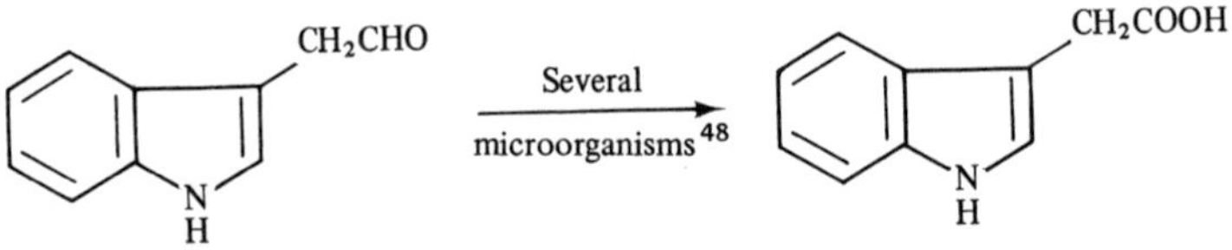

Indole-3-acetaldehyde

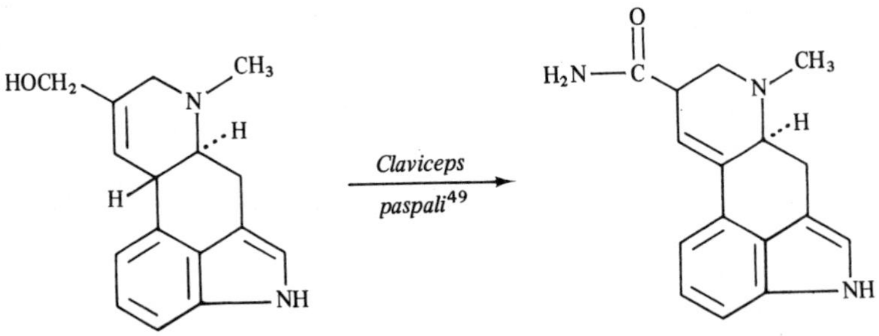

Elymoclavine → Lysergic amide (isolated)

(2) <u>Secondary Alcohol → Ketone</u>.   In the microbial oxygenation of a
methylene group, the product may be either an alcohol or a ketone.
In our earlier discussion of such oxidations (Chapter 1), we have not
distinguished between these reactions.   It is worth noting at this
point that the ketonic products undoubtedly arise as a consequence of
microbial oxidation of the initially formed alcohol.   Such examples
are not presented here, but should be sought in the earlier section.

It is also worth noting that most, if not all, oxidations of alcohols to ketones (or, for that matter, of alcohols to aldehydes to acids, as discussed above) are reversible. Much work, done primarily from the point of view of the reductive process, has afforded valuable insights into the stereochemical course of these reactions, into the structural characteristics of the substrate that have an influence, and into other aspects.[50] The potential value of using this information for oxidative reactions should not be overlooked, although its detailed discussion does not fall within the scope of this book.

(a) Hydroxyl Not on a Ring (includes most sugars). There are a number of examples of the microbial oxidation of simple alcohols to ketones[51,52] but this area is of no practical interest to the synthetic organic chemist. If, however, the molecule in which the hydroxyl group is situated is more complex, thus presenting more problems of selective reaction with chemical reagents, it may be worth considering the microbial reagent. This point has been especially well illustrated with sugars and related substances. Rather than to list the many examples, only a few with yields are shown. They have been formulated in standard carbohydrate bar style for brevity and to focus attention on the pertinent oxidation reaction.

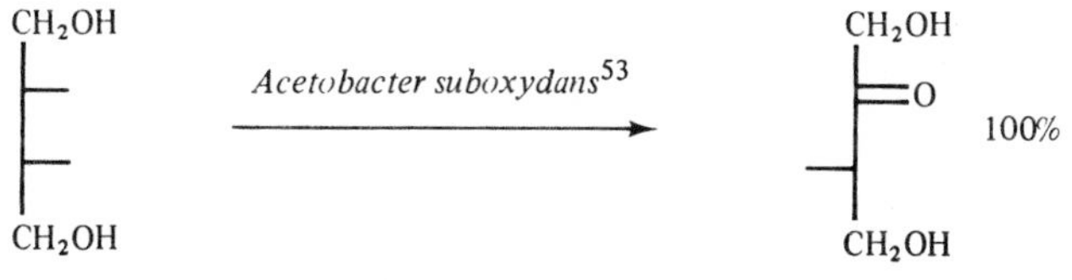

meso-Erythritol   L-Erythrulose

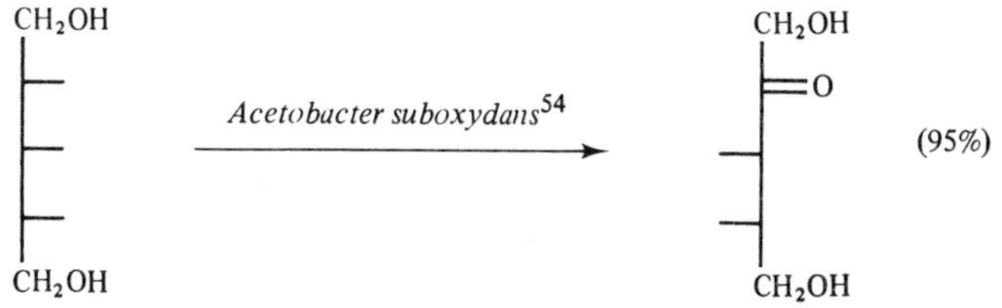

Adonitol   L-Adonulose

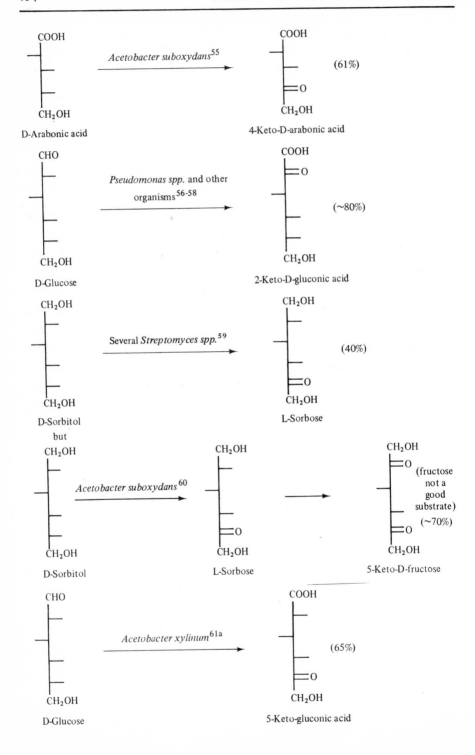

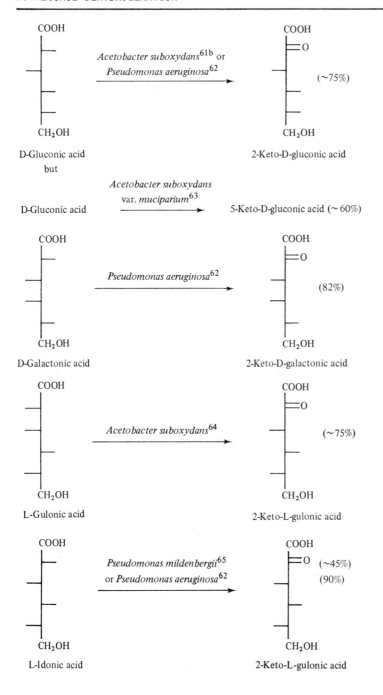

D-Gluconic acid
but

Acetobacter suboxydans
var. muciparium[63]

D-Gluconic acid ⟶ 5-Keto-D-gluconic acid (~60%)

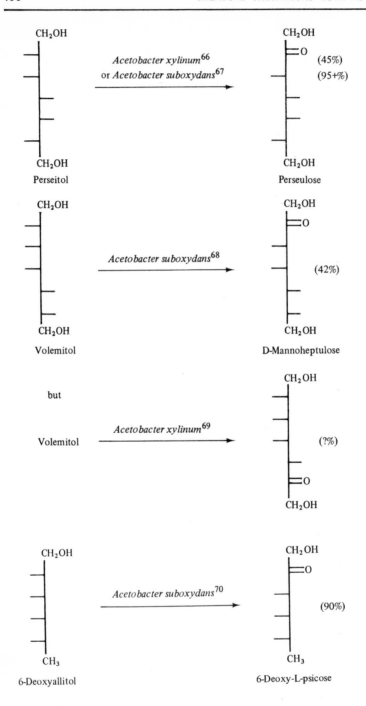

Perseitol   *Acetobacter xylinum*[66] or *Acetobacter suboxydans*[67]   Perseulose   (45%) (95+%)

Volemitol   *Acetobacter suboxydans*[68]   D-Mannoheptulose   (42%)

but

Volemitol   *Acetobacter xylinum*[69]   (?%)

6-Deoxyallitol   *Acetobacter suboxydans*[70]   6-Deoxy-L-psicose   (90%)

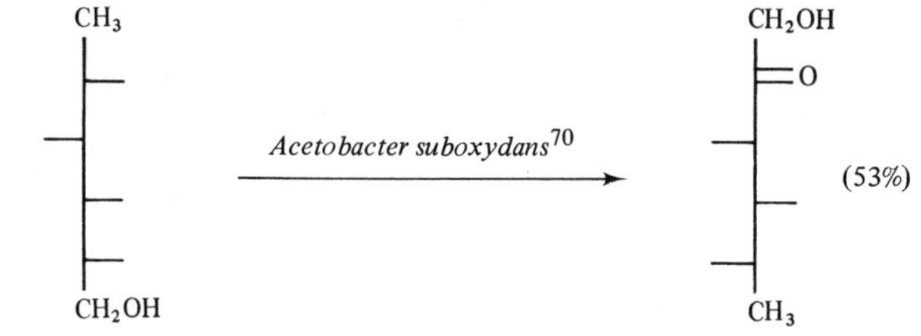

CH₃ ... *Acetobacter suboxydans*[70] ... CH₂OH, O ... (53%) ... CH₃

1-Deoxy-D-sorbitol   6-Deoxy-L-sorbose

Oxidations of aminoalcohols to aminoketones are known, but yield information is lacking.[71]   In the case of the isolated enzyme system from *Escherichia coli*, 1-amino-2-propanol is oxidized to aminoacetone.[72]   Other aminoalcohols also are oxidized:

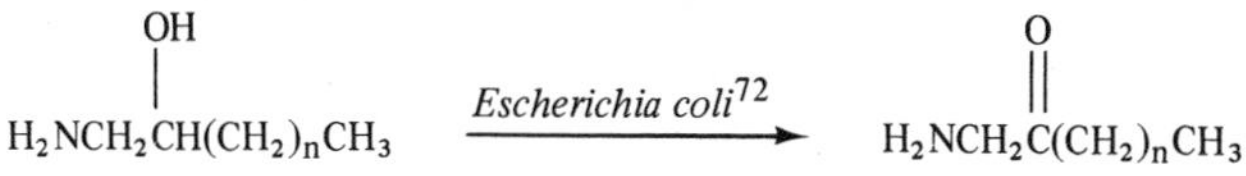

$$\text{H}_2\text{NCH}_2\underset{\underset{\text{OH}}{|}}{\text{CH}}(\text{CH}_2)_n\text{CH}_3 \xrightarrow{\textit{Escherichia coli}^{72}} \text{H}_2\text{NCH}_2\underset{\underset{\text{O}}{\|}}{\text{C}}(\text{CH}_2)_n\text{CH}_3$$

n = 0, 1

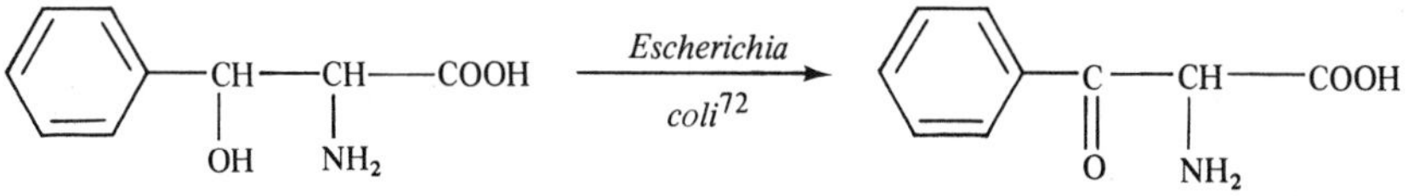

Carnitine has been oxidized with cell-free extracts from *Pseudomonas aeruginosa*[73]:

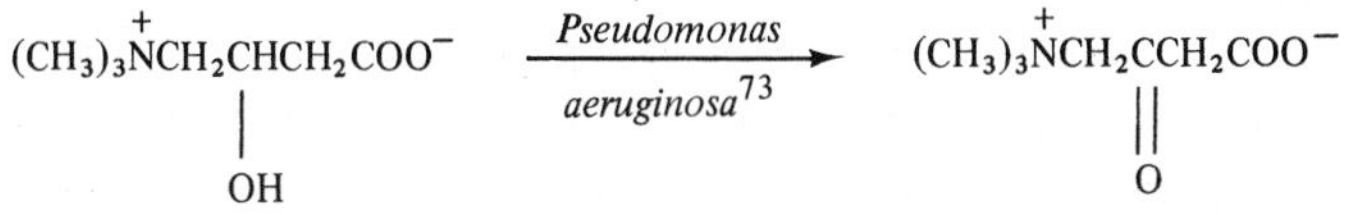

$$(\text{CH}_3)_3\overset{+}{\text{N}}\text{CH}_2\underset{\underset{\text{OH}}{|}}{\text{CH}}\text{CH}_2\text{COO}^- \xrightarrow[\textit{aeruginosa}^{73}]{\textit{Pseudomonas}} (\text{CH}_3)_3\overset{+}{\text{N}}\text{CH}_2\underset{\underset{\text{O}}{\|}}{\text{C}}\text{CH}_2\text{COO}^-$$

Carnitine

In the case of chloramphenicol (with several microorganisms), the over-all reaction is more complex[74]:

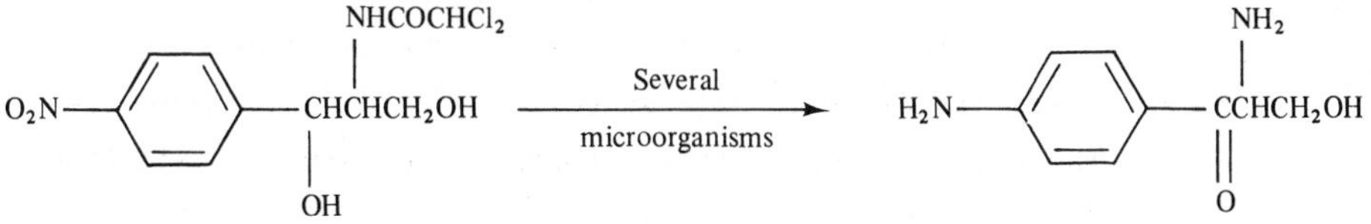

Chloramphenicol

The oxidation of *erythro-* (but not *threo-*) ethylmalic acid to α-ketovaleric acid by an enzyme preparation from *Ps. aeruginosa* suffers from the fact that it is easier to prepare the product than the substrate by chemical means,[75] while the oxidation

$$
\begin{array}{c}
\text{COOH} \\
| \\
\text{HCCH}_2\text{CH}_3 \\
| \\
\text{HCOH} \\
| \\
\text{COOH}
\end{array}
\quad \xrightarrow{\textit{Pseudomonas aeruginosa}^{75}} \quad
\text{CH}_3(\text{CH}_2)_2 \overset{\overset{\displaystyle O}{\|}}{—} \text{C} — \text{COOH}
$$

Ethylmalic acid

of isohydroxycamphoric acid to isoketocamphoric acid by a *Diphtheroid* is so deeply embedded in the degradation sequence of camphor that synthetic utility is doubtful.[76]

$$
\begin{array}{c}
\text{CH}_3 \\
| \\
\text{CHOH} \\
| \\
\text{H}_3\text{CCCH}_3 \\
| \\
\text{HOOCCH}_2\text{CHCH}_2\text{COOH}
\end{array}
\quad \xrightarrow{\textit{Diphtheroid}^{76}} \quad
\begin{array}{c}
\text{CH}_3 \\
| \\
\text{C}=\text{O} \\
| \\
\text{H}_3\text{CCCH}_3 \\
| \\
\text{HOOCCH}_2\text{CHCH}_2\text{COOH}
\end{array}
$$

Isohydroxycamphoric acid   Isoketocamphoric acid

(b) Hydroxyl on a Ring. The known microbiological oxidations of simple cyclic alcohols, such as cyclohexanol to cyclohexanone[77] will be of little interest to the synthetic chemist. Only in the case of the more complex molecules can the microbial method offer certain advantages.

The oxidations of cyclic polyols (cyclitols) by *Acetobacter suboxydans* have been studied extensively, especially by Posternak and his co-workers over the past several decades, and the selectivity of the organism for axially oriented hydroxyls has been firmly established. Early attempts[78,79] to define more precisely the role of nearby substituents in conferring oxidative selectivity have not been substantiated.[80] The bulk of the work has been carried out in a Warburg apparatus, but a few reactions that have yielded tangible products (ordinarily as a derivative) will suffice as illustrative examples.   (In these formulas, the solid lines represent hydrogen atoms.)

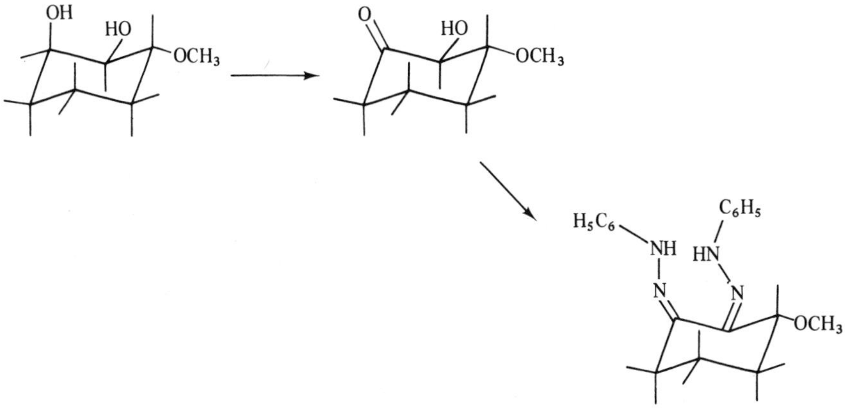

$(54\%)^{81}$

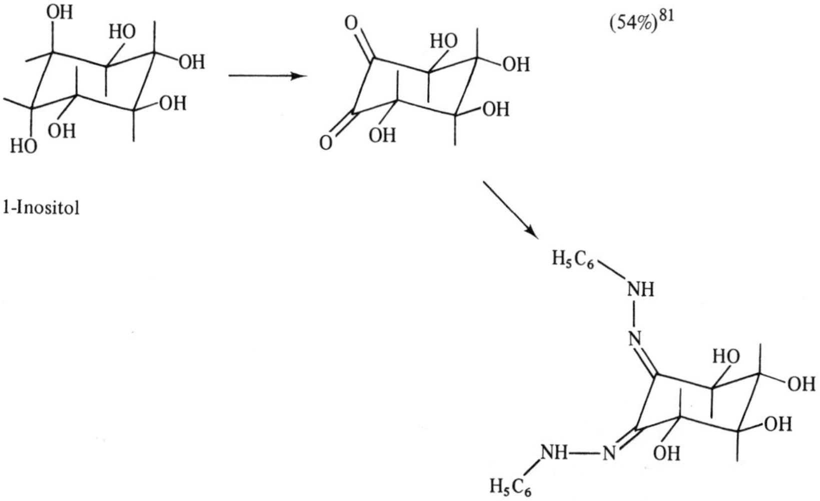

1-Inositol

$(29\%)^{78}$

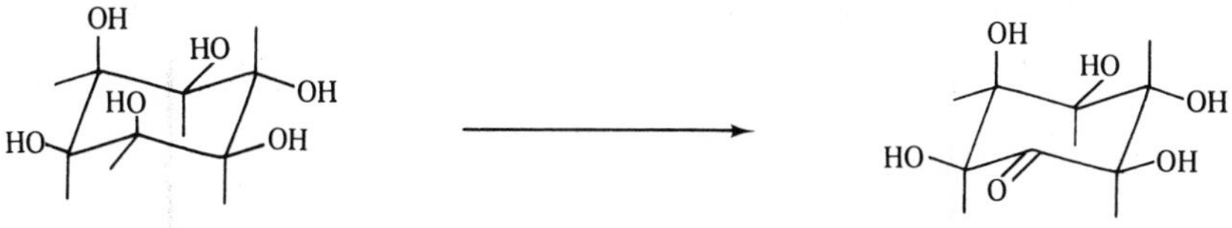

1-*epi*-Inositol

1-*epi*-*meso*-Inosose

$(68\%)^{82}$

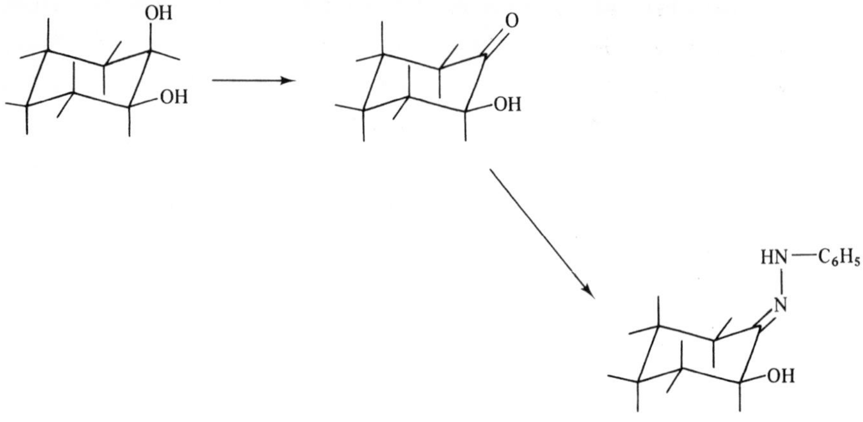

$(76\%)^{83}$

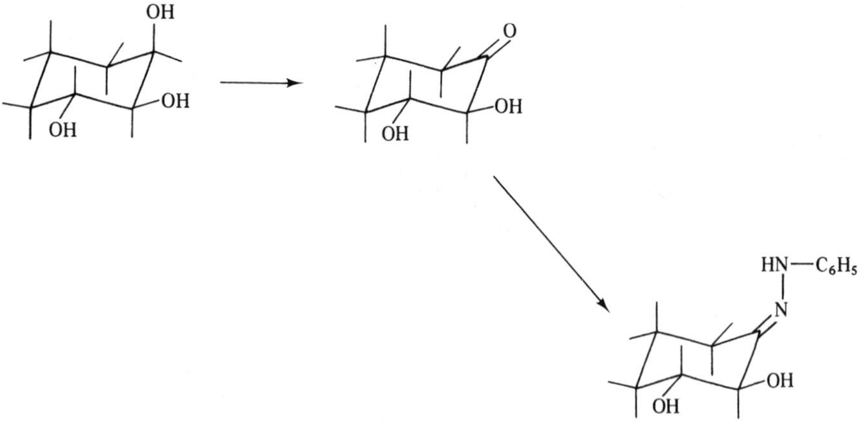

$(?\%)^{84}$

Aside from the steroids, the application of microbial oxidation to terpenoid alcohols has not been extensive. The investigations of Gunsalus and co-workers into the degradation of camphor have shown that such reactions are involved in the overall pathway, and an enzyme system from *Pseudomonas putida* has

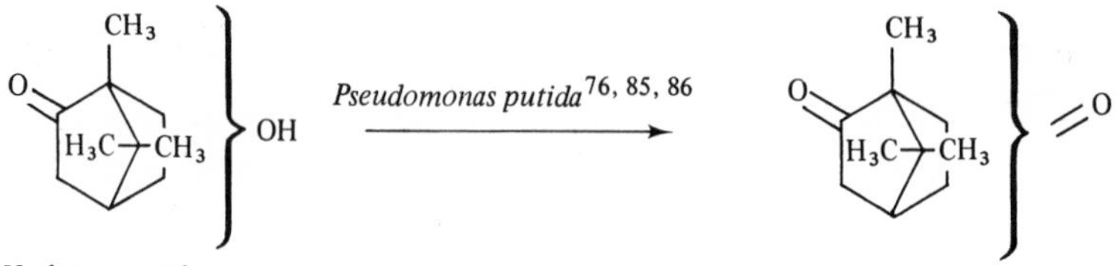

Hydroxycamphors

been shown to act on a variety of cyclic alcohols in addition to
5- or 6-hydroxycamphor. Similar oxidations have been seen as
part of the microbial oxidative patterns of several terpenoid
substances[87-89] but in the cases of the kessane-2β,7- or 2β,8α-
diols derived from α-kessyl alcohol, these prove to be fairly

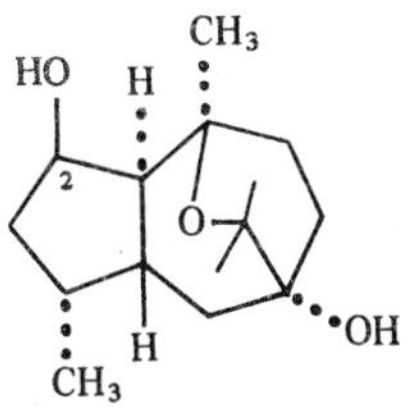

Kessane-2β, 7-diol 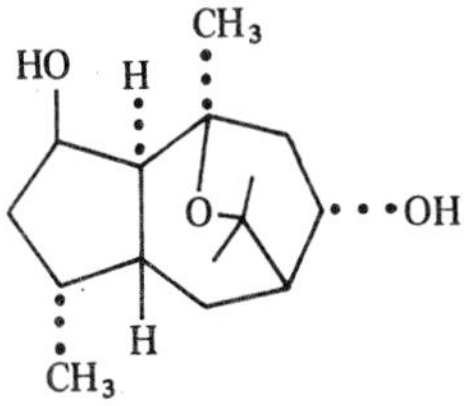   Kessane-2β, 8-diol

resistant to microbial oxidation.[90]   The interesting *Nocardia
restrictus* transformation of the steroid fragment, while com-

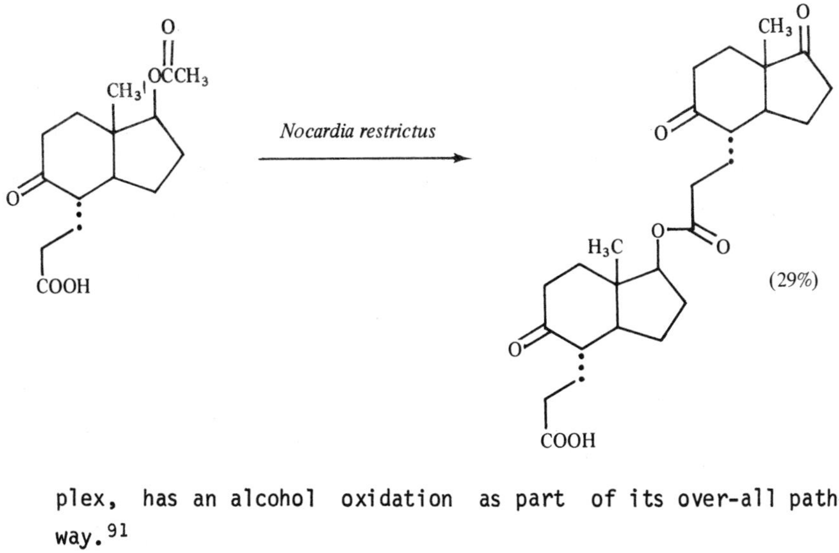

Nocardia restrictus

(29%)

plex, has an alcohol oxidation as part of its over-all pathway.[91]

Steroids represent a major area in which alcohol → ketone
oxidations have been used for synthetic purposes, although in
most cases chemical oxidation is the preferred method. The examples are intended more to suggest possible uses of a microbial
method for other substrates than to illustrate useful transformations *per se*.

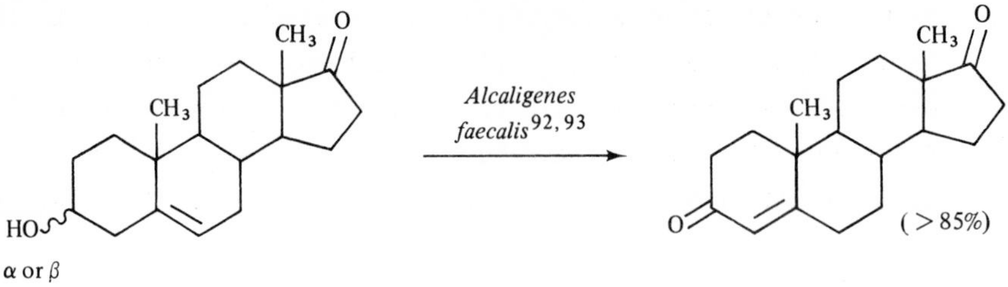
CH3
O
CH3
HO
α or β
Alcaligenes faecalis92, 93
CH3
O
CH3
O
(> 85%)

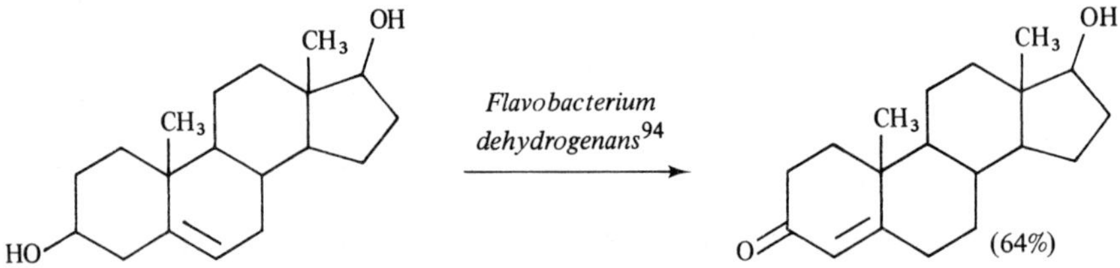
CH3
OH
CH3
HO
Flavobacterium dehydrogenans94
CH3
OH
CH3
O
(64%)

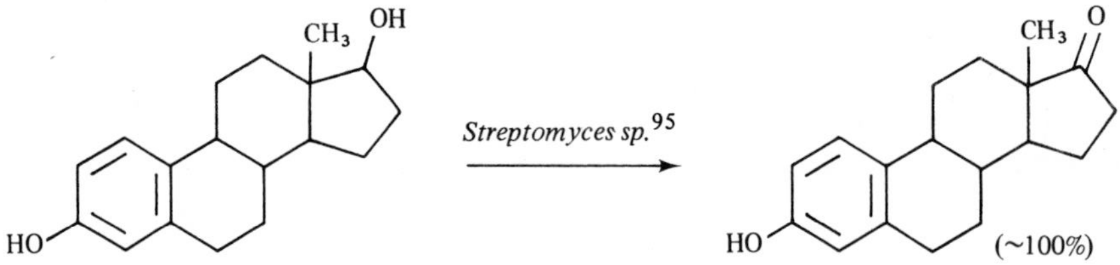
CH3
OH
HO
Streptomyces sp.95
CH3
O
HO
(~100%)

Estradiol

Estrone

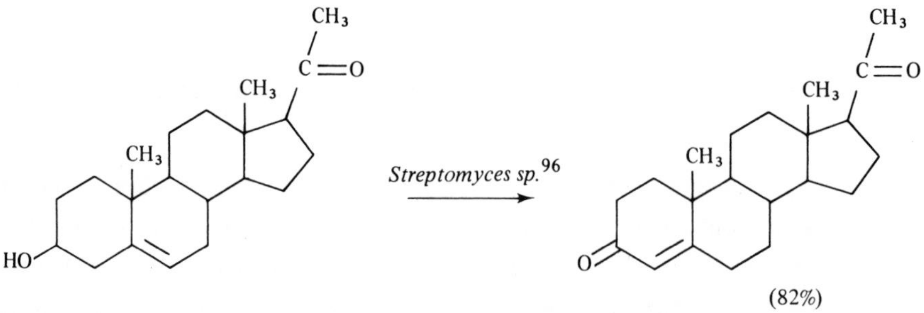
CH3
C=O
CH3
CH3
HO
Streptomyces sp.96
CH3
C=O
CH3
CH3
O
(82%)

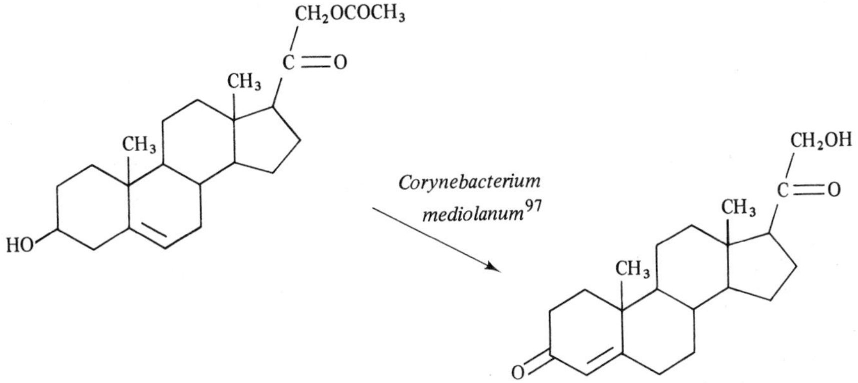
CH₂OCOCH₃
C=O
CH₃
CH₃
HO
Corynebacterium mediolanum[97]
CH₂OH
C=O
CH₃
CH₃
O

(34%)

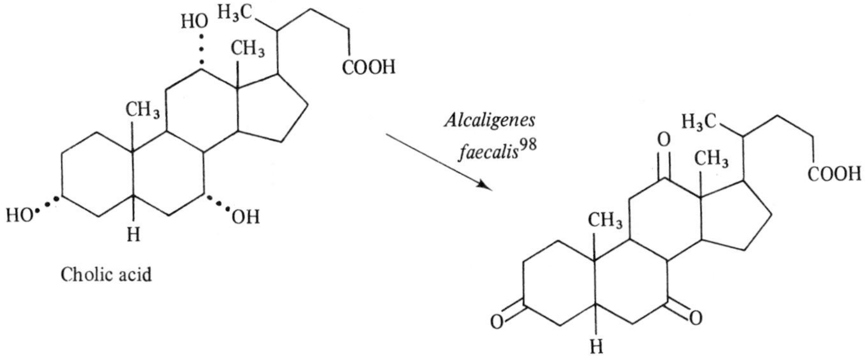
HO
H₃C
CH₃
COOH
CH₃
HO
OH
H
Cholic acid
Alcaligenes faecalis[98]
H₃C
O
CH₃
COOH
CH₃
O
O
H

(83%)

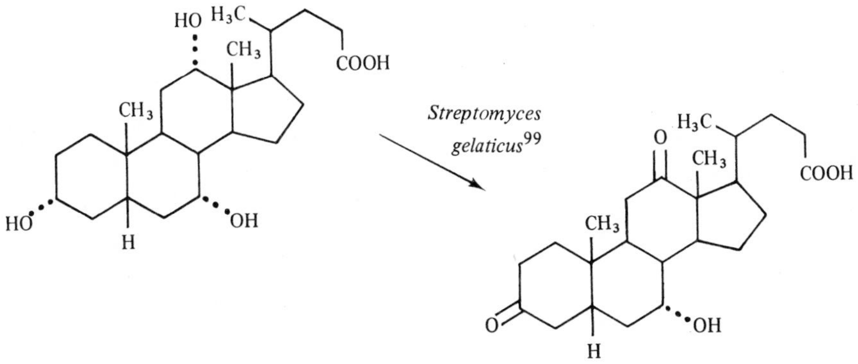
HO
H₃C
CH₃
COOH
CH₃
HO
OH
H
Streptomyces gelaticus[99]
H₃C
O
CH₃
COOH
CH₃
O
OH
H

(6.7%)

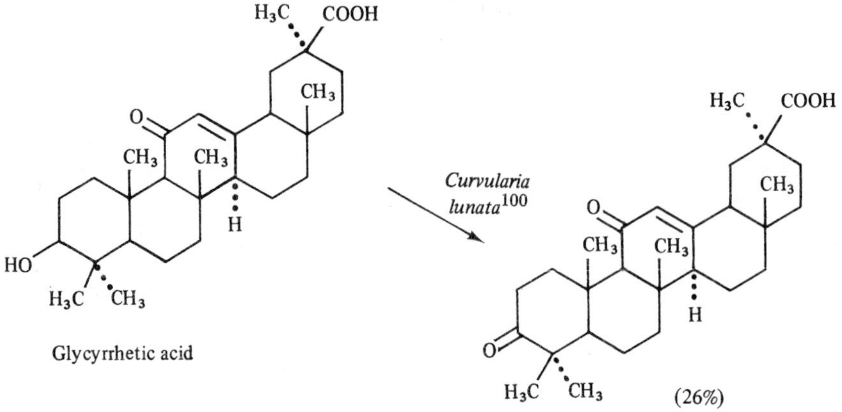

Glycyrrhetic acid

(26%)

Oxidation of existing or of newly formed hydroxyl groups often occurs concomitantly with other reactions in the microbial transformation of steroids, for example:

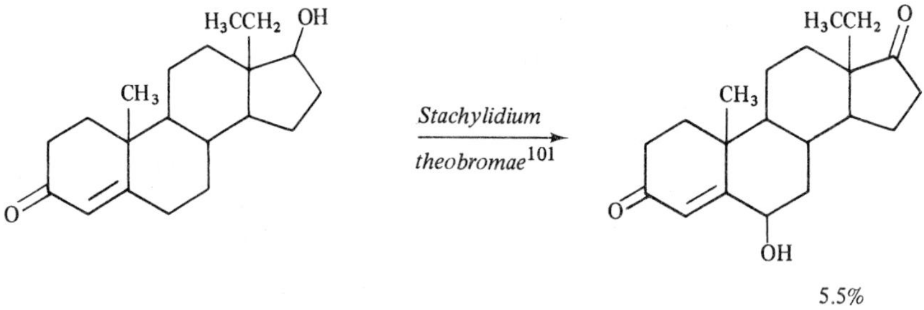

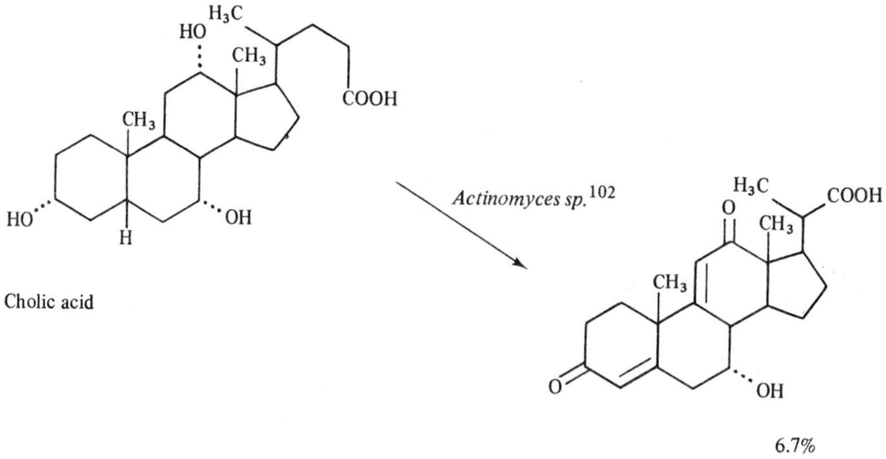

5.5%

Cholic acid

6.7%

REFERENCES

1. L. E. Casida, Jr., *Industrial Microbiology*, Wiley, New York, 1968, pp. 333-334.

2. E. Azoulay and M. T. Heydeman, *Biochim. Biophys. Acta 73*, 1 (1963).

3. V. F. Kazimirova and N. V. Novotel'nov, *Tr. Leningrad. Tekhnol. Inst. Kholod. Prom. 14*, 194 (1956); *Chem. Abstr. 51*, 11460 (1957).

4. J. C. Senez and E. Azoulay, *Biochim. Biophys. Acta 47*, 307 (1961).

5. E. Azoulay and J. C. Senez, *Ann. Inst. Pasteur 98*, 868 (1960).

6. R. I. Leavitt, *U. S. Patent* 3,326,772 (June 20, 1967).

7. W. B. Jakoby, *J. Biol. Chem. 232*, 75 (1958).

8. A. C. van der Linden and C. G. R. Huybregtse, *Abstr. Amer. Chem. Soc. 154*, Q5 (1967).

9. L. L. Wallen, F. H. Stodola, and R. W. Jackson, *Type Reactions in Fermentation Chemistry*, United States Department of Agriculture, 1959, Reactions 287-290.

10. Reference 9, Reactions 310-339.

11. Reference 1, pp. 344-347.

12. Reference 9, Reactions 291-309.

13. H. Lyr, *Enzymologia 24*, 69 (1962).

14. A. K. Williams and R. G. Eagon, *J. Bacteriol. 77*, 167 (1959).

15. Reference 9, Reactions 60-66.

16. Reference 9, Reactions 67-71.

17. T. Asai, K. Aida, and Y. Ueno, *J. Agr. Chem. Soc. Jap. 26*, 625 (1952).

18. E. Masuo and Y. Nozaki, *Ann. Rept. Shionogi Res. Lab. 6*, 110 (1956).

19. F. Challenger, V. Subramanian, and T. K. Walker, *Nature 119*, 674 (1927).

20. C. T. Goodhue and E. E. Snell, *Biochemistry 5*, 393 (1966).

21. V. Nurmikko, E. Salo, H. Hakola, K. Makinen, and E. E. Snell, *Biochemistry 5*, 399 (1966).

22. C. T. Goodhue and E. E. Snell, *Biochemistry 5*, 403 (1966).

23. C. T. Goodhue and J. R. Schaeffer, *3rd International Fermentation Symposium*, September, 1968, Abstr. G2-7.

24. T. Hayaishi, K. Hirano, H. Ueda, and C. Tatsumi, *Agr. Biol. Chem. 31*, A21 (1967).

25. W. Seubert and E. Fass, *Biochem. Z. 341*, 35 (1964).

26. N. R. Ballal, P. K. Bhattacharyya, and P. N. Rangachari, *Biochem. Biophys. Res. Commun. 29*, 275 (1967).

27. N. R. Ballal, P. K. Bhattacharyya, and P. N. Rangachari, *Biochem. Biophys. Res. Commun. 23*, 473 (1966).

28. N. R. Ballal, P. K. Bhattacharyya, and P. N. Rangachari, *Indian J. Biochem. 5*, 1 (1968).

29. B. E. Cross, R. H. B. Galt, and J. R. Hanson, *J. Chem. Soc.*, 295 (1964).

30. B. E. Cross and K. Norton, *Chem. Commun.*, 535 (1965).

31. P. Hodge, E. R. H. Jones, and G. Lowe, *J. Chem. Soc. (C)*, 1216 (1966).

32. J. D. Bu'Lock in *Comparative Phytochemistry* (T. Swain, ed.), Academic, New York, 1966, pp. 79-95.

33. C. E. Castro and E. W. Bartnicki, *Biochim. Biophys. Acta 100*, 384 (1965).

34. T. Yura and H. J. Vogel, *Biochim. Biophys. Acta 24*, 648 (1957).

35. T. Yura and H. J. Vogel, *J. Biol. Chem. 234*, 339 (1959).

36. A. F. Calvert and V. W. Rodwell, *J. Biol. Chem. 241*, 409 (1966).

37. R. Shaw, *Nature 209*, 1369 (1966).

38. R. I. Leavitt, *J. Gen. Microbiol. 49*, 411 (1967).

39. M. Katagiri, S. Takemori, K. Nakazawa, H. Suzuki, and K. Akagi, *Biochim. Biophys. Acta 139*, 173 (1967).

40. M. Kitagawa, *J. Biochem. 43*, 553 (1956).

41. M. Merle and R. H. Baum, *Abstr. Amer. Chem. Soc.*, No. 158, MICR 47 (1969).

42. H. Kosmol, K. Kieslich, and H. Gibian, *Justus Liebigs Ann. Chem. 711*, 42 (1968).

43. S. Fujise, *Biochem. Z. 236*, 241 (1931).

44. H. D. Dakin, *J. Biol. Chem. 18*, 91 (1914).

45. D. Gross, A. Feige, A. Zureck, and H. R. Schütte, *European J. Biochem. 4*, 28 (1968).

46. J. Youatt, *Australian J. Exp. Biol. Med. Sci. 40*, 191 (1962); *Chem. Abstr. 57*, 12997 (1962).

47. W. J. Robbins and E. C. Lathrop, *Soil Sci. 7*, 475 (1919).

48. K.-W. Glombitza, *Experientia 23*, 101 (1967).

49. K. Mothes, K. Winkler, D. Gröger, H. G. Floss, U. Mothes, and F. Weygand, *Tetrahedron Lett.*, 933 (1962).

50. See R. Bentley, *Molecular Asymmetry in Biology*, vol. II, Academic, New York, 1970, Chapter 1.

51. Reference 9, Reactions 1-3, 26-28.

52. D. A. Klein and F. A. Henning, *Appl. Microbiol. 17*, 676 (1969).

53. E. I. Fulmer and L. A. Underkofler, *Iowa State College J. Sci. 21*, 251 (1947); *Chem. Abstr. 41*, 6299 (1947).

54. T. Reichstein, *Helv. Chim. Acta 17*, 996 (1934).

55. J. Liebster, M. Kulhanek, and M. Tadra, *Chem. Listy 47*, 1075 (1953).

56. J. J. Stubbs, L. B. Lockwood, E. T. Roe, B. Tabenkin, and G. E. Ward, *Ind. Eng. Chem. 32*, 1626 (1940).

57. L. B. Lockwood, B. Tabenkin, and G. E. Ward, *J. Bacteriol. 42*, 51 (1941).

58. Y. Ikeda, *J. Agr. Chem. Soc. Jap. 24*, 56 (1950).

59. E. Masuo and E. Kondo, *Ann. Rept. Shionogi Res. Lab. 6*, 115 (1956).

60. K. Sato, Y. Yamada, K. Aida, and T. Uemura, *Agr. Biol. Chem. 31*, 877 (1967).

61. a. K. Bernhauer and K. Schön, *Z. Physiol. Chem. 180*, 232 (1929).

    b. K. Bernhauer and H. Knobloch, *Naturwiss. 26*, 819 (1938).

62. M. Kulhanek, *Chem. Listy 47*, 1071 (1953).

63. K. Bernhauer and H. Knobloch, *Biochem. Z. 303*, 308 (1940).

64. B. E. Gray, *U. S. Patent* 2,421,612 (June 3, 1947).

65. B. E. Gray, *U. S. Patent* 2,421,611 (June 3, 1947).

66. G. Bertrand, *Compt. Rend. 147*, 201 (1908).

67. R. M. Hann, E. B. Tilden, and C. S. Hudson, *J. Amer. Chem. Soc. 60*, 1201 (1938).

68. L. C. Stewart, N. K. Richtmyer, and C. S. Hudson, *J. Amer. Chem. Soc. 71*, 3532 (1949).

69. E. L. Dulaney, W. J. McAleer, M. Koslowski, E. D. Stapley, and J. Jaglom, *Appl. Microbiol. 3*, 336 (1955).

70. H. Kaufmann and T. Reichstein, *Helv. Chim. Acta 50*, 2280 (1967).

71. J. M. Turner, *Biochem. J. 98*, 7P (1966).

72. J. M. Turner, *Biochem. J. 104*, 112 (1967).

73. H. Aurich, H. P. Kleber, and W. D. Schöpp, *Biochim. Biophys. Acta 139*, 507 (1967).

74. G. N. Smith and C. S. Worrel, *Arch. Biochem. 28*, 232 (1950).

75. R. Rabin, I. I. Salamon, A. S. Bleiweis, J. Carlin, and S. J. Ajl, *Biochemistry 6*, 377 (1968).

76. P. J. Chapman, G. Meerman, I. C. Gunsalus, R. Srinivasan, and K. L. Rinehart, Jr., *J. Amer. Chem. Soc. 88*, 618 (1966).

77. J. Ooyama and J. W. Foster, *Antonie van Leeuwenhoek 31*, 45 (1965).

78. B. Magasanik and E. Chargaff, *J. Biol. Chem. 174*, 173 (1948).

79. B. Magasanik, R. E. Franzl, and E. Chargaff, *J. Amer. Chem. Soc. 74*, 2618 (1952).

80. Th. Posternak, A. Rapin, and A. L. Haenni, *Helv. Chim. Acta 40*, 1594 (1957).

81. P. Barbezat, D. Reymond, and Th. Posternak, *Helv. Chim. Acta 50*, 1811 (1967).

82. Th. Posternak, *Helv. Chim. Acta 29*, 1991 (1946).

83. Th. Posternak, D. Reymond, and H. Friedli, *Helv. Chim. Acta 38*, 205 (1955).

84. Th. Posternak and F. Ravenna, *Helv. Chim. Acta 30*, 441 (1947).

85. W. H. Bradshaw, H. E. Conrad, E. J. Corey, I. C. Gunsalus, and D. Lednicer, *J. Amer. Chem. Soc. 81*, 5507 (1959).

86. I. C. Gunsalus, P. J. Chapman, and J. F. Kuo, *Biochem. Biophys. Res. Commun. 18*, 924 (1965).

87. H. Hikino, K. Aota, Y. Tokuoka, and T. Takemoto, *Chem. Pharm. Bull. 16*, 1088 (1968).

88. H. Hikino, S. Nabetani, and T. Takemoto, *Yakugaku Zasshi 89*, 809 (1969).

89. H. Hikino, T. Kohama, and T. Takemoto, *Chem. Pharm. Bull. 17*, 1659 (1969).

90. H. Hikino, Y. Tokuoka, Y. Hikino, and T. Takemoto, *Tetrahedron 24*, 3147 (1968).

91. E. Kondo, B. Stein, and C. J. Sih, *Biochim. Biophys. Acta 176*, 135 (1969).

92. L. Mamoli and A. Vercellone, *Chem. Ber. 71*, 1686 (1938).

93. H. B. Hughes and L. H. Schmidt, *Proc. Soc. Exptl. Biol. Med. 51*, 162 (1942).

94. A. Ercoli, *Z. Physiol. Chem. 270*, 266 (1941).

95. M. Welsch and C. Heusghem, *Compt. Rend. Soc. Biol. 142*, 1074 (1948).

96. A. Ercoli, *Boll. Sci. Fac. Chim. Ind. Bologna 18*, 1279 (1940); *Chem. Abstr. 38*, 1540 (1944).

97. L. Mamoli, *Chem. Ber. 72*, 1863 (1939).

98. L. H. Schmidt, H. B. Hughes, M. H. Green, and E. Cooper, *J. Biol. Chem. 145*, 229 (1942).

99. S. Hayakawa, Y. Saburi, and H. Teraoka, *Proc. Japan Acad. 32*, 519 (1956); *Chem. Abstr. 51*, 2111 (1957).

100. L. Canonica, G. Jommi, V. M. Pagnoni, F. Pelizzoni, B. M. Ranzi, and C. Scolastico, *Gazz. Chim. Ital. 96*, 820 (1966).

101. K. Kieslich, H. D. Berndt, R. Wiechert, U. Kerb, G. Schulz, and H. J. Koch, *Justus Liebigs Ann. Chem. 726*, 161 (1969).

102. S. Hayakawa, *Proc. Japan Acad. 30*, 133 (1954); *Chem. Abstr. 50*, 388 (1956).

Chapter 10

OXIDATION OF AMINO TO NITRO

The synthetic organic chemist usually thinks in terms of the reduction of a nitro compound to an amine, rather than the oxidation of
an amine to a nitro compound, largely because of the complexity of
the latter reaction, and its tendency to produce highly colored,
tarry materials instead of clean, homogeneous products. As one of
nature's more unusual and less well publicized excursions into the
exotic, several microorganisms are able to oxidize amino groups of
selected substrates to the corresponding nitro compounds under very
mild conditions and may be considered to be the reagent of choice in
certain circumstances.

Since several microorganisms produce natural products containing
the nitro group, it might be expected that they can oxidize simple
amino-containing substances. For example, *Streptomyces thioluteus*,
which produces the nitro compound aureothin, oxidizes *p*-aminobenzoic

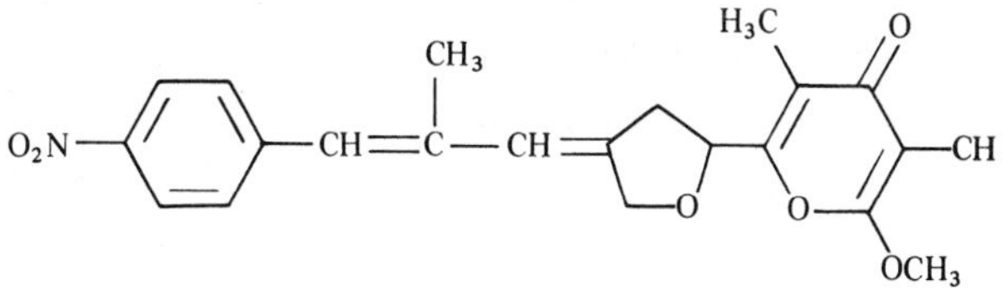

Aureothin

to *p*-nitrobenzoic acid.[1,2] When washed *S. thioluteus* cells are used,
the assayed yield reached about 85% after two hours.

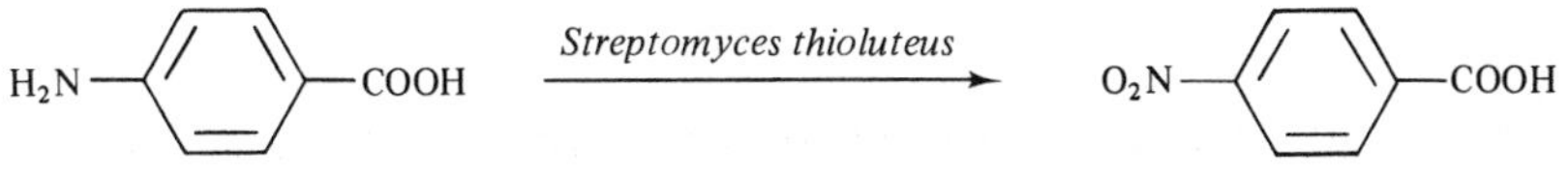

p-Aminobenzoic acid

Several other compounds analogous to p-aminobenzoic acid are also oxidized by S. thioluteus. p-Aminophenylacetic acid, p-amino-L-phenylalanine, and p-aminoacetophenone afford the corresponding nitro compound, although at substantially slower rates than for the p-aminobenzoic acid, while p-aminobenzaldehyde gives a mixture of p-nitrobenzoic acid, p-nitrobenzaldehyde, and p-nitrobenzyl alcohol. Since a number of closely related substances fail to undergo oxidation, there would seem to be considerable substrate specificity for p-amino substituted substrates.

Based on the observations that (1) the incubation of p-dimethylaminobenzaldehyde with S. thioluteus gives p-dimethylaminobenzaldehyde N-oxide, and (2) that the oxidation of p-aminobenzoate to p-nitrobenzoate in an ^{18}O-enriched atmosphere leads to ^{18}O incorporation into the nitro group (two ^{18}O per $-NO_2$),[3] the mechanism for the reaction $-C-NH_2 \rightarrow -C-NO_2$ has been proposed to be:

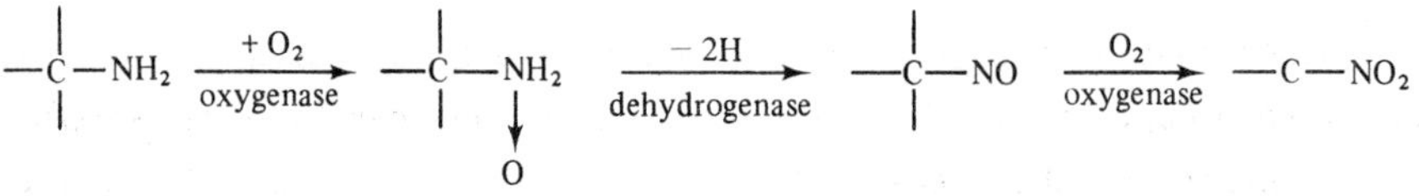

The postulation of the intermediate nitroso compound, made plausible by the evidence that the Streptomyces-produced pigment ferroverdin contains a nitroso group,[4] would seem to offer some additional

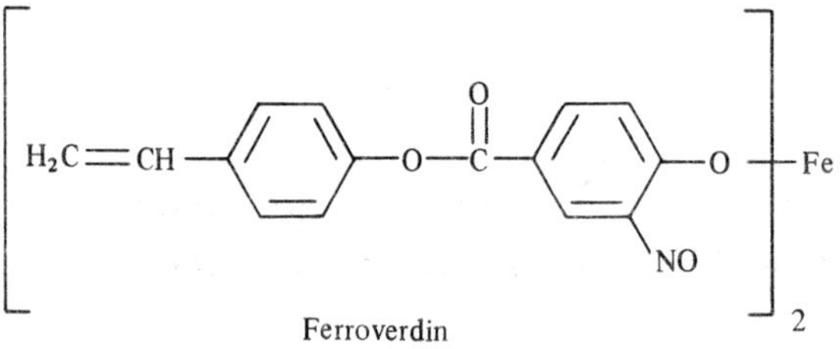

Ferroverdin

possibilities for the study of aromatic amine oxidations as useful synthetic reactions.

In the course of studies of the biosynthesis of the antibiotic azomycin (2-nitroimidazole), it has been shown that the antibiotic-

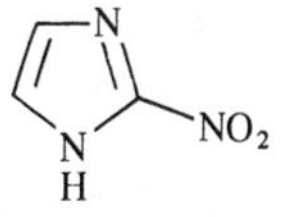

Azomycin

producing microorganism (*Streptomyces* species strain LE/3342) is able to oxidize several aminoimidazoles to nitroimidazoles.[5,6]   2-Amino-imidazole itself is converted into azomycin, apparently in about 50% yield. (The yield cannot be determined accurately, owing to the stimulation of *de novo* production of the antibiotic by the 2-aminoimidazole.) It has been speculated, without any experimental support, that an aminoimidazole riboside is first formed, that this undergoes oxidation to the nitro compound, and that the ribose moiety then is cleaved off.   Results for a series of alkyl-substituted aminoimidazoles are indicated in Table 1, the product in each case being the corresponding nitroimidazole.

Table 1. Oxidation of Aminoimidazoles to Nitroimidazoles

Substrate	Yield, %
2-Amino-4(5)-methylimidazole	25
2-Amino-4(5)-ethylimidazole	36
2-Amino-4(5)-propylimidazole	29
2-Amino-4(5)-isopropylimidazole	25
2-Amino-4(5)-butylimidazole	3
2-Amino-4,5-dimethylimidazole	8
2-Amino-1-methylimidazole	-
2-Amino-4(5)-phenylimidazole	-

Aminoimidazoles represent a structure type that is somewhat more amenable than average to chemical oxidative transformation of the amino to the nitro group. Azomycin has been synthesized (nonmicrobiologically) in 48.5% yield by chemical oxidation of 2-aminoimidazole, but when the 4(5)-methyl compound is oxidized in the same way, the product yield is only 4.5%, in contrast to the 25% obtained with the microorganism.[7] On the other hand, chemical oxidation of 1-methyl-2-aminoimidazole succeeds where the microbiological method fails.

Studies of the biosynthesis of the antibiotic chloramphenicol by a *Streptomyces* species have shown that the nitro group arises from an amino precursor,[8] and it seems reasonable that several other natural-

$$
\begin{array}{ccc}
\text{O} \quad \text{CH}_2\text{OH} & & \text{O} \quad \text{CH}_2\text{OH} \\
\| \quad | & & \| \quad | \\
\text{Cl}_2\text{CHCNHCH} & & \text{Cl}_2\text{CHCNHCH} \\
| & & | \\
\text{HCOH} & \xrightarrow{\textit{Streptomyces sp.}} & \text{HCOH} \\
\end{array}
$$

Chloramphenicol

ly occurring nitro compounds may also arise biosynthetically *via* oxidation of an amino group. In the case of 3-nitropropionic acid (hiptagenic acid, bovinocidin), evidence has been presented[9-15] that its

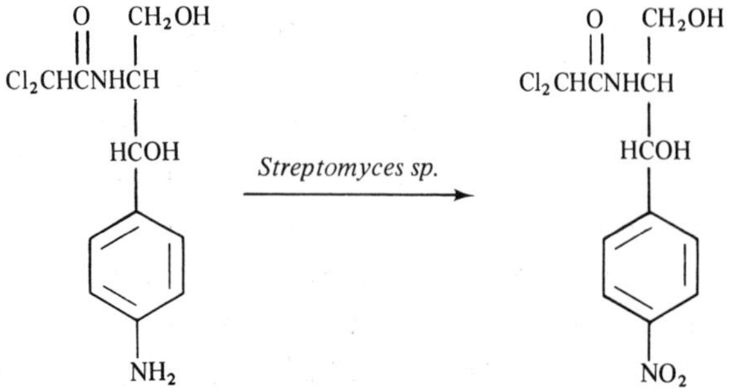

$$
\begin{array}{c}
\text{COOH} \\
| \\
\text{H}_2\text{N}-\text{CHCH}_2\text{COOH} \longrightarrow \text{O}_2\text{NCH}=\text{CHCOOH} \longrightarrow \text{O}_2\text{N}-\text{CH}_2\text{CH}_2\text{COOH}
\end{array}
$$

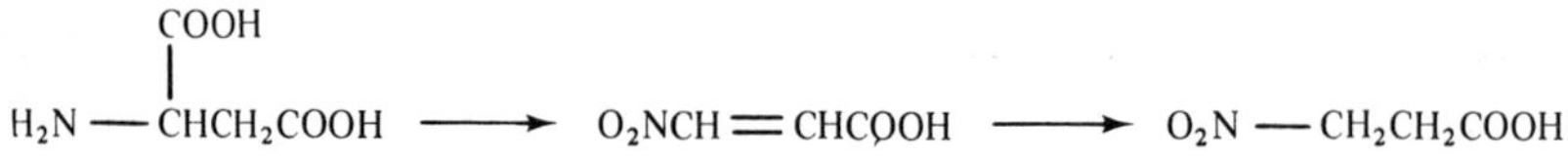

Aspartic acid   3-Nitropropionic acid

biosynthesis occurs from aspartic acid *via* β-nitroacrylic acid, but the problem is not conclusively resolved.

Pyrrolnitrin apparently is formed when the microbiological cleavage of tryptophane by *Pseudomonas aureofaciens* is followed by a variety of reactions including amino group oxidation and ring halo-

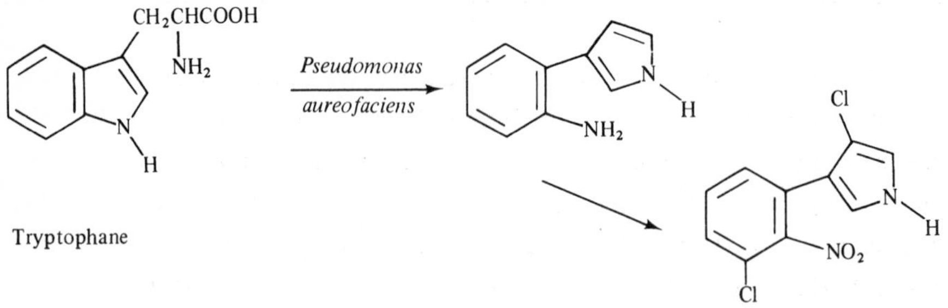

Tryptophane

Pyrrolnitrin

genation.[16,17] Interestingly, tryptophanes substituted at the 5,6- or 7-position give rise to correspondingly substituted pyrrolnitrins,[18] thus suggesting that this complex series of transformations may have an important synthetic utility.

The aristolochic acids,[19] 2-hydroxy-3-nitrophenylacetic acid,[20] and 1-amino-2-nitrocyclopentanecarboxylic acid,[21] as well as those

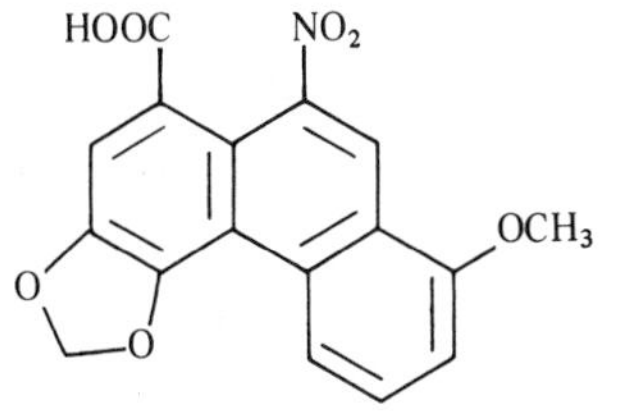

Aristolochic acid I

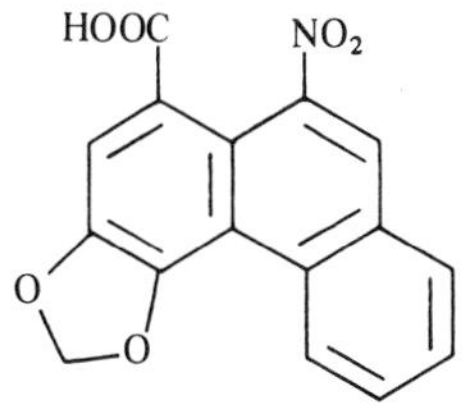

Aristolochic acid II

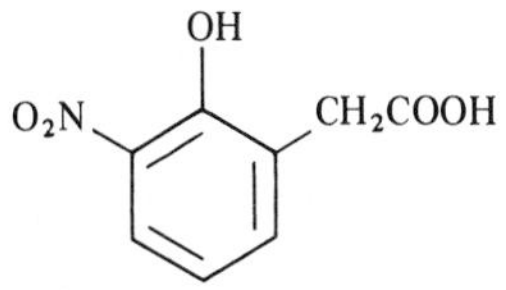

2-Hydroxy-3-nitro-
phenylacetic acid

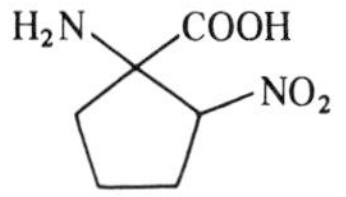

1-Amino-2-nitro-
cyclopentanecarboxylic acid

polypeptide antibiotics (rufomycin,[22] ilamycin[23]) that contain nitro-tyrosine, remain to be explored with regard to the possibility that they, too, arise by the oxidative transformation of amino to nitro.

REFERENCES

1. S. Kawai, K. Kobayashi, T. Oshima, and F. Egami, *Arch. Biochem. Biophys. 112*, 537 (1965).

2. S. Kawai, T. Oshima, and F. Egami, *Biochim. Biophys. Acta 97*, 391 (1965).

3. S. Kawai, T. Oshima, and F. Egami, *Biochim. Biophys. Acta 104*, 316 (1965).

4. A. Ballio, H. Bertholdt, A. Carilli, E. B. Chain, V. DiVittorio, A. Tonolo, and L. Vero-Barcellona, *Proc. Roy. Soc. (London) 158B*, 43 (1963).

5. G. C. Lancini, E. Lazzari, and G. Sartori, *J. Antibiot. 21*, 387 (1968).

6. G. C. Lancini, D. Kluepfel, E. Lazzari, and G. Sartori, *Biochim. Biophys. Acta 130*, 37 (1966).

7. A. G. Beaman, W. Tantz, T. Gabriel, O. Keller, V. Toome, and R. Duschinsky, *Antimicrob. Ag. Chemother.*, 469 (1965).

8. R. C. McGrath, L. C. Vining, F. Sala, and D. W. S. Westlake, *Can. J. Biochem. 46*, 587 (1968).

9. P. D. Shaw and D. Gottlieb, in *Biogenesis of Antibiotic Substances* (Z. Vanek and Z. Hostalek, eds.), Academic, New York, 1965, pp. 261-269.

10. P. D. Shaw and N. Wang, *J. Bacteriol. 88*, 1629 (1964).

11. S. Gatenbeck and B. Forsgren, *Acta Chem. Scand. 18*, 1750 (1964).

12. J. H. Birkenshaw and A. M. L. Dryland, *Biochem. J. 93*, 478 (1964).

13. J. W. Hylin and H. Matsumoto, *Arch. Biochem. Biophys. 93*, 542 (1960).

14. A. J. Birch, B. J. McLoughlin, H. Smith, and J. Winter, *Chem. Ind.*, 840 (1960).

15. P. D. Shaw and J. A. McCloskey, *Biochemistry 6*, 2247 (1967).

16. D. H. Lively, M. Gorman, M. E. Haney, and J. A. Ware *Antimicrob. Ag. Chemother.*, 462 (1966).

17. R. P. Elander, R. L. Hamill, M. Gorman, and J. Mabe, *Abstr. Amer. Chem. Soc. 154*, Q42 (1967).

18. R. L. Hamill, R. P. Elander, M. Gorman, and D. H. Lively, *Abstr. Amer. Chem. Soc. 154*, Q41 (1967).

19. a. M. Pailer, L. Belohlav, and E. Simonitsch, *Monatsh. Chem. 87*, 249 (1956).

    b. M. Pailer and A. Schleppnik, *Monatsh. Chem. 88*, 367 (1957).

20. M. Isono, *J. Agr. Chem. Soc. Jap. 28*, 562 (1954).

21. P. W. Brian, G. W. Elson, H. G. Hemming, and M. Radley, *Nature 207*, 998 (1965).

22. T. Takita, K. Oki, Y. Okami, K. Maeda, and H. Umezawa, *J. Antibiot. Ser. A 15*, 46 (1962).

23. T. Takita, H. Naganawa, K. Maeda, and H. Umezawa, *J. Antibiot. Ser. A 17*, 129 (1964).

Chapter 11

OXIDATION OF AMINO TO HYDROXL

Depending upon the substrate used, as well as other factors, the microbiologically mediated replacement of an amino group by oxygen can result in a phenol, an alcohol, an aldehyde, or an acid. The biochemical mechanisms and pathways for these transformations are several and relatively unrelated. There is no shortage of descriptions of these processes in the biochemical literature, where they appear under the headings "transamination," "oxidative deamination," "deamination," "amino acid oxidases," and "dehydrases." The reader is referred to any good biochemistry textbook for an introductory overview and for citations of additional source material.

The following discussion reflects the paucity of examples that could be considered useful for synthetic work, and is not to be construed as an exhaustive or balanced discussion of what has long been a purely biochemical research area.

Aromatic and Heteroaromatic Amino. The replacement of an aromatic amino group by a hydroxyl has unfortunately been designated "deamination" by biochemists. Mechanistically, the reaction (at least in the case of pyrimidines and purines) is hydrolytic, but the net result may be viewed formally as an oxidation.

Although the transformation of adenosine to inosine by microbial enzymes has been known and studied for decades, the intent has rarely, if ever, been to obtain useful quantities of the product since this is usually more readily accomplished by chemical means. The examples are cited only to suggest that in the case of a given molecule too sensitive to permit chemical "deamination," the microbiological method may merit careful consideration.

217

The adenosine deaminase enzyme from *Aspergillus oryzae* that converts adenosine to inosine also brings about the conversion of the "unnatural" substrate 3-isoadenosine to 3-isoinosine.[1] The same overall reaction can be carried out chemically, but the lability of the

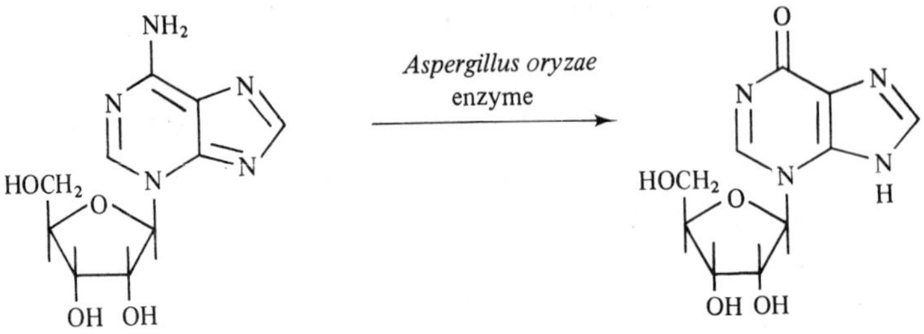

3-Isoadenosine   3-Isoinosine

glycosidic bond restricts the applicability of some methods. The enzymatic method is thus potentially attractive although it has not actually been used preparatively.

A variety of microorganisms able to attach ribose to the adenosine isomer 4-amino-1H-pyrazolo[3,4-d]pyrimidine, but not to 4-hydroxy-1H-pyrazolo[3,4-d]pyrimidine also give the riboside of the latter compound (up to 35% yield).[2] This must arise by the sequence shown.

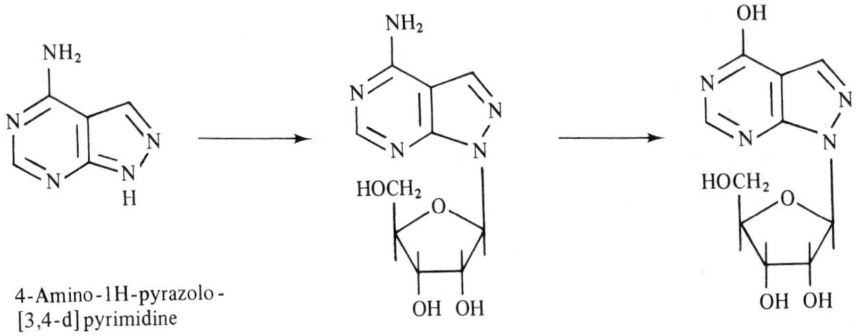

4-Amino-1H-pyrazolo-
[3,4-d]pyrimidine

The interesting nucleoside cordycepin (3'-deoxyadenosine) is oxidized to 3'-deoxyinosine in 44% yield by *Escherichia coli*[3] and is

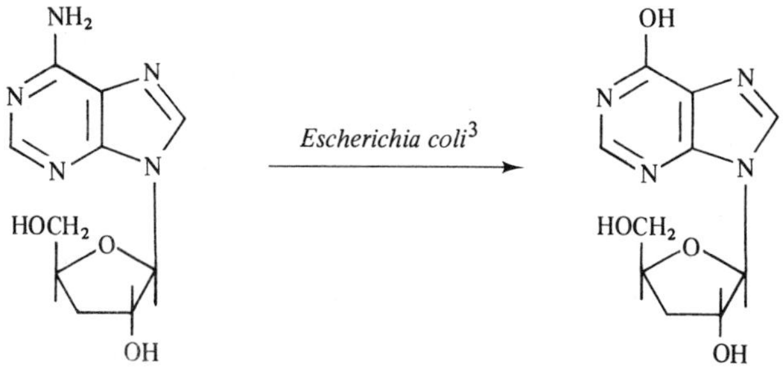

Cordycepin

deaminated by a purified adenosine deaminase from *Aspergillus ory-*
zae.[4]   This deamination may occur chemically as well, but the conditions are not clearly defined.[3,4]   The antibiotic formycin (7-amino-
3β-D-ribofuranosyl-1H-pyrazolo[4,3-d]pyrimidine) is converted to for-

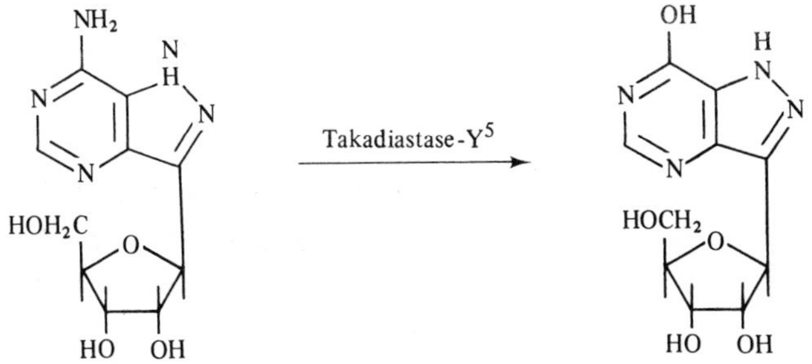

Formycin   Formycin B

mycin B by a variety of organisms or their enzymes.[5] When commercial
Takadiastase-Y (from *A. oryzae*) is used, the yield of crystalline
formycin B is 74%.

The purified adenosine deaminase from *A. oryzae* deaminates a
variety of other nucleosides and nucleotides, but some molecules do
not undergo reaction. It has been suggested[4] that unusual structural
features, such as the deazapurine moiety, as in tubercidin, or the

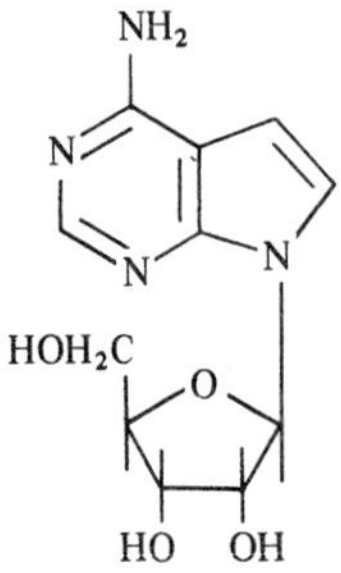

Tubercidin 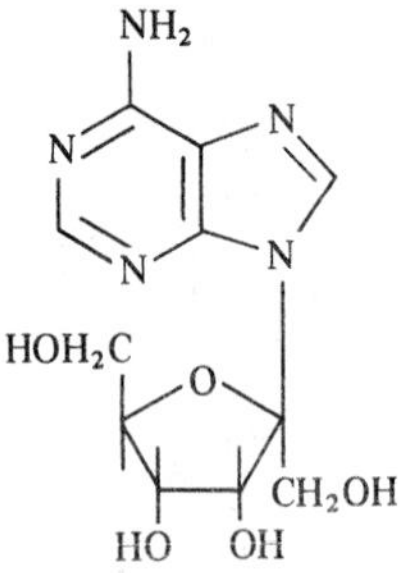   Psicofuranine

carbinol group at the 1'-position, as in psicofuranine, are responsible for preventing deamination of these resistant molecules. The
ability of other microorganisms, particularly actinomycetes and bacteria, to carry out these deaminations has been examined.[6]

Pyrithiamine, a thiamine-antagonist, undergoes oxidation to oxypyrithiamine by a *Staphylococcus aureus* mutant or by an enzyme preparation derived from this organism.[7]

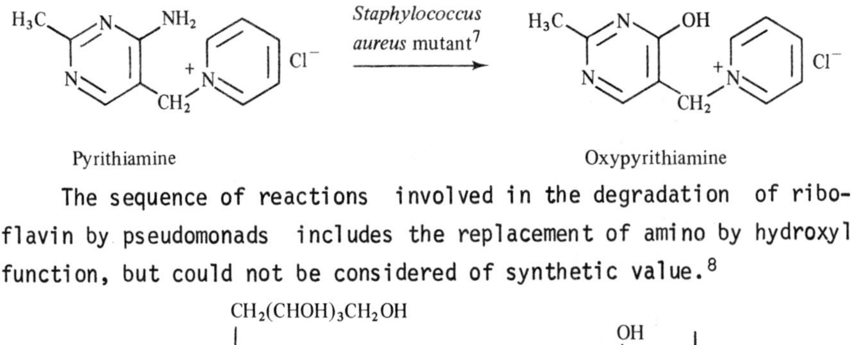

Pyrithiamine   Oxypyrithiamine

The sequence of reactions involved in the degradation of riboflavin by pseudomonads includes the replacement of amino by hydroxyl
function, but could not be considered of synthetic value.[8]

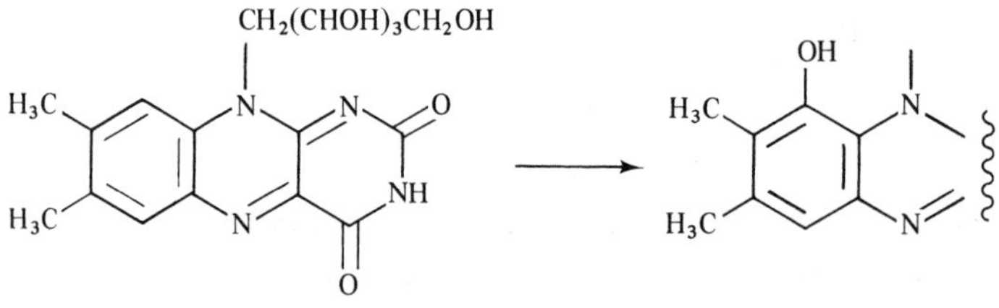

Riboflavin

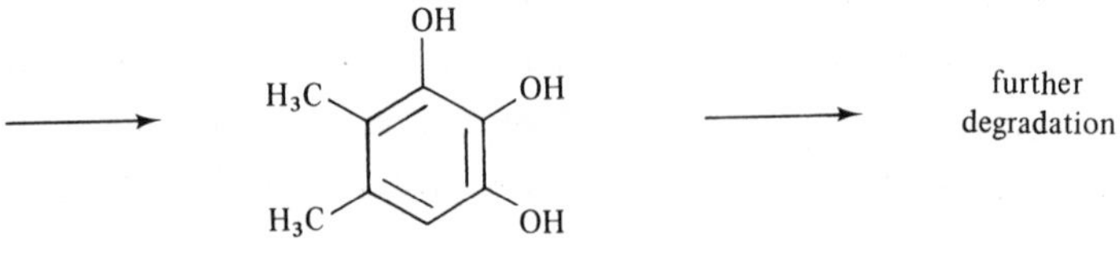

further
degradation

Similarly, this reaction type is involved in the degradation of the herbicide 3,5-dinitro-*o*-cresol (DNOC) by a pseudomonad,[9] and in the degradation of 5-hydroxyanthranilic acid by *Nocardia opaca*,[10] but

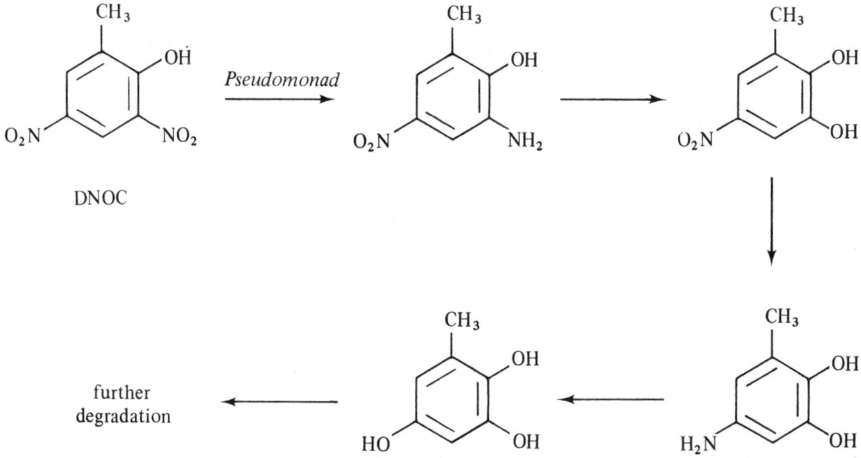

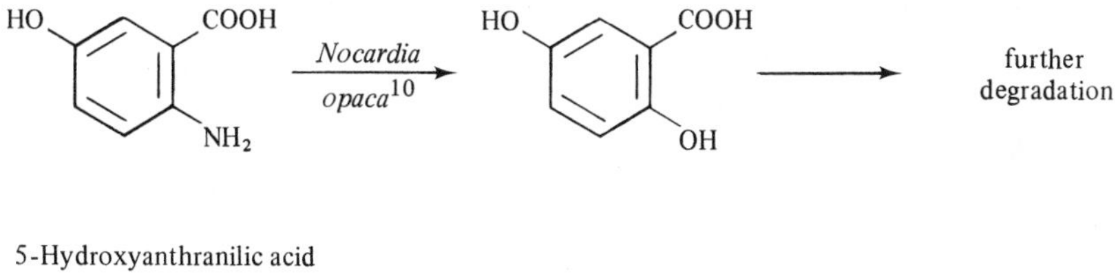

DNOC

5-Hydroxyanthranilic acid

again, no synthetic value can be attributed to these sequences. However, these examples may suggest the applicability of this reaction type to other substrates under circumstances where synthetic value could be established.

Aliphatic or Araliphatic Amino. A large number of primary amines and diamines are oxidized by an amine dehydrogenase from a *Pseudomonas* species to give the corresponding aldehydes (not isolated).[11] The following substrates are oxidized at rates sufficiently high to suggest some potential utility to the chemist: methylamine, ethyl-

amine, propylamine, butylamine, pentylamine, hexylamine, 1,2-diamino-
ethane, 1,3-diaminopropane, 1,4-diaminobutane (putrescine), 1,5-di-
aminopentane, 1,6-diaminohexane, spermine, spermidine, ethanolamine,
histamine, and L-ornithine.   In the case of the diamines, it is not
clear whether both amino groups are oxidized, but the abundance of

$$H_2NCH_2CH_2CH_2CH_2NH_2 \xrightarrow{\text{\textit{Pseudomonas} species}} H_2NCH_2CH_2CH_2CHO$$
Putrescine

other work[12-14] with putrescine suggests that an aminoaldehyde is
formed.

The *Pseudomonas* enzyme fails to oxidize secondary, tertiary, or
aromatic amines, and a number of primary amines are oxidized only at
exceedingly slow rates. Interestingly, 1,2-diaminopropane is not oxidized. These facts suggest another utility to the chemist, namely to
selectively alter individual components of a mixture of isomers or
homologs, in order to simplify a complex mixture and thereby perhaps
to facilitate the isolation of a desired component.

The metabolism of tryptamine by *Hygrophorus conicus* involves
more than one oxidative reaction type, but the conversion of the primary amino group to carboxyl is included.[15]

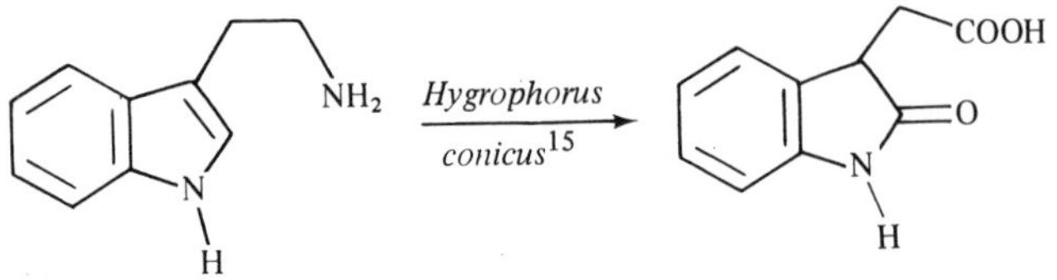

Tryptamine   (43%)

The conversion of histamine to imidazoleacetic acid in 60-75%
(assay) yield by an *Achromobacter* species (that converts isoamylamine

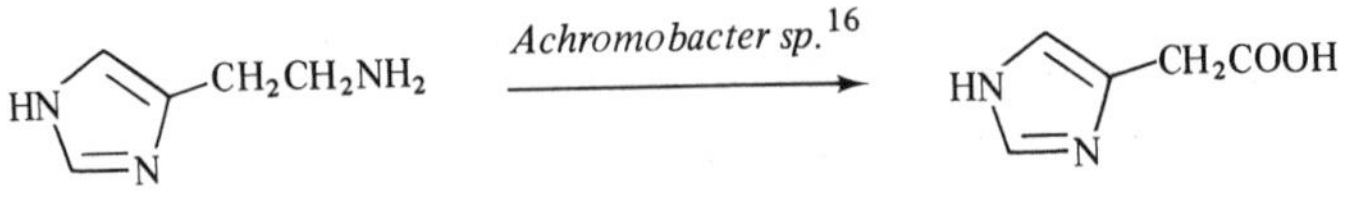

Histamine

to isovaleric acid in 63% yield) is arrested at this oxidation stage by the inclusion of arsenite in the medium.[16]

Tyramine oxidase, the enzyme from *Sarcina lutea* responsible for the oxidation of tyramine to p-hydroxyphenylacetaldehyde, has been isolated and crystallized.[17] This enzyme also oxidizes dopamine, but

HO—⟨⟩—$CH_2CH_2NH_2$ $\xrightarrow[lutea]{Sarcina}$ HO—⟨⟩—CH_2CHO

(HO)   (HO)

Tyramine (Dopamine)

not a variety of other monoamines. Although the existing data[18,19] refer to assays only, the availability of a pure enzyme would appear to enhance the potential synthetic utility of these reactions.

The product S-ethylmercaptolactic acid has been isolated from the oxidation of S-ethyl-L-cysteine by *Saccharomyces cerevisiae*, and

CH_3CH_2S—CH_2—CH—COOH $\xrightarrow[cerevisiae^{20}]{Saccharomyces}$ CH_3CH_2—S—CH_2—CH—COOH

| |
NH$_2$ OH

S-Ethyl-L-cysteine

the organism is reported to oxidize S-methyl-L-cysteine, L-methionine, and L-ethionine as well.[20]

Arthrobacter globiformis degrades aminoacetone completely, for use as a nutrient. If the washed cells are incubated with this substrate in the presence of iodoacetate (which blocks further degradation), a 70% conversion (assay) to pyruvaldehyde occurs.[21]

$$CH_3\overset{\overset{O}{\|}}{C}—CH_2NH_2 \xrightarrow[globiformis^{21}]{Arthrobacter} CH_3\overset{\overset{O}{\|}}{C}—CHO$$

Aminoacetone

The preparation of ^{14}C-labeled kynurenic acid from ^{14}C-L-tryptophane is a useful biosynthetic process, in view of the specificity of isotope labeling that it affords. The reaction sequence involves at one point the oxidation of an α-amino acid to an (unisolated) α-keto-acid, which spontaneously reacts further.[22]

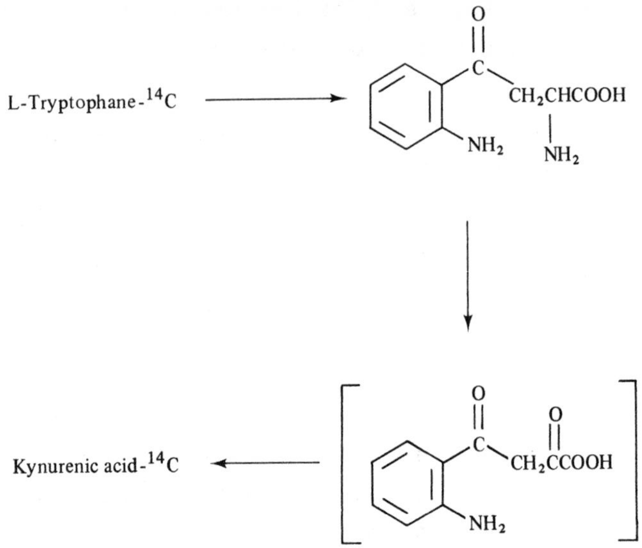

The microbiological degradation of nicotine and of nornicotine involves __two__ amine oxidations, although the two reactions have seldom been carried out individually. The degradation of nicotine by *Achrom-obacter nicotinophagum* (which attacks only this substrate and not a

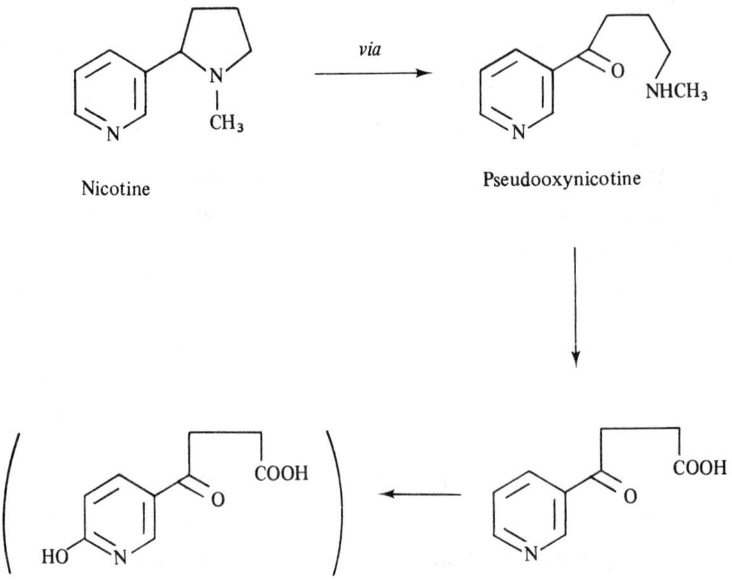

host of closely related ones) gives 4-(3-pyridyl)-4-oxobutyric acid
in 32% yield,[23] while the same reaction with a *Pseudomonas* strain[24]
gives a 20% yield.   (The *Achromobacter nicotinaphagum* also gives 4-
[6-hydroxy-3-pyridyl]-4-oxobutyric acid from nicotine).   In the case
of nornicotine, using a *Pseudomonas* species, the yield of 6-hydroxy-
nicotinoylpropionic acid is 38%.[26]

   Oxidation of the aminosteroid conessine by *Gloeosporium cyclaminis* or by *Hypomyces haematococcus* results in the replacement of the

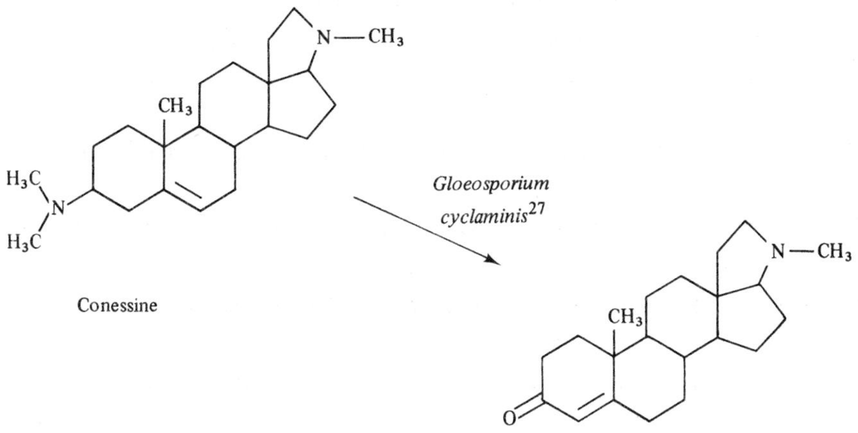

Conessine

*Gloeosporium
cyclaminis*[27]

3-dimethylamino-Δ^5-structure by a 3-keto-Δ^4-structure in 36% yield
(*Gloeosporium cyclaminis*).[27]

REFERENCES

1. R. Wolfenden, T. K. Sharpless, I. S. Ragade, and N. J. Leonard,
    J. Amer. Chem. Soc. 88, 185 (1966).

2. P. J. Curtis and D. R. Thomas, *Biochem. J. 82*, 381 (1962).

3. M. E. Herr and H. C. Murray, unpublished results.

4. Y. Fukagawa, T. Sawa, T. Takeuchi, and H. Umezawa, *J. Antibiot.
    Ser. A 18*, 191 (1965)

5. H. Umezawa, T. Sawa, Y. Fukagawa, G. Koyama, M. Murase, M.
    Hamada, and T. Takeuchi, *J. Antibiotic. Ser. A 18*, 178 (1965).

6. T. Sawa, Y. Fukagawa, I. Homma, T. Takeuchi, and H. Umezawa, *J.
    Antibiot. Ser. A 20*, 317 (1967).

7. A. K. Sinha and G. C. Chatterjee, *Biochem. J. 107*, 165 (1968).

8. P. Z. Smyrniotis, H. T. Miles, and E. R. Stadtman, *J. Amer. Chem. Soc. 80*, 2541 (1958).

9. M. S. Tewfik and W. C. Evans, *Biochem. J. 99*, 31P (1966).

10. R. B. Cain, *Antonie van Leeuwenhoek 34*, 417 (1968).

11. R. R. Eady and P. J. Large, *Biochem. J. 106*, 245 (1968).

12. R. Michaels and K-H. Kim, *Biochim. Biophys. Acta 115*, 59 (1966).

13. K-H. Kim and T. T. Tchen, *Biochem. Biophys. Res. Commun. 9*, 99 (1962).

14. K-H. Kim, *J. Biol. Chem. 239*, 783 (1964).

15. D. J. Siehr, *J. Amer. Chem. Soc. 83*, 2401 (1961).

16. K. Satake, S. Ando, and H. Fujita, *J. Biochemistry (Tokyo) 40*, 299 (1953).

17. H. Yamada, T. Uwajima, H. Kumagai, M. Watanabe, and K. Ogata, *Biochem. Biophys. Res. Commun. 27*, 350 (1967).

18. H. Yamada, T. Uwajima, H. Kumagai, M. Watanabe, and K. Ogata, *Agr. Biol. Chem. 31*, 890 (1967).

19. H. Yamada, H. Kumagai, T. Uwajima, and K. Ogata, *Agr. Biol. Chem. 31*, 897 (1967).

20. G. A. Maw and C. M. Coyne, *Arch. Biochem. Biophys. 117*, 499 (1966).

21. M. L. Green and J. B. Lewis, *Biochem. J. 106*, 267 (1968).

22. T. Tanaka and E. J. Behrman, *Anal. Biochem. 1*, 181 (1960).

23. J. W. Hylin, *Arch. Biochem. Biophys. 83*, 528 (1959).

24. E. Wada and K. Yamasaki, *J. Amer. Chem. Soc. 76*, 155 (1954).

25. J. W. Hylin, *J. Bacteriol. 76*, 36 (1958).

26. E. Wada, *Arch. Biochem. Biophys. 64*, 244 (1956).

27. J. deFlines, A. F. Marx, E. P. van der Waard, and D. van der Sijde, *Tetrahedron Lett.*, 1256 (1962).

General Reference

E. F. Gale, *Biochem. J. 36*, 64 (1942).

Chapter 12

SULFUR OXYGENATION

The microbial oxygenation of thioethers to sulfoxides and sulfones has been studied sparingly. With several important exceptions, these reactions appear to proceed in a manner similar to their chemical counterparts in the limited number of cases that have been examined. Thus, microbial oxidation of sulfur to the sulfoxide level in some cases leads to small amounts of sulfone as well, just as is observed in many of the chemical reactions used to carry out this transformation. The amounts of contaminating sulfone are small, however, and this problem does not appear serious in most cases. Yields of sulfoxides have been low in comparison to yields realized by chemical methods; these undoubtedly could be improved in most cases by studying the effect of changes in the many variables affecting a microbial reaction. There are, however, two distinct benefits apparent at the present time to be gained from use of the "microbial reagents" in the oxidation of thioethers to sulfoxides. The first of these is that of selective oxygenation in the presence of other labile functional groups, typical in general of microbial reactions. The second, and perhaps most important advantage to be gained by these methods, is in the stereospecific oxygenation of sulfur, leading in the cases studied to optical purities ranging from 5 to 94%. This valuable feature of these reactions was apparent in the first report of a microbial oxidation of sulfur, the oxygenation of biotin to biotin (-)-sulfoxide by *Aspergillus niger*[1-3] in which only the (-)-form of the product was isolated. However, the yield of this material was minuscule, 2 parts per million occurring in the dry material of the filtrate of the mold.

The closely related antibiotics lincomycin and clindamycin are both oxidized to the corresponding sulfoxides, the latter in about 20% yield by *Streptomyces armentosus*.[4,5]

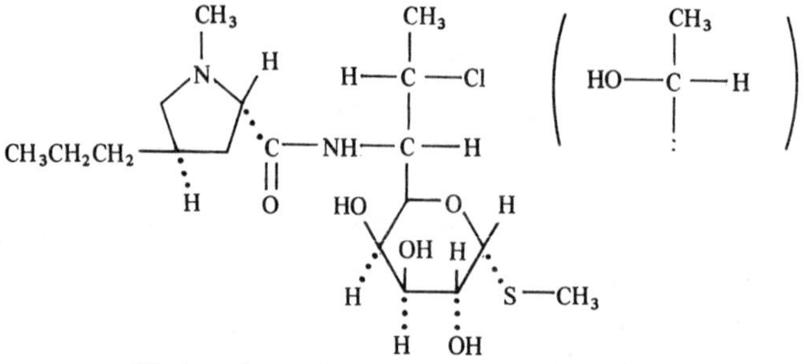

Clindamycin   (Lincomycin)

Steroidal methylthioethers undergo microbial oxygenation on sulfur. Fermentation of 17β-acetoxy-7α-methylthioandrost-4-en-3-one with
Calonectria decora gives 17β-hydroxy-7α-methylsulfinylandrost-4-en-3-
one (34%).[6] Comparison of the 17β-acetoxy derivative with both enan-

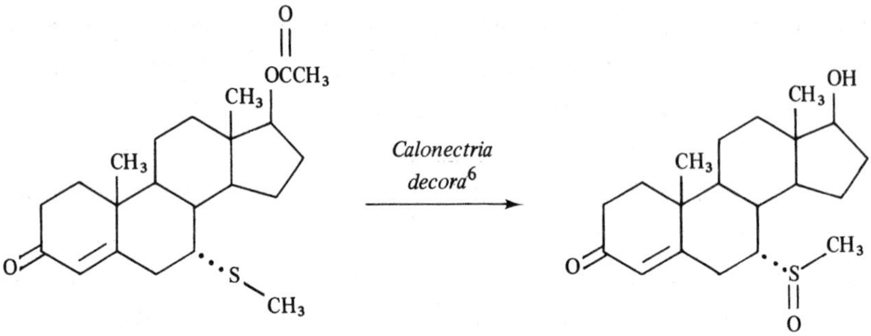

17β-Acetoxy-7α-methylthio-
androst4-en-3-one

tiomeric sulfoxides prepared by chemical synthesis leads to the conclusion that the microbial product is formed stereospecifically in
this reaction. Fermentation of either 17α- or 17β-methylthioandrost-
4-en-3-one with *Rhizopus stolonifer* gives the corresponding 17α- or
17β-methylsulfinylandrost-4-en-3-ones.[7] In the case of the 17α-

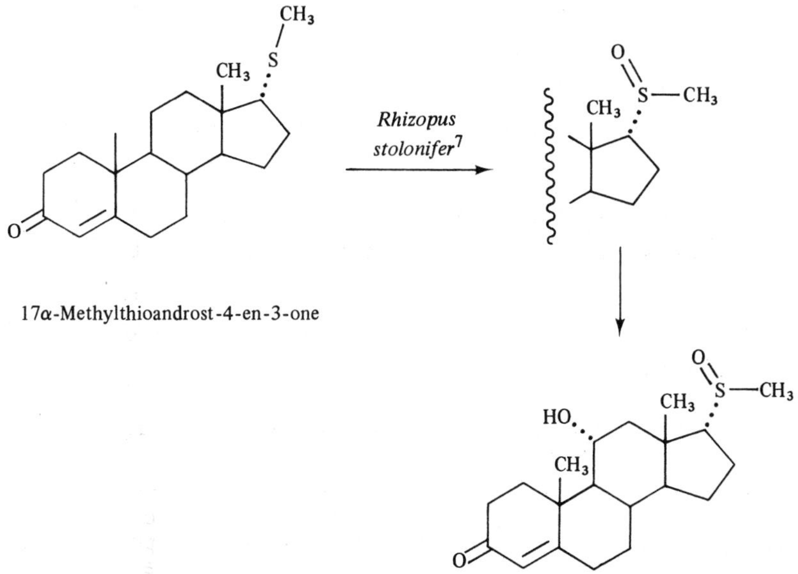

17α-Methylthioandrost-4-en-3-one

methylthioether substrate, isomeric (and presumably epimeric) sulfoxides are obtained in a ratio of 7.5 to 1, of which one undergoes further reaction at C-11, giving 11α-hydroxy-17α-methylsulfinylandrost-4-en-3-one. The stereospecificity observed in these reactions led Dodson, Newman, and Tsuchiya[8] to investigate the possibility of preparing optically active sulfoxides from optically inactive, unsymmetrical sulfides. In this they were successful, finding that phenyl benzyl sulfide is oxygenated by *Aspergillus niger* to a mixture of

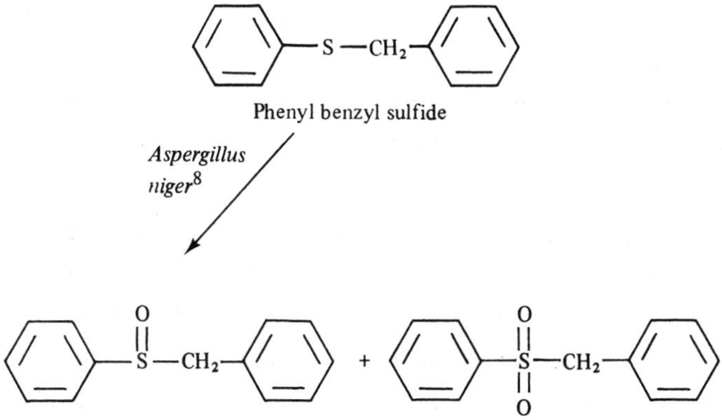

Phenyl benzyl sulfide

phenyl benzyl sulfoxide (23%) and phenyl benzyl sulfone (9%). The sulfoxide is optically active, with an optical purity of 18%. A second sulfide, methyl β-naphthyl sulfide, is also oxygenated with some stereoselectivity to the sulfoxide (together with some sulfone) by *A. niger*.[8]

The above observation of stereoselective oxygenation of sulfides has been extended to a series of compounds having aromatic, benzylic and aliphatic substituents on sulfur.[9,10] The sulfoxides, obtained in yields up to 65% from this series of substrates, have optical purities ranging from 5 to 94% (Table 1).

Table 1. Oxidation of Sulfides to Sulfoxides with *Aspergillus niger*

Sulfide (A-S-B)	Yield, % (optical purity)[a] from growing culture	Yield, % (optical purity) from acetone powder
p-$CH_3C_6H_4$-S-$tert$-Bu	24 (94)	25 (100)
p-$CH_3C_6H_4$-S-CH_2-p-$C_6H_4CH_3$	11 (88)	4 (94)
$C_6H_5CH_2$S-$tert$-Bu	24 (77)	65 (91)
p-$CH_3C_6H_4CH_2$-S-$tert$-Bu	--	49 (78)
p-$CH_3C_6H_4$-S-$CH_2C_6H_5$	19 (82)	3 (81)
p-$CH_3C_6H_4$-S-i-Pr	--	12 (69)
p-$CH_3C_6H_4$-S-CH_3	48 (32)	0
p-$CH_3C_6H_4$-S-n-Bu	--	14 (33)
$C_6H_5CH_2$-S-CH_3	18 (46)	0
$C_6H_5CH_2$-S-C_6H_5	9 (5)	14 (27)
p-$tert$-BuC_6H_4$-S-$CH_2C_6H_5$	0	8 (13)

[a]Based on highest known rotations.

From the knowledge of the absolute configurations and optical purities of the products, it is possible to attempt a rough correlation between the stereoselectivity of the oxidation and the substrate structure. Additional examples will be needed to refine this correlation further, but it now appears that the stereoselectivity is such that A will generally have more aromatic character than B in the sulfoxide products (*i.e.*, A = *p*-methylphenyl → benzyl → *tert.*-butyl = B).[10]

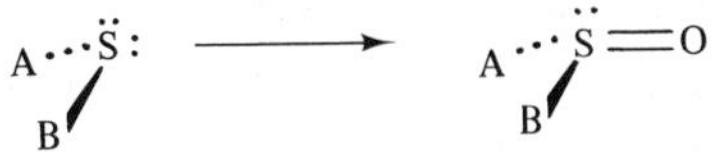

$$A \cdots \ddot{S} : \quad\longrightarrow\quad A \cdots \ddot{S} = O$$

(with B attached below each S)

The above oxygenations of sulfides to sulfoxides may also be carried out with acetone powders of *A. niger*.[10]   (The preparation of an acetone powder of a culture represents the first step in one common approach to the isolation and purification of enzymes.   In this case the oxygenating enzyme is being sought. The cells of the microorganism are mechanically ruptured and the portion that is insoluble in cold acetone is obtained in the form of a powder following removal of the acetone. The substrate is then shaken with the acetone powder in water or a suitable buffer solution.)   The sulfides are generally oxygenated to sulfoxides in yields roughly the same as those obtained by the use of whole cultures.   In addition the optical purity of the sulfoxides obtained from the acetone powder procedure have higher degrees of optical purity and less contaminating sulfone.

Since stereoselectivity also occurs in the oxygenation of sulfoxides to sulfones by *A. niger*, it may be inferred that the stereoselectivity of the organism for the oxygenation of <u>sulfide</u> to <u>sulfoxide</u> is not the only cause of the optical activity found in the sulfoxides.[11]

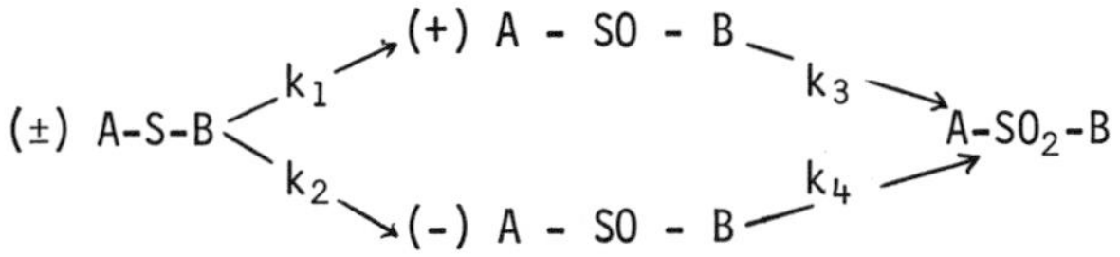

$$
(\pm)\ A\text{-}S\text{-}B
\begin{array}{c}
\xrightarrow{k_1} (+)\ A - SO - B \xrightarrow{k_3} \\
\searrow_{k_2} (-)\ A - SO - B \xrightarrow{k_4}
\end{array}
A\text{-}SO_2\text{-}B
$$

In other words, in the above scheme, the rate constants k_3 and k_4 have an effect on the observed optical activity of the product sulfoxide as do the rate constants k_1 and k_2.

REFERENCES

1. L. D. Wright and E. L. Cresson, *J. Amer. Chem. Soc. 76*, 4156 (1954).
2. L. D. Wright, E. L. Cresson, J. Valiant, D. E. Wolf, and K. Folkers, *J. Amer. Chem. Soc. 76*, 4160 (1954).

3. L. D. Wright, E. L. Cresson, J. Valiant, D. E. Wolf, and K. Folkers, *J. Amer. Chem. Soc.* *76*, 4163 (1954).

4. A. D. Argoudelis and D. J. Mason, *J. Antibiotics* *22*, 289 (1969).

5. A. D. Argoudelis, J. H. Coats, D. J. Mason, and O. K. Sebek, *J. Antibiotics* *22*, 309 (1969).

6. C. E. Holmlund, K. J. Sax, B. E. Nielson, R. E. Hartman, R. H. Evans, Jr., and R. H. Blank, *J. Org. Chem.* *27*, 1468 (1962).

7. R. M. Dodson and P. B. Sollman, *U. S. Patent* 2,999,101 (September 5, 1961).

8. R. M. Dodson, N. Newman, and H. M. Tsuchiya, *J. Org. Chem.* *27*, 2707 (1962).

9. B. J. Auret, D. R. Boyd, and H. B. Henbest, *Chem. Commun.*, 66 (1966).

10. B. J. Auret, D. R. Boyd, H. B. Henbest, and S. Ross, *J. Chem. Soc. (C)*, 2371 (1968).

11. B. J. Auret, D. R. Boyd, and H. B. Henbest, *J. Chem. Soc. (C)*, 2374 (1968).

Chapter 13

OTHER MICROBIAL OXIDATIONS

After all of the major manifestations of chemical oxidations
with microorganisms have been more or less neatly cataloged, there
remains an embarrassing residuum of miscellaneous reactions that must
somehow be dealt with. Some of these reactions represent composite
sequences of several of the reaction types discussed earlier. Others,
such as oxidative decarboxylation or N-oxide formation, seem to be
representable as a single step but are not numerous enough to warrant
separate chapters. Still others, such as N-demethylation, although
formally representable as reductions, undoubtedly proceed by an oxidative pathway. A number of these reactions that appear to have some
potential synthetic utility are presented here. The reader is cau-
tioned that some examples, particularly of composite reactions, have
already found their way into earlier discussions.

Oxidative Decarboxylation. The conversion of L-lysine to 5-
aminovaleric acid by *Pseudomonas fluorescens*[1,2] in 65% yield (assay)
involves an oxygenative decarboxylation of the substrate, giving an
intermediate aminoamide that undergoes further hydrolysis to a δ-

$$
\begin{array}{ccc}
CH_3NH_2 & CH_2NH_2 & \\
| & | & CH_2NH_2 \\
(CH_2)_3 \xrightarrow[\textit{fluorescens}]{\textit{Pseudomonas}} (CH_2)_3 \longrightarrow | \\
| & | & (CH_2)_3 \\
CHNH_2 & C\text{-}NH_2 & | \\
| & \| & COOH \\
COOH & O & \\
\end{array}
$$

L-Lysine

aminoacid. The oxygenase responsible for the first step may be isolated and used to obtain the amide.[1]

Analogous oxidative decarboxylations of aromatic substrates are
found in the conversion of carboxyl groups of salicylic acid and certain related compounds to hydroxyl groups.[3-6]

233

Salicylic acid

An unusual pathway to the formation of p-hydroxyaniline by *Myco-bacterium tuberculosis* (and several other *Mycobacterium* species) in-volves the conversion of p-aminobenzyl alcohol to p-hydroxyaniline.[7]

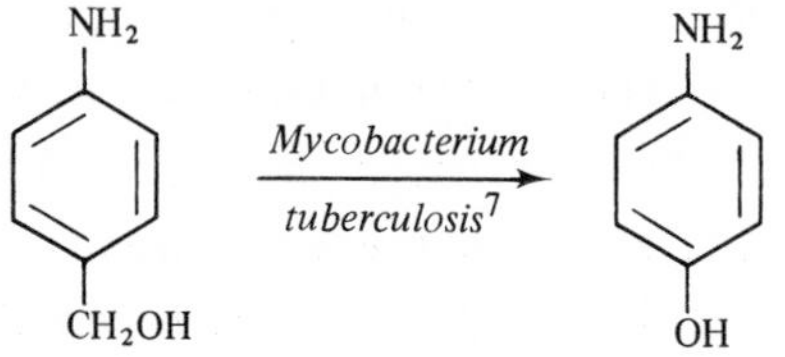

p-Aminobenzyl alcohol

The p-aminobenzyl alcohol is derived from p-aminobenzoic acid. The suggestion has been made that this alcohol is an intermediate in oxy-genations of aniline.[7]

N-Dealkylation. A frequently encountered microbial reaction, which certainly involves oxidative reactions in some (and probably all) cases, is the cleavage of alkyl amines. The cleavage of nicotine and nicotine derivatives by *Arthrobacter oxydans*[8-10] or by a *Pseudo-*

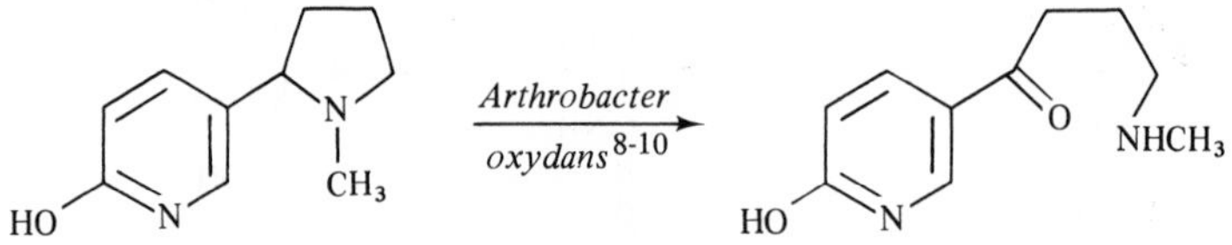

6-Hydroxynicotine

monas species[11,12] results in the isolation of keto-amines, which can be seen to arise from an oxidative cleavage of the pyrrolidine ring. In the former case, a cell-free enzyme preparation has been ob-tained.[10]

Other examples of N-dealkylations are found in several N-demethylation reactions in which the oxidation state of the alkyl group is not easily determined. Several examples illustrate this point. Tropine is demethylated by *Arthrobacter atropini*, giving tropigenin.[13]

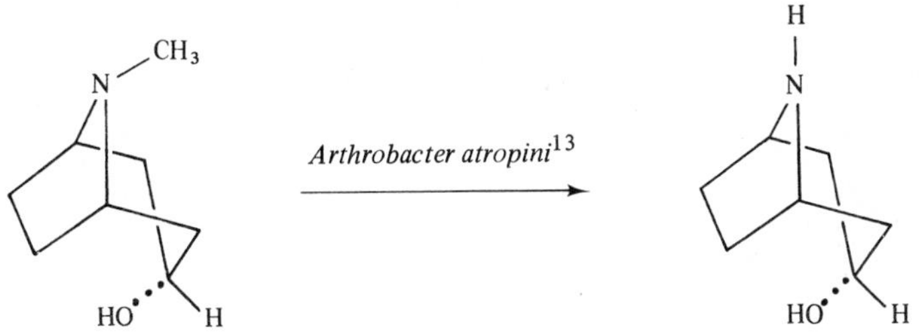

Tropine

Diazepam is demethylated by *Pellicularia filamentosa*[14] and by *Sporotrichum sulfurescens*[15] (12%).   This reaction parallels that found in

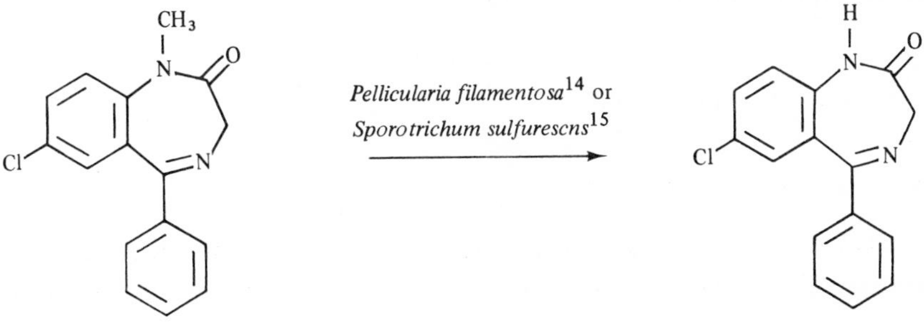

Diazepam

mammalian metabolism of this minor tranquilizer. The antibiotic clindamycin is converted to an N-demethyl compound (10%) by the action of *Streptomyces punipalus*,[16] but this reaction can be done in a superior

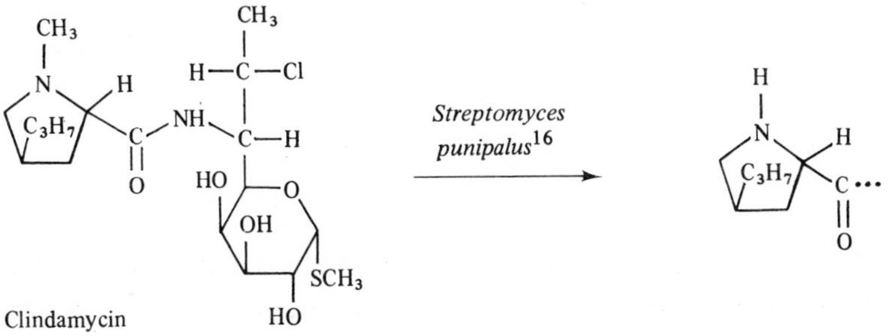

Clindamycin

manner by a catalytic demethylation.[17]   Finally, 6,14-ethenotetrahy-
drothebaine alkaloids are demethylated (3-10%) by *Cunninghamella ech-
inulata*[18]; a reaction that is difficult to achieve by chemical means

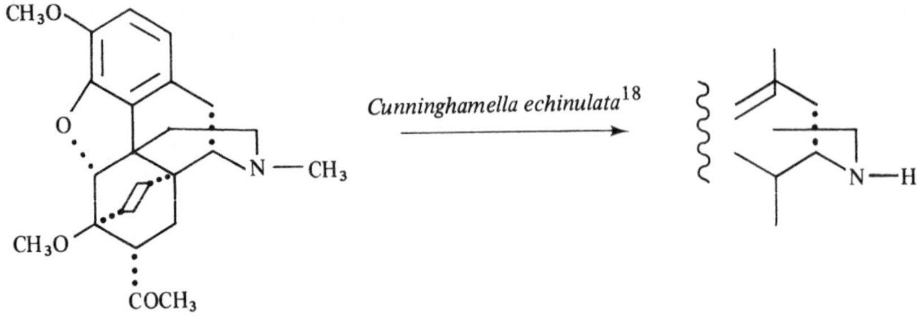

without affecting the other portions of the molecule.

Another N-alkyl cleavage reaction, catalyzed by an enzyme preparation from a *Pseudomonas* species, is the conversion of α-amino-β-
pyrazolylpropionic acid to pyrazole.[19]

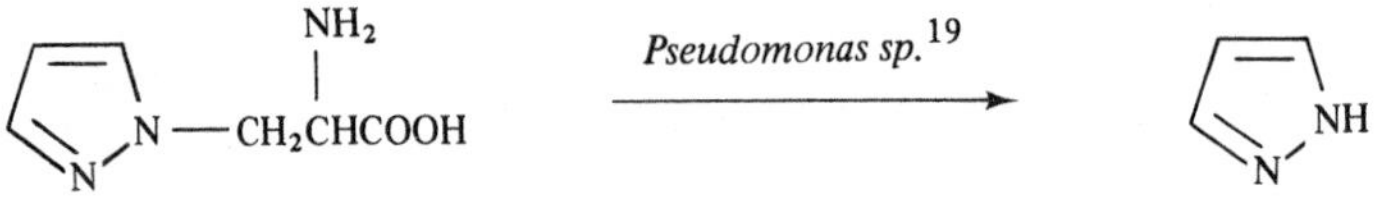

α-Amino-β-pyrazolylpropionic acid

Related to the above amine dealkylations is the cleavage of a
sulfide bond of lincomycin by *Streptomyces lincolnensis*.[20]   (The related sulfur oxygenation reaction is discussed in Chapter 12.)

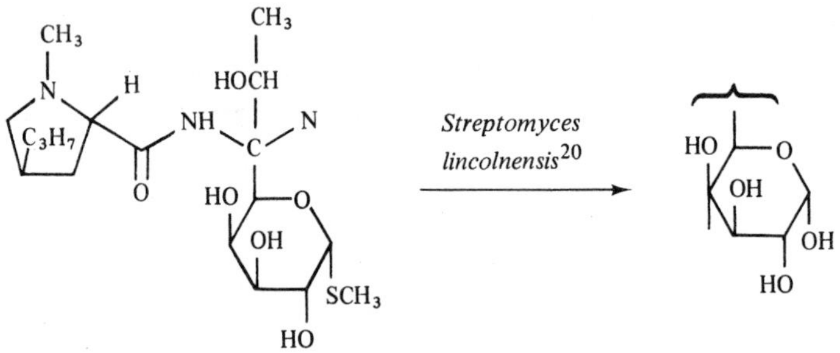

Lincomycin

Oxidative reactions, including dehydrogenation (Chapter 8), are
involved in the conversion of L-pipecolic acid to L-α-aminoadipic
acid by a *Pseudomonas* species.[21]

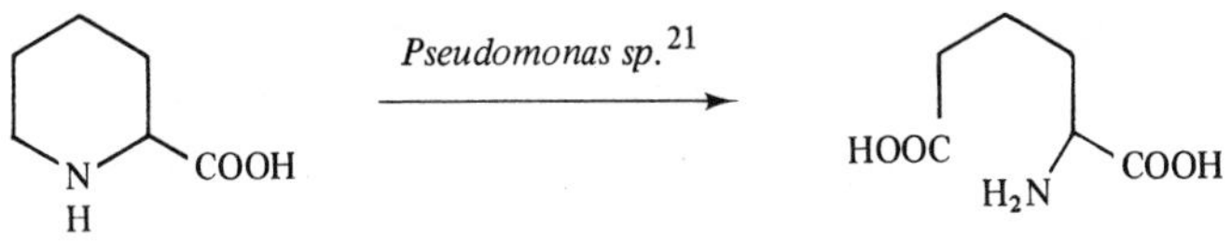

L-Pipecolic acid

N-Oxide Formation. Another reaction occurring occasionally with amines is oxidation to amine oxides. Three examples are given:

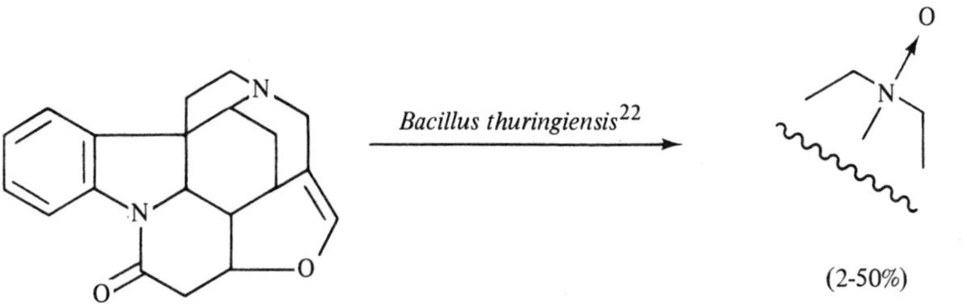

(2-50%)

Strychnine

Diazepam

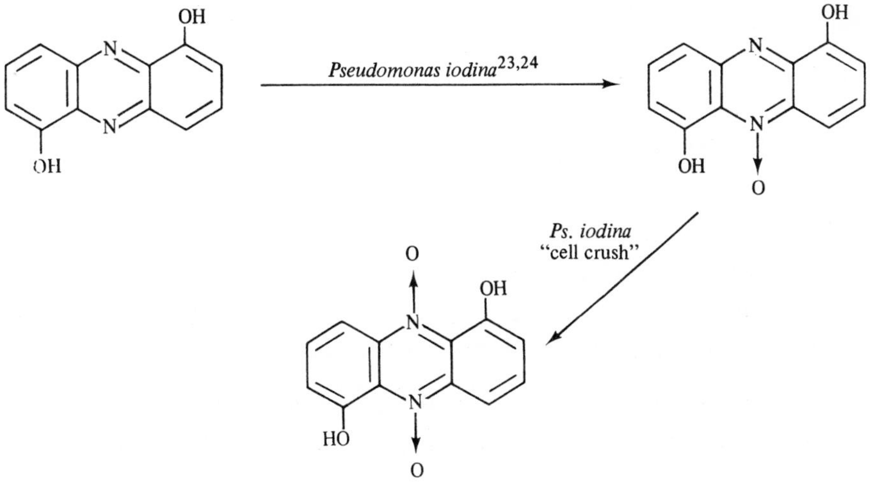

<u>Composite Reactions</u>. Among the numerous works of the Indian
group led by P. K. Bhattacharyya is a description of the conversion
of α-santalene to *tere*-santalol (*A*) by *Aspergillus niger*.[25] *tere*-San-

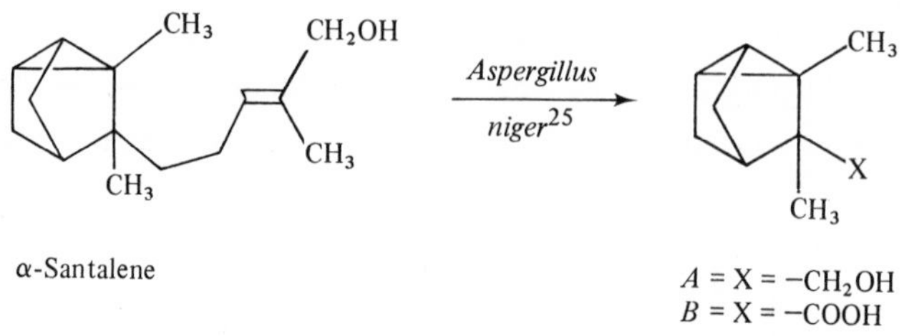

α-Santalene

A = X = −CH₂OH
B = X = −COOH

talic acid (*B*) represents the major product from this bioconversion
and likely arises from further oxidation of *A*.

A similar loss of several carbons is seen in the conversion of
the 3-acetate of cholest-5-ene-3β,19-diol to estrone by a soil microorganism CSD-10 in 72% yield,[26] (and in similar reactions of closely

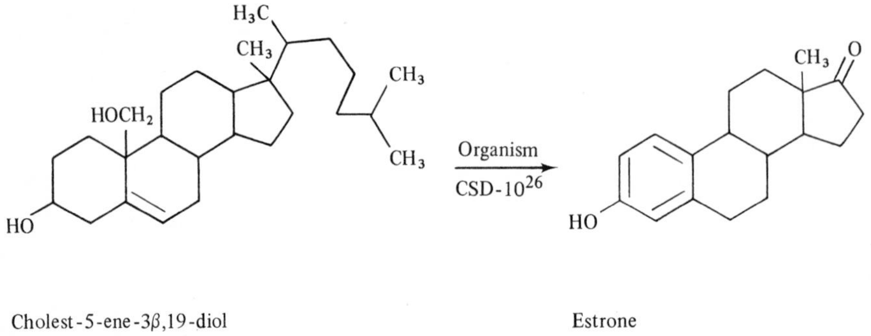

Cholest-5-ene-3β,19-diol   Estrone

related substrates with *Nocardia* and *Mycobacteria* species[27]).   The
reaction involves β-oxidation of intermediate fatty acids[28,29] and
this is probably also true for the α-santalene degradation.

The degradation of diosgenone to 16-oxygenated-1,4-androstadien-
3-ones by *Fusarium solani*[30] probably proceeds *via* hydrolysis, alcohol

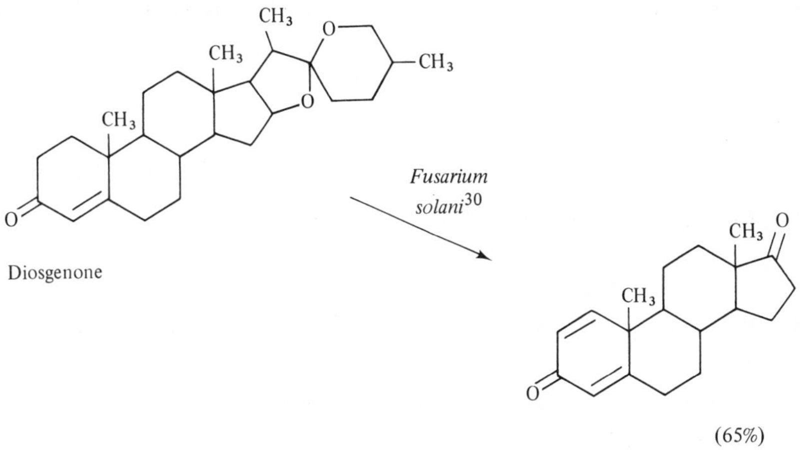

Diosgenone

Fusarium
solani[30]

(65%)

dehydrogenation, microbial Baeyer-Villiger oxidation, alkyl dehydrogenation, reverse aldol, and other reactions.

A remarkable feature of microbial oxygenation reactions is the
rarity of structural rearrangement accompanying the oxygenation reaction. Thus, the oxygen is generally introduced into the molecule with
no change in the remainder of the structure, with the possible exception of a few allylic shifts in the cases of olefin oxygenations. It
is of interest therefore, that Bhattacharyya and co-workers find
products having camphane and menthane skeletons in the products of α-
and β-pinene oxygenations by a *Pseudomonas* species.[31,32] Thus, from
α-pinene, borneol and oleuropeic acid are found to be among the prod-

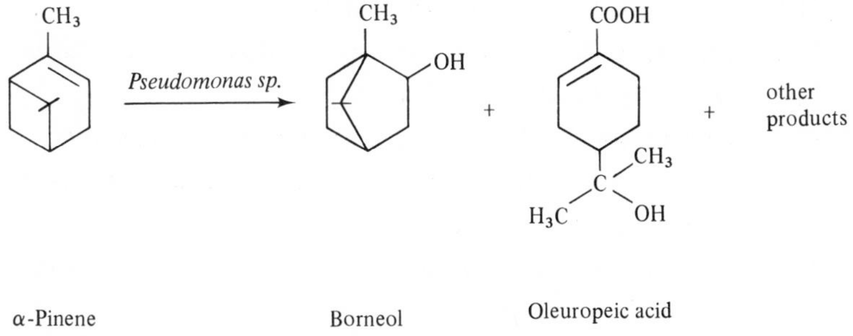

Pseudomonas sp.

+   +   other
products

α-Pinene   Borneol   Oleuropeic acid

ucts representing the aforementioned structural types. To explain
the formation of these and similar products arising from the fermentations, Bhattacharyya has proposed formation of intermediate proto-

nated species that undergo carbonium ion rearrangements   in the classical chemical sense.

REFERENCES

1. H. Takeda and O. Hayaishi, *J. Biol. Chem. 241*, 2733 (1966).
2. H. Hagino and K. Nakayama, *Agr. Biol. Chem. 31*, A21 (1967).
3. S. Yamamoto, M. Katagiri, H. Maeno, and O. Hayaishi, *J. Biol. Chem. 240*, 3408 (1965).
4. M. Katagiri, H. Maeno, S. Yamamoto, and O. Hayaishi, *J. Biol. Chem. 240*, 3414 (1965).
5. M. Katagiri, S. Yamamoto, and O. Hayaishi, *J. Biol. Chem. 237*, PC2413 (1962).
6. M. Katagiri, S. Takemori, K. Suzuki, and H. Yasuda, *J. Biol. Chem. 241*, 5675 (1966).
7. N. H. Sloane, K. G. Untch, and A. W. Johnson, *Biochim. Biophys. Acta 78*, 588 (1963).
8. L. I. Hochstein and S. C. Rittenberg, *J. Biol. Chem. 235*, 795 (1960).
9. F. A. Gries, K. Decker, and M. Bruhmuller, *Z. Physiol. Chem. 325*, 229 (1961).
10. K. Decker and H. Bleeg, *Biochim. Biophys. Acta 105*, 313 (1965).
11. E. Wada, *Arch. Biochem. Biophys. 72*, 145 (1957).
12. E. Wada and K. Yamasaki, *J. Amer. Chem. Soc. 76*, 155 (1954).
13. J. Kaczkowski and M. Mozejko-Toczko, *Acta Microbiol. Pol. 9*, 173 (1960); *Chem. Abstr. 56*, 5197d (1962).
14. G. Greenspan, H. W. Ruelius, and H. E. Alburn, *U. S. Patent* 3,453,179 (July 1, 1969).
15. R. A. Johnson and H. C. Murray, unpublished observation.
16. A. D. Argoudelis, J. H. Coats, D. J. Mason, and O. K. Sebek, *J. Antibiotics 22*, 309 (1969).
17. R. D. Birkenmeyer and L. Dolak, *Tetrahedron Lett. 58*, 5049 (1970).
18. L. A. Mitscher, W. W. Andres, G. O. Morton, and E. L. Patterson, *Experientia 24*, 133 (1968).
19. M. Takeshita, Y. Nishizuka, and O. Hayaishi, *Biochim. Biophys. Acta 48*, 409 (1961).

20. A. D. Argoudelis and D. J. Mason, *J. Antibiotics 22*, 289 (1969).

21. D. R. Rao and V. W. Rodwell, *J. Biol. Chem. 237*, 2232 (1962).

22. P. Bellet and D. Gerard, *Ann. Pharm. Franc. 20*, 928 (1962).

23. N. N. Gerber and M. P. Lechevalier, *Biochemistry 4*, 176 (1965).

24. H. A. Lechevalier, in *Biogenesis of Antibiotic Substances* (Z. Vanek and Zd. Hostalek, eds.), Academic, New York, 1965.

25. B. R. Prema and P. K. Bhattacharyya, *Appl. Microbiol. 10*, 529 (1962).

26. C. J. Sih, S. S. Lee, Y. Y. Tsong, K. C. Wang, and F. N. Chang, *J. Amer. Chem. Soc. 87*, 2765 (1965).

27. R. Deghenghi, S. Rakhit, K. Singh, C. Vezina, and C. J. Sih, *Steroids 10*, 313 (1967).

28. C. J. Sih, K. C. Wang, and H. H. Tai, *J. Amer. Chem. Soc. 89*, 1956 (1967).

29. C. J. Sih, H. H. Tai, and Y. Y. Tsong, *J. Amer. Chem. Soc. 89*, 1957 (1967).

30. E. Kondo and T. Mitsugi, *J. Amer. Chem. Soc. 88*, 4737 (1966).

31. O. P. Shukla, M. N. Moholay, and P. K. Bhattacharyya, *Indian J. Biochem. 5*, 79 (1968).

32. O. P. Shukla and P. K. Bhattacharyya, *Indian J. Biochem. 5*, 92 (1968).

Chapter 14

PRACTICAL EXPERIMENTAL METHODS

INTRODUCTION

The organic chemist who lacks microbiological training or experience faces methodology problems that he is unable to define and
whose scope he cannot grasp. If his only objective in using a microbiological oxidation is to push through rapidly to the end result
(Does the molecule oxidize, or doesn't it? What product is formed?),
it would be most expeditious for him to enlist the participation of
an interested and well-equipped microbiologist. In this way, the necessary probing reactions will consume little time on the part of either the chemist or microbiologist and can lead rapidly to a decision
regarding scale-up of the process. Whether the same type of participation can be applied to scale-up work depends on the magnitude of
scale-up desired and on the size and nature of the facilities available.

It is conceivable that such a collaborative effort would satisfy
the immediate objective of the organic chemist, but it would probably
leave him deficient in appreciation of microbiological methodology,
some of which he may wish to acquire for personal use in his own laboratory. The best way for him to acquire a rapid exposure to the
techniques and equipment of microbiology is to spend a few days in a
laboratory where work of this type is being actively pursued and
where he can be given direct instruction. Even the simplest techniques may require several written pages for their accurate exposi-
tion, thereby conveying to the reader an unwarranted impression of
their complexity. In contrast, a moment or two of demonstration by an
expert, and a bit of guidance for the first attempts of the neophyte,
can result in a practical grasp of a method that carries with it a
self-confidence not often acquired from a text. Beyond this, the use

243

of microorganisms has a "feel" (and smell) that cannot be conveyed by
the written word.   (The organic chemist   need only think back to the
transition from his own first days in a laboratory to realize that
the doubts inspired by the newly presented tools of his trade by now
have been transformed into an indefinable "feel," familiarity, and
self-confidence.) The following discussion of methodology has thus
been kept to a bare minimum, and is intended merely to call attention
to several important facets that are sometimes taken for granted in a
microbiologically oriented text. Since essentially all of the microbiological reactions described in this book have been done under aerobic conditions, the following discussion refers entirely to fermentations of this type.

EQUIPMENT

   General.   Any reasonably well-equipped chemical laboratory will
contain much of the equipment needed for fermentation work.   Erlenmeyer flasks, beakers, graduated cylinders, funnels, a Bunsen burner
or two, and a balance are common and indispensable.   A pH meter of
modest accuracy will be needed frequently.   In addition to these, it
will be necessary to acquire asbestos gloves for handling hot equipment, and some of the equipment items described below.

   Fermentation in Flasks.   At the present, highly empirical state
of the art of microbiological oxidations, a chemist will wish to make
several probing experiments with a minimal commitment of his proposed
substrate.   Such probing is most commonly carried out by incubating
a few milligrams of compound with a microorganism grown in an Erlenmeyer flask.   As a rule of thumb, the volume of the contents of a
flask should be no more than about one-fifth the total volume of the
flask. A 500-ml Erlenmeyer flask containing 100 ml of culture slurry
is suitable for transformation of up to 20 mg of most substrates.

   Fermentations in flasks must be agitated in some manner to ensure aeration of the microorganism. Many types of mechanical shakers
are available, at all levels of sophistication of added instrumenta-
tion. Given a room or cupboard in which the temperature can be main-

tained at about 25-30°C, the simplest and least expensive rotary action shaker will serve nicely and will present the fewest maintenance
problems.   If a suitably thermo-regulated room is not available, an
incubator-shaker or a water-bath shaker may be utilized.   The former
will prove more useful than the latter; unfortunately, the lowest-
priced incubator-shaker costs about twice as much as the lowest-
priced water-bath shaker, and about three to four times as much as a
simple rotary shaker without temperature control. Since an Erlenmeyer
flask size of 500 ml offers maximum versatility, shakers should be
equipped to accommodate this size.

Preparation for carrying out fermentations in flasks requires
that the flask that will contain the (culture) medium be sterilized.
Five-hundred milliliter flasks may be sterilized in an ordinary pressure cooker at 15 psig <u>steam</u> (not steam plus air) pressure for about
20 min.   This will result in an internal temperature of about 121°C.
Additional pressure will be required at higher altitudes to achieve
this temperature.   For larger flasks, or for larger fermentors of
other types (see below), an autoclave will be required.   Autoclaves
are available in a variety of models, with various degrees of sophistication and elegance, but almost all are "fixed installation" equipment, requiring piped-in steam and water as well as a waste water
drain system. (The larger scale fermentors that include the capability for steam sterilization within their over-all structure are not
likely to find their way into a chemical laboratory because of their
size and cost.)

<u>Fermentation in Fermentors</u>.   Fermentors (other than Erlenmeyer
flasks) can be obtained in all sizes, from the smallest bench-top
glass model to the giant industrial fermentor. While fermentors offer
many conveniences in the scale-up of microbial reactions, they may
require additional study of the fermentation conditions before results are obtained analogous to those seen in the Erlenmeyer flask.
This is a consequence (in part) of the mechanical differences between
these two systems. Fermentations in stirred vessels may be better or
worse (in terms of yield, nature of product, specificity of reaction,
etc.) than those in shaken flasks.   The chemist who contemplates the

acquisition of fermentors beyond the shaken-flask stage should proceed only after extensive consultation with his microbiological colleagues, since the cost of even the simplest equipment is high, and there are a number of related problems (agitation rate, aeration rate, air sterilization, foam control, etc.) whose discussion is beyond the scope of this book.

THE MICROORGANISM

Acquisition.   To use microorganisms as chemical reagents, they must first be brought into the laboratory, and the equipment for their handling must be available. Pure cultures ($i.e.$, a single species uncontaminated with other species) of microorganisms may be acquired by purchase from a suitable agency, such as the American Type Culture Collection (ATCC) or the Centraalbureau voor Schimmelcultures (CBS). These and several others are cited (with addresses) at the end of the chapter.   Alternatively, a culture may sometimes be obtained from an investigator who has reported on its use. In this case, less reliance can be placed on the purity of the culture. In either case, the organism should be requested as a culture on a solid culture medium in a tube, and it will be necessary to make transfers to ($i.e.$, grow more organism in) additional tubes or to other containers in order to obtain sufficient material for further use.

Maintenance and Growth.   In order to transfer the culture to other containers it is necessary to have ready suitable nutrient preparations.   Any appropriate mixture of nutrients is called a culture medium, and it may be liquid or may be solidified (gelled) by the inclusion of a gelling agent, such as agar-agar.   In everyday microbiological jargon, the term "medium" generally refers to a liquid culture medium and "agar" (whether or not it actually contains agar-agar) to a solid culture medium. Fortunately for both the chemist and the microbiologist, many culture media are commercially available (see Appendix for sources), both as solid media (in culture tubes or on culture plates [Petri dishes]) or as dry mixes (suitable for making up to liquid media by addition of water, followed by ster-

ilization).   The ATCC catalog contains examples of typical media and
their components   as well as an indication of which medium to select
for a particular culture.

   Storage.   A culture may frequently be stored for a reasonable
period of time (three to six months) by growing it on an agar medium
in a culture tube, which is then placed in a refrigerator. (There are
several techniques for longer term storage, but these are best reser-
ved to the microbiologist.   The chemist whose culture accidentally
dies out can always reorder his culture from the culture collection.)

   Transfer.   The transfer of a culture from one solid medium to
another is required for both the storage of the culture and the sav-
ing of "seed" for a later fermentation, while transfer to a liquid
medium is required in order to obtain the larger quantities of culture needed to start a fermentation. The former is done with the use
of an inoculating loop or needle that is sterilized by heating to
redness in a burner flame. The loop must be cool during the transfer,
however. (One way to cool a hot loop is to touch it briefly to the
bare agar in the tube.) Using the cool loop, a few cells of the culture are scraped from the one medium onto the loop and then rubbed
off onto the other medium.   If the two media are in culture tubes or
flasks, the caps or plugs are removed aseptically and the mouths of
the tubes or flasks heated briefly in a burner flame both before and
after the transfer to kill contaminating organisms lurking there and
to keep air currents moving out of the tube. All of these operations
should be carried out as rapidly as possible in a clean environment
where there is a minimum of air currents in order to avoid accidental
contamination from air-borne microorganisms.   The newly inoculated
culture must be allowed to grow by incubating it at an appropriate
temperature, usually for several days, before it can be used or
stored.

   When a culture is transferred to a liquid medium in order to
carry out a fermentation, the inoculum may consist of vegetative
(growing) cells or of spores (resting cells, somewhat analogous to
seeds).   Vegetative cells must usually be scraped from the culture

tube into the sterile liquid medium in a suitably stoppered culture flask, using an inoculating loop as described above. Alternatively, sterile water may be added to the culture tube, and the culture then agitated with a sterile loop, following which the slurry is trans- ferred (by sterile pipet or by pouring) into the recipient flask or fermentor. All aseptic handling precautions, including the flaming of mouths of containers, must be observed. Transfer of a slurry is more commonly carried out using organisms that sporulate freely, since the spores are easily dislodged, frequently without the use of the loop, and may be pipeted into several flasks at a reasonably uniform inocu- lation rate. · Most fermentations proceed best (and with minimum risk of contamination) if the inoculum is "heavy," *i.e.*, a large number of cells are used to start the growth in a new flask or fermentor. It may thus be necessary to proceed through one or more scale-up inocu- lations to arrive at the desired amount of the organism suitable for heavy inoculation. A culture of vigorously growing cells amounting to 5-10% of the fermentation volume is usually adequate.

After the outgrowth of the inoculum in the growth medium has reached the desired stage (for most organisms, this will be about 24- 48 hours), the cells may most simply be used for fermentation by add- ing the substrate and continuing the incubation. Alternatively, the cells may first be harvested by centrifugation or filtration, washed to remove remaining medium ingredients, and then used as a suspension in water or a buffer solution, or processed further to give various types of enzyme preparations.

Dangerous Microorganisms (Pathogens). A number of microorganisms are classified as pathogenic (*i.e.*, causative agent for human dis- ease) by the Federal Government and may be supplied only to highly qualified users. It is probably wise for the organic chemist to avoid entirely the use of this type of organism unless he is able to ar- range for its use under the careful supervision of and in the facili- ties of a *bona fide* microbiologist. Other organisms, classified as plant pathogens, may be shipped into certain areas only after permis- sion is obtained from the State Plant Quarantine Office or State Ag- riculture Department. Since such permission would ordinarily be given only to highly qualified users, organisms falling into this category should probably also be avoided if possible. Whether a given culture falls into these categories can easily be ascertained from the supplier (in the event that this is a culture collection).

When cultures (pathogenic or otherwise) are no longer needed, they should be killed by autoclaving or by bactericidal treatment (*e.g.*, treatment with a 5% phenol solution for a few hours at room temperature) before disposal.

"Easy" Organisms. Microorganisms show great variability in such properties as growth, sporulation, vigor, resistance to contamination, and other attributes that can, for our purposes, be lumped together somewhat subjectively to generate an "ease of handling" scale. The chemist who is just beginning to work in this area will probably want to start with some of the "easy" organisms from the following list: *Acetobacter suboxydans, Aspergillus niger, Calonectria decora, Corynebacterium simplex, Nocardia restrictus,* most *Pseudomonas* species, *Rhizopus nigricans, Saccharomyces cerevisiae,* and *Sporotrichum sulfurescens.*

MISCELLANEOUS OPERATIONAL NOTES

Is the Culture Ready to Use? Before a chemist commits his precious substrate to the fermentation flask, he will want to have some reasonable assurance that the culture he is growing there is all that it should be. His level of sophistication as a microbiologist, and his access to analytical equipment will be the limiting factors on the accuracy of his determination. Ideally, he will check the pH of the beer and observe a specimen microscopically. More probably, he will look at and smell the beer to make his judgment.

Bacterial and yeast growth in culture will generally result in a more or less uniformly cloudy suspension of cells, not unlike thin cream of chicken soup in appearance. The pH of such cultures at the end of their most vigorous growth (at which time most of a limiting nutrient in the medium has been consumed) may be high (pH 8) or low (pH 3), depending upon the buffering capacity and nature of the medium ingredients.   Media with complex (protein) nitrogen sources such as corn steep solids will generally result in higher pH values than media made up of ammonia, glucose, and other simple ingredients. The odor may range from pleasantly yeasty (some yeasts) or pleasantly earthy (some *Streptomyces* species) to various degrees of mild unpleasantness (rotten cabbage odor, for example).

Molds (fungi), on the other hand, most often will result in thick mats or marble-sized balls of growth ranging in color from yellow or pink through brown to black.   A flask lacking such luxuriant growth would in all probability be suspect and best not used. The molds offer a range of odors, most of which are not unpleasant, and some of which can be as downright appetizing as freshly baked bread. A contaminated flask will probably smell putrid! The pH of mold fermentations is a consequence of the same factors discussed above for bacteria and yeasts.

How is Substrate Added?   Given a properly grown, uncontaminated organism, the substrate may be introduced in several ways.   Unless special facilities are available, the flask will have to be opened, the substrate added, and the flask closed again, all under aseptic conditions to avoid contamination (although the hazards are fewer at this stage than they are early in the growth of the culture, owing to nutrient depletion and the competitive advantage of the "fully grown" culture).   A liquid, or in some cases, a solid substrate (if finely divided) may be added as is, but more frequently in solution.   Ordinarily a solid substrate need not be sterilized -- a fortunate circumstance, since many molecules would not be stable under autoclave conditions. However, if·the substrate is added as an aqueous solution, sterile water should be used to minimize the contamination hazard.

A substrate not soluble in water is most often dissolved in a water-miscible organic solvent.   The solvent must be used sparingly, in order to avoid inhibitory effects on the microorganism.   Solvents that have been used successfully include ethanol, acetone, dimethylsulfoxide, and N,N-dimethylformamide (DMF).   Of these, the last has been the most useful in our experience and does not appear to be in-

hibitory when used at 0.5% of the beer volume. DMF need not be sterilized, but it may be useful to add 5 g of phenol to each liter of
DMF to kill any lurking microbial contaminants.

Although the addition of a water-insoluble substrate dissolved
in, say, DMF, usually results in adequate dispersion, this is not invariably the case. The substrate will then fail to undergo oxidation.
The addition of a dispersing agent, such as Ultrawet 30-DS (Arco
Chemical Co., Chicago, Illinois) to the flask will frequently convert
failure to success.   For a first attempt, an amount equal to about
0.25% of the beer volume should be added. This material need not be
sterilized.

Use of Spores for Oxidative Reactions. All of the preceding comments relate to the use of a vegetative culture, used in the flask in
which it has been brought to maximum growth from a small inoculum of
spores or vegetative cells. Most of the comments are also relevant to
the use of washed cultures. These two techniques have been by far the
most widely used, but the more recent introduction[1] of the technique
of fermentation using certain types of spores in a non-nutritive environment (*i.e.*, where they will not germinate and proliferate) may
eventually represent the most desirable approach, at least from the
chemist's point of view.   It is conceivable that in the future pack-
ages of spores of a variety of microorganisms may be commercially
available and that these may be purchased like any chemical reagent,
kept on the shelf (or in the refrigerator) until needed, and then
shaken with the candidate substrate with only minimal attention to
aseptic handling.

Where is the Product?   The duration of the incubation of substrate with microorganism may be arbitrarily set as a certain length
of time (*e.g.*, 24, 48, or 72 hr) following the addition of substrate
or, better, by aseptic withdrawal of small samples from the fermentation, which can be analyzed (*e.g.*, by extraction and thin-layer
chromatography) for extent of completion of reaction. The fermentation may then be stopped by chilling the flasks or fermentor in a
cold room (refrigerator). At this point, one is left with a potful of

beer from which to isolate the product(s).   As an initial operation
the cells of the microorganism and the clear beer may be separated by
either filtration (often facilitated by the use of Celite filter aid)
or centrifugation.   Extraction of the clear beer (methylene chloride
is recommended as the first choice of solvent) and examination of the
extract will show if the products reside here. It should not be surprising to expect water-insoluble products to be found here!

Isolation from a clear beer is not different from any standard
chemical workup. On the other hand, isolation from the organism (mat,
cake, mycelium, and pellet are terms often used to describe this solid material) may require some ingenuity, particularly if rupture of
the cells is necessary, but extraction with acetone (or other suitable solvent) may be satisfactory. Unfortunately, solvent extraction
of many microorganism cells leads to extraction of sterols and other
lipids, which may complicate purification work, and the chemist is
urged to use such supernatural or spiritual forces as are available
to him in order to ensure emergence of product into the clear beer.

All equipment that has been used for handling living microorganisms should be autoclaved. If this is not feasible because of equipment limitations, it should be soaked in a disinfectant (we recommend
the use of 5% aqueous phenol) for an hour or so.

Filter cakes should also be treated, and the use of 5% phenol
may be the easiest way, but a solvent-extracted beer is unlikely to
contain viable cells and may be discarded without further treatment.

REFERENCES

1. K. Singh, S. N. Sehgal, and C. Vezina, *Appl. Microbiol. 16*, 393
    (1968).

General References

G. Smith, *An Introduction to Industrial Mycology*, 4th ed., Arnold,
    London, 1954, pp. 226-272.
    Chapter XI. Laboratory Equipment and Technique.

W. B. Sarles, W. C. Frazier, J. B. Wilson, and S. G. Knight, *Microbiology*, 2nd ed., Harper & Brothers, New York, 1956, pp. 127-145.
Chapter 11. Culture Media.
Chapter 12. Sterilization.
Chapter 13. Isolation of Pure Cultures.
B. D. Davis, *Microbiology*, Hoeber Medical Division, Harper & Row, New York, 1967, pp. 336-353.
Chapter 11. Sterilization and Disinfection.
L. E. Casida, Jr., *Industrial Microbiology*, Wiley, New York, 1968, pp. 117-143.
Chapter 7. Fermentation Media.
Chapter 8. Inoculum Preparation.
Chapter 9. Scale-up of Fermentations.
L. L. Wallen, F. H. Stodola, and R. W. Jackson, *Type Reactions in Fermentation Chemistry*, U. S. Department of Agriculture, Publication No. ARS-71-13, 1959, pp. 1-11.
Introduction. Tools and Methods of Fermentology.
C. M. Christensen, *The Molds and Man*, McGraw-Hill, New York, 1965 (paperback edition), pp. 226-247.
Chapter 12. Experiments with Fungi.
The American Biology Teacher, volume 30, no. 6 (August 1968). A special issue dedicated to microbiology, available upon request from Difco Laboratories, Detroit, Michigan 48201. Highly recommended.

Appendix to PRACTICAL EXPERIMENTAL METHODS

1. Microbial Culture Sources
    -- American Type Culture Collection (ATCC)
        12301 Parklawn Drive
        Rockville, Maryland   20852
        The ATCC Catalog should be acquired, since it contains much valuable information about culture ordering and handling, and about culture media.
    -- Centraalbureau voor Schimmelcultures
        Baarn (molds, actinomycetes) or Delft (yeasts), The Netherlands

-- Northern Utilization Research and Development Division (NRRL)
   Agricultural Research Service
   U. S. Department of Agriculture
   Peoria, Illinois
-- Commonwealth Mycological Institute (CMI) (fungi only)
   Kew, Surrey, England
-- National Collection of Industrial Bacteria (NCIB) (bacteria
   only)
   Torrey Research Station
   Aberdeen, Scotland

2. Fermentation Media Suppliers
   -- Difco Laboratories
      Detroit, Michigan 48201
   -- BBL, Division of BioQuest
      P. O. Box 175
      Cockeysville, Maryland 21030

3. Fermentation Equipment Suppliers

   a. Glassware or Plastic Ware
      -- Belco Glass, Inc.
         340 Edrudo Road
         Vineland, New Jersey 08360
      -- Falcon Plastics
         Division of B-D Laboratories, Inc.
         5500 West 83rd Street
         Los Angeles, California 90045
      -- Corning Glass Works
         Corning, New York 14830

   b. Autoclaves
      -- American Sterilizer Company
         Erie, Pennsylvania
      -- Arthur H. Thomas Company
         P. O. Box 779
         Philadelphia, Pennsylvania 19105

-- Wilmot Castle Company
Rochester, New York

c. Shakers, Fermentors, Incubators, etc.
-- New Brunswick Scientific Co., Inc.
1130 Somerset Street
New Brunswick, New Jersey
-- Scientific Products
Division of American Hospital Supply Corp.
1210 Leon Place
Evanston, Illinois 60201
-- Chemapec, Inc.
1 Newark Street
Hoboken, New Jersey 07030
-- The Virtis Company, Inc.
Gardiner, New York 12525
-- Biotech, Inc.
12221 Parklawn Drive
Rockville, Maryland 20852
-- Fermentation Design, Inc.
726 North Graham Street
Allentown, Pennsylvania 18103

GLOSSARY

Aerobic: A procedure (growth, biotransformation, fermentation, etc.) carried out in the presence of air or oxygen.

Agar (Agar-agar): (1) A polysaccharide from seaweed, commonly used as a base (gelling agent) for solid media. (2) A solid medium (jargon).

Aseptic: Procedures that minimize accidental entry of undesired organisms into a culture or fermentation.

Autoclave: A device in which equipment and media are sterilized at elevated temperature.

Bacteria: A large group of single-celled microorganisms.

Beer: The result of the growth of a microorganism in a liquid culture (removal of the organism gives a "clear beer").

Contamination: The presence of undesired organisms in a culture or a fermentation.

Culture: A population of microorganisms confined in an environment in which viability is retained.

Fungus: A mold. A type of filamentous plant lacking chlorophyll, frequently involved in the decay of dead organic matter.

Fermentation: The transformation of one molecule type (substrate) to another (product) by microorganism enzymes. (An older, more restrictive definition is now used only by older, more restrictive microbiologists.)

Germination: The "sprouting" of spores to produce a proliferation of vegetative growth.

Incubation: (1) Keeping a culture under a given set of conditions for a defined time. (2) Allowing a culture to act upon a substrate under defined conditions.

Inoculate: The deliberate introduction of microorganisms (usually of a single species) into a culture medium.

Inoculum: The microorganism(s) used to inoculate.

Medium (Culture Medium): (1) An appropriate mixture of nutrients capable of supporting the growth and multiplication of a microorganism. (2) A liquid culture medium (jargon).

Mold: see "Fungus"

Nutrient: A material that the microorganism uses as food or growth stimulant. (Nutrient substrates may give no isolatable products.)

Pseudomonad: A *Pseudomonas* species, or closely related microorganism, usually not yet fully classified taxonomically.

Pure Culture: A single species, uncontaminated with other species.

Species: A taxonomically distinct kind of microorganism. Individual cells within a species may differ in biochemical properties. [Abbreviated sp. (singular) or spp. (plural).]

Spores: Reproductive bodies, of one or more cells, at rest (nongrowing) until introduced into a nutrient medium. The "seeds" of microorganisms.

Sporulation: The production of spores by a vegetative growth of a microorganism.

Sterilize: To kill all living organisms.

Strain: A pure culture descended from a single individual of a species and thus presumably more biochemically homogeneous than the aggregate of all individuals of a species.

Substrate: The material or compound to be acted on by an organism or an enzyme to produce a product chemically related to the substrate. Preferably not a nutrient.

Substrate Specificity: A semiquantitative measure of the preference of an organism (enzyme) for a substrate.

Transfer: The introduction of a small amount of an organism into a virgin nutrient environment, in order to increase the supply of the organism.

Vegetative: Microorganism cells that are, or have recently been, growing and multiplying.

Yeast: A type of fungus that has single cells (as opposed to molds, which are filamentous).

AUTHOR INDEX

Underlined numbers give the page on which the complete reference is
listed.

A

Abbott, B. J., 179(34), <u>183</u>
Agurell, S., 97(33), <u>110</u>
Ahearn, G. P., 157(5), <u>162</u>
Aida, K., 98(37,39), 99(41), <u>110</u>,
  187(17), 194(60), <u>205</u>, <u>207</u>
Ajl, S. J., 198(75), <u>208</u>
Akagi, K., 191(39), <u>206</u>
Akhtar, M., 175(17), 176(21), <u>183</u>
Alburn, H. E., 107(74), 108(74),
  <u>112</u>, 235(14), 237(14), <u>240</u>
Ali Khan, M. Y., 39(107), <u>80</u>
Alsche, W. J., 161(26), <u>163</u>
Ambrus, G., 15(56), 16(56),
  54(122), <u>78</u>, <u>81</u>
Amici, A. M., 9(44), <u>77</u>
Anderson, H. V., 9(38), 59(38),
  60(38), <u>77</u>
Ando, S., 222(16), <u>226</u>
Andres, W. W., 35(93), <u>80</u>,
  107(73), <u>112</u>, 133(33,34), <u>140</u>,
  236(18), <u>240</u>
Aota, K., 28(78), <u>79</u>, 93(23), <u>109</u>,
  201(87), <u>208</u>
Arai, Y., 55(124), <u>81</u>
Archer, S., 101(52), 102(52,53,54),
  <u>110</u>, <u>111</u>
Argoudelis, A. D., 227(4,5), <u>232</u>,
  235(16), 236(20), <u>240</u>, <u>241</u>
Arima, K., 135(44), <u>141</u>, 172(16),
  <u>182</u>
Arnold, N., 122(40), <u>125</u>
Asai, T., 98(37,38,39), <u>110</u>,
  187(17), <u>205</u>
Auret, B. J., 230(9,10,11),
  231(10), <u>232</u>
Aurich, H., 197(73), <u>208</u>
Azoulay, E., 179(31,32), <u>183</u>,
  186(2,4,5,11), <u>205</u>

B

Bachrach, U., 182(43), 183(44),
  <u>184</u>

Baggi, G., 149(39), <u>154</u>
Bal, M., 172(6), <u>182</u>
Ballal, N. R., 189(26,27,28), <u>206</u>
Ballio, A., 212(4), <u>215</u>
Barbezat, P., 199(81), <u>208</u>
Bartnicki, E. W., 190(33), <u>206</u>
Basso, L. V., 183(45), <u>184</u>
Baum, R. H., 191(41), <u>206</u>
Beaman, A. G., 214(7), <u>216</u>
Beck, D., 107(72), <u>111</u>
Behrman, E. J., 151(52), <u>154</u>,
  224(22), <u>226</u>
Beliveau, J., 97(34), <u>110</u>
Bellet, P., 237(22), <u>241</u>
Belohlav, L., 215(19a), <u>216</u>
Benedict, R. G., 113(1), <u>123</u>
Bentley, R., 193(50), <u>207</u>
Berberian, D. A., 101(52),
  102(52,53), <u>110</u>, <u>111</u>
Berger, J., 89(11,12), <u>108</u>
Bergmann, F., 138(57), <u>142</u>
Berndt, H. D., 204(101), <u>209</u>
Bernhauer, K., 194(61a),
  195(61b,63), <u>207</u>
Bertholdt, H., 212(4), <u>215</u>
Bertrand, G., 196(66), <u>207</u>
Bhattacharyya, P. K., 89(13,14),
  90(16), 91(17,18), 92(18,19,20),
  93(21,22), 100(42,43), 101(18),
  <u>108</u>, <u>109</u>, <u>110</u>, 116(12,13), <u>124</u>,
  189(26,27,28), <u>206</u>, 238(25),
  239(31,32), <u>241</u>
Biaggi, C., 149(43), <u>154</u>
Bianchi, M. L., 9(44), <u>77</u>
Biellmann, J. F., 29(79,80), <u>79</u>
Birch, A. J., 214(14), <u>216</u>
Birkenmeyer, R. D., 236(17), <u>240</u>
Birkenshaw, J. H., 214(12), <u>216</u>
Blackwood, A. C., 139(60), <u>142</u>
Blank, R. H., 228(6), <u>232</u>
Bleeg, H., 136(45), <u>141</u>, 234(10),
  <u>240</u>
Bleiweiss, A. S., 198(75), <u>208</u>
Bloch, K., 177(24,25), 178(27),
  <u>183</u>